Physics: An Integrated Approach
Volume II, 2nd Ed.

Eric J. Page

CRYSTAL
MA

2nd Edition 2018 – Eric J. Page

ISBN: 1724764047
ISBN-13: 978-1724764041

BRIEF CONTENTS

Table of Contents

List of Tables

ACKNOWLEDGEMENTS

This book would not have been possible without the amazing support of my family. Thank you all for putting up with all the late nights and weekends preparing for class, writing papers and grant proposals and working on this book. Without you, none of this would be possible Thanks to all the members of the USCGA Physics Section for suggestions and edits: Lorraine Allen, Maria Aristizabal, Paul Blodgett, David Harris, Royce James, Briana Jewczyn, Scott Jones, Rich Paolino, Brooke Stutzman, James Toomey, and Kyle Young. Finally, many thanks to Greg Severn and David Devine for inspiration and so many conversations about how to explain and relate physics – the pillars of this book.

ABOUT THE AUTHOR

Eric J. Page earned a doctorate in Physics from the University of Rochester, in 2005. He then worked at Allegheny College in Meadville, PA as a Visiting Assistant Professor before taking a tenure-track position for five years at the University of San Diego. Dr. Page is currently the Associate Dean for Academic Affairs and an Associate Professor of Physics at the United States Coast Guard Academy. Dr. Page's research is primarily in the Science of Learning where he studies how cognitive biases and heuristics (mental errors and shortcuts, respectively) affect student learning. Dr. Page is also the author of the forthcoming books "Why don't they get it? How colleges students' brains fool them and what we can do about it" and "Learning in College: A primer for students."

PREFACE

Why a new physics textbook when there are literally hundreds currently available in the market? There are two important answers to this question (although there are actually many answers). First, physics texts are designed to be very broad, and none before this one were designed for the Coast Guard Academy. Although we strive to provide a solid introduction to physics to give you all the tools you need to succeed in an ever evolving technological society, our service-focused program need to have slightly different focus than other books provide. As Coast Guard officers, no matter what your major, you will be in an environment the ability to quickly analyze technical information will set you apart from the others. The physics sequence at CGA must provide you with the basic, fundamental tools needed to complete such tasks.

Second, no textbook out there adequately integrates over thirty years of research in cognitive science and physics education on how students learn. To be fair, there are several recent texts that are based on the results of physics education research. Some of these texts present information in a modern order that is more conducive to learning; some provide a scaffold to help students understand where the readings are going; some focus more heavily on conceptual building blocks in addition to the traditional plug-and-chug mentality. This book does build on these ideas; however, no other physics textbook utilizes the general results from cognitive science of how students learn and integrate material in a different way.

Education in the sciences is traditionally based on *massed practice*: A topic is introduced, you spend time discussing it and watching/doing a few example problems, complete a few homework problems, get tested on the ideas and then generally forget about it until the final exam. "Cramming" works well for massed practice. In fact, although many college students will outwardly say that cramming isn't good, the vast majority do it, and do well on standard tests because of it. You feel like you are learning, and you have all (or at least some) of the information available to you for the exam. And then you forget…most of it. This is called *performance versus learning*: you may perform well on the exam, but the information/ideas have not become part of your long-term ethos. Forgetting after *performance* can be so powerful that when presented with a similar problem just a few days later, you may feel like you've never seen the entire topic before. Sound familiar?

Decades of careful research into how people learn have suggested several ways to improve upon the standards of massed practice. However, each of these introduce what are known as *desirable difficulties* – a feeling that something is harder or even impossible to learn initially, but yet leads to far less forgetting in the long term. Another way to say this is that *learning is hard*. If it seems easy, there is good chance you either fooled yourself into thinking you understand it and/or you have left yourself open to forgetting.

Here are some of the modern techniques that have proven to be extremely powerful:

1) *Spacing*: Returning to topics more than once on a given pattern. Spacing is usually done is smaller chunks when information is first introduced and then the intervals can increase over time.

2) *Interleaving*: Bouncing back and forth between topics, sometimes in ways that may not initially make sense. This allows your brain to spread out knowledge so that it is not compartmentalized and hard to recall.

3) *Testing*: It is extremely easy to convince yourself that you understand something when someone does it for you (examples and demonstrations). In fact, the less you know, the easier it can be to convince yourself you understand (this is called the Dunning-Kruger effect). The best way to avoid this is to test yourself and be tested. Testing yourself is a far more powerful study technique than reviewing material you have read previously.

4) *Generating*: Having students generate questions/problems/scenarios is very powerful because it forces the brain to integrate more information than simply solving problems. This is also why tutoring/teaching is a powerful way to make sure you understand a subject because you have to generate answers to others questions.

5) ***Less Organization***: This one surprises people and is often met with consternation by seasoned instructors. The less organized learning materials/worksheets/tests are, the more conducive they are to long-term learning. It appears that with less organization, the brain is forced to do its own organizing and analyzing, making learning deeper and forgetting less effective.

I've attempted to build a number of these techniques into the textbook and accompanying class worksheets. I won't claim that this textbook is the end-all-be-all for integrating these ideas together; however, I do assure you that while writing this text, the important results of cognitive science have been at the forefront of my thoughts, right along with how to best present physics content based on my interactions with students over the last decade. I hope *Physics: An Integrated Approach* helps aid your learning in physics and I am *always* open to comments to help improve this book.

Eric J. Page
May, 2018
(Revised from the 1st edition of 2014)

201 – Introduction to Waves

Wave phenomena surround us in our everyday lives: we see ripples after dropping a pebble into a pond (water waves), we hear a baby crying in the next room (sound waves) and we see cars pass by while driving on the highway (electromagnetic waves). In this unit, we start to understand how to describe waves and describe some of the most important basic features such as frequency and wavelength.

Integration of Ideas

The idea of simple harmonic oscillations
The relationship between force, displacement and work

The Bare Essentials

- Mechanical waves come in two types, longitudinal and transverse.

- The wave equation describes how waves can be made of *any* shape.

The Wave Equation

$$0 = \frac{\partial^2 f}{\partial x^2} - \frac{1}{v^2}\frac{\partial^2 f}{\partial t^2}$$

Description – Any function, f, which is a solution to the wave equation represents a function of x and t that varies regularly in space and time and represents a wave that moves through space with a phase speed v.

- Sinusoidal waves – amplitude and phase

Sinusoidal Waves

$$f(x,t) = A\sin(\omega t - kx)$$

Description – This equation gives the displacement from equilibrium of a point on a sinusoidal wave with amplitude A, angular frequency, ω, and angular wave number, k at position x and time t.

Note 1: $\omega = 2\pi f$, $k = \frac{2\pi}{\lambda}$

Note 2: For a transverse wave, this displacement will be measured perpendicularly from equilibrium and for a longitudinal wave the displacement is measured parallel to the wave from equilibrium.

- The direction of propagation of a sinusoidal wave is along the axis given in the phase of the sine function and is *opposite* of the sign between the kx and ωt terms.

Some Important Relationships
Sinusoidal Waves

Angular Frequency, $\omega = 2\pi f$

Angular Wavenumber, $k = \frac{2\pi}{\lambda}$

Phase Speed, $v_p = \frac{\lambda}{T}$
  (moves one wavelength in one period)

$f = \frac{1}{T}$, frequency is cycles per second, period is time per cycle.

Using these relationships, the phase speed determined in a number of ways:

$$v_p = \frac{\omega}{k} = \frac{\lambda}{T} = \lambda f$$

- Complex waves are a combination of many frequencies

Fourier Theorem

$$f(x) = \sum_{n=0}^{\infty} A_n \sin(k_n x) + B_n \cos(k_n x)$$

Description – Fourier's Theorem describes how a wave of any shape can be made as the combination of sine and cosine waves of different frequencies.

201-1: The Physics concept of a wave

Consider: *What exactly is a wave?*

OUR SENSES ARE CONSTANTLY BOMBARDED by waves every second of every day without us even realizing this is occurring. So, what is a wave? In physics-speak, a wave is any disturbance that propagates in space and time. This is a very general description that doesn't really connect waves to our surroundings, so, what are some of the waves that we encounter every day? We all know about water waves like those produced by dropping a pebble so that it falls on the glassy surface of still water. Many of you may also know that sound is a wave – the propagation of pressure differences in air. Light is also a wave, known as an electromagnetic wave that is created by either objects at a given temperature (e.g. the sun) or by accelerating charged particles (e.g. electrons in your cell phone's antenna).

As far as we know, all waves can be placed into one of three categories, mechanical waves, electromagnetic waves and matter waves:

- *Mechanical Waves* – These constitute many of the waves that we are familiar with, including water waves, sound waves and earthquake (seismic) waves. Mechanical waves are formed by the displacement of some *medium*, which is a background material that acts as a substance for the wave to travel in. Mechanical waves will be the focus of the next couple of chapters.
- *Electromagnetic Waves* – Electromagnetic waves (EM) also surround us constantly and include light, ultraviolet rays, radio waves, microwaves, infrared radiation and gamma rays. EM waves do not require a medium to travel through and can therefore travel through empty space. We will study EM waves in unit 207.
- *Matter Waves* – Not known until the early parts of the 20th century, matter waves are a result of particle-wave duality, one of the tenants of quantum mechanics that states that all particles (such as protons and electrons) also possess wave properties. We will study matter waves in unit 222.

For now, we are going to focus in on mechanical waves, although many of the properties we discuss will translate over to the two other types of waves as well. As noted above, a mechanical wave is formed by some disturbance in a medium – the medium could be a stretched string (wave on a string), air (sound waves) or the earth (seismic waves). All mechanical waves *require* a medium (some background substance) for the wave to travel in.

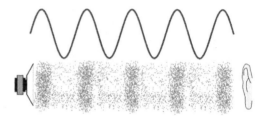

Figure 201-1 shows two important types of waves. The top part of a figure shows a *transverse wave*, where the motion of the particles is perpendicular to the direction of propagation of the wave. Such a wave can be created by holding a string taut and creating pulses by moving your hand up and down. The pulse you create will be vertical, but the pulse will travel horizontally. An important part of creating such a wave is that the particles in the string are connected to each other. When you move your hand up and down, the bonds between the part of the string you are holding and the next part of the string create a force that causes the next part of the string to go up and down, and then the next part and so on.

Figure 201-1. (top) transverse wave on a string, where the string oscillates perpendicular to the motion of the wave. (bottom) Longitudinal sound wave with areas of high density and low density along the length of the wave. Note that in both cases the wave is moving to the right in the figure.

It is also possible for a disturbance to be propagated along the same axis as the disturbance itself. This type of wave motion is called a *longitudinal wave* and is pictured in the bottom part of Figure 201-1. The figure shows a *sound wave* (which are longitudinal in fluids). As the wave passes, some areas of air are compressed and some areas are decompressed. The compressed areas of air have a higher pressure than the areas around them, which causes them to try and decompress. This compression and decompression acts as the restoring force for a longitudinal wave and causes the wave to propagate through the air in the same direction as the compression and decompression.

Another important concept when discussing waves is the *restoring force*. When the medium is disturbed producing a wave, there is some force that tries to bring it back to its equilibrium position. In the discussion of the transverse wave above, I described how the hand moving up and then down was connected to pieces later down the string because of the forces between molecules – this is the restoring force in this case. When the molecules move up, there is a force that wants to bring them back down to where they started – the equilibrium position. In a longitudinal wave a similar thing happens – as the material is compressed by the initial disturbance, the compressed area has a higher pressure which means it now pushes on the material farther down the line. In fact, longitudinal waves are also called pressure waves, because it is this pressure difference that causes the wave to propagate.

Neglecting surfaces, solids can support both transverse and longitudinal waves, but fluids can only support longitudinal (pressure) waves. This is because whereas a transverse dislocation in a solid has a restoring force because of the molecular interactions, a fluid does not (fluid molecules can slide by each other). Both phases support longitudinal waves because if an area of high pressure is created, the molecules will push back on their surroundings to equalize the pressure. This is important because in air and water, sound waves are only longitudinal. However, in solids, sound waves can have both transverse and longitudinal components, such as in earthquake (seismic) waves.

201-2: The Wave Equation

Consider: *Is there a mathematical way of knowing if a function represents a wave?*

How do we know if a disturbance is capable of creating waves supported by a medium? It turns out that there is a differential equation, known as the **wave equation**, which can tell us just that. Although the wave equation can be derived by looking at forces on a stretched string and then assuming that the result extends to all waves, I'd like to give a very intuitive derivation of the equation. Consider Figure 201-3 that represents both a downward and upward pulse similar to Figure 201-1. No matter what the actual shape of the curve, we could describe it by some function f. The horizontal line through the center of Figure 201-3 is meant to represent the **equilibrium position** of the particles, i.e., where the string would be if given no pulse. What we want to do is relate how curved a section of the Figure is to the acceleration of that same section. So, first consider near the equilibrium line – near equilibrium the string is pretty straight, that is to say that the concavity of the string is small. Since those same points are near the equilibrium position, the force on that section and therefore the acceleration is also near zero. Consider now a point near the top or bottom of the string.

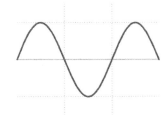

Figure 201-3. A sinusoidal "pulse" with positive and negative parts.

These areas are very curved (have a large concavity) and are far away from the equilibrium position and therefore have a large force and a large acceleration. Generalizing these two situations, it is not hard to assume that the concavity of a section of the string is proportional to the acceleration of that section as well.

Now for the math: As we discussed in Physics I the acceleration of a particle is given by the second derivative of the particle's position with respect to time. Also, if you remember from math, the concavity of a shape is given by the second derivative with respect to position. So, by saying that the concavity is proportional to the acceleration, we are mathematically saying

$$\frac{\partial^2 f}{\partial x^2} \propto \frac{\partial^2 f}{\partial t^2}, \tag{201-1}$$

Remember that although I'm talking about this as though the wave is on a string, one of our basic properties of waves is that there is a restoring force trying to bring the particles back to equilibrium, so our discussion holds for all waves.

What about the constant of proportionality? What is it? Let's think about this in terms of dimensional analysis. Since we are talking about particles moving, the function f should have units of distance. Therefore, the second derivative with respect to position would have dimensions of m/m^2, and the second derivative with respect to time would have units of m/s^2. In order to convert m/s^2 to m/m^2, we have to multiply the right side by something with units of $s^2/m^2 = (s/m)^2$. Although these units may seem weird, $(s/m)^2$ is the unit for inverse speed squared. In fact, the constant of proportionality *is* the speed of the wave, more generally known as the **phase speed** of the wave. If we call this speed v and plug it into the equation above we have the general (one-dimensional) wave equation.

The Wave Equation

$$0 = \frac{\partial^2 f}{\partial x^2} - \frac{1}{v^2}\frac{\partial^2 f}{\partial t^2} \tag{201-2}$$

Description: Any function, f, which is a solution to the wave equation represents a function of x and t that varies regularly in space and time and represents a wave that moves through space with a phase speed v.

The wave equation is a *second order linear partial differential equation*. Although intimidating, it is not hard to determine whether a function fits the equation as shown in the example below. Remember, if a given function is a solution to the wave equation it represents a wave and if not, it doesn't. Such equations are the backbone of advanced physics, although we only touch the surface here. If you would like to see a full mathematical derivation of the wave equation, you can find it on the internet at many places including http://hyperphysics.phy-astr.gsu.edu/hbase/waves/waveq.html.

Example 201-1: Weird wave

For what value(s) of c is the function

$$f(x,t) = e^{x-ct}$$

a solution to the wave equation?

Solution:

This is a direct application of the wave equation (Eq. 201-2). In order to solve for c, we must plug the function $f(x,t)$ into the wave equation. I will first compute each term and then combine.

Remember that when taking a partial derivative, you treat every other variable except the variable of interest (the variable in the denominator of the derivative) as a constant.

Also, we must remember to use the chain rule. The derivative of the exponential function is just the function itself, but we must then multiply by the derivative of the argument of the exponential.

Here, I start with the derivative with respect to x:

$$\frac{\partial f}{\partial x} = e^{x-ct}\frac{\partial}{\partial x}(x-ct) = e^{x-ct}.$$

Therefore,

$$\frac{\partial^2 f}{\partial x^2} = e^{x-ct}.$$

As for the derivative with respect to t:

$$\frac{\partial f}{\partial t} = e^{x-ct}\frac{\partial}{\partial t}(x-ct) = -ce^{x-ct}$$

Therefore,

$$\frac{\partial^2 f}{\partial t^2} = c^2 e^{x-ct}.$$

We can now relate these via the wave equation,

$$0 = \frac{\partial^2 f}{\partial x^2} - \frac{1}{v^2}\frac{\partial^2 f}{\partial t^2},$$

leading to

$$0 = e^{x-ct} - \frac{1}{v^2}c^2 e^{x-ct}.$$

Note that each term has the same exponential term, so this can be canceled:

$$0 = 1 - \frac{1}{v^2}c^2.$$

Finally, this equation can be solved for c:

$$c = v.$$

So, we can see that our $f(x,t)$ is a solution to the wave equation as long as the constant is the speed of the wave.

201-3: Sinusoidal Waves

Consider: *How do we describe basic types of waves?*

One of the most important functions that satisfies the wave equation is

$$f(x,t) = A\,sin(kx - \omega t). \tag{201-3}$$

Since the sine function repeats itself indefinitely, this equation represents a *periodic wave*. These sinusoidal waves are very fundamental, and it may appear at first glance that their use is very limited. However, it turns out that waves of any shape can be created by the superposition of sinusoidal waves through a process known as Fourier synthesis. We will study Fourier's theorem in section 201-4 below; however, for now it is just important to realize that the basic sinusoidal wave is a component of all waves. Therefore, understanding how sinusoidal waves work gives us a strong foothold for how real waves behave.

Since the sinusoidal wave is a function of both position and time, we can visualize it in both ways as can be seen in Figure 201-4. The top part of Figure 201-4 represents a picture, or snapshot, of a sinusoidal wave at a particular time. The maximum displacement of the wave from equilibrium is called

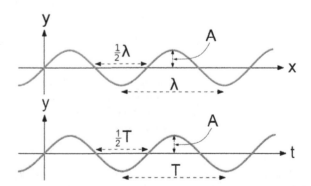

Figure 201-4. Sinusoidal waves as a function of distance (top) and of time (bottom). The definitions of wavelength, period and amplitude are shown in the graphs.

4

the *amplitude*, A, and the distance between peaks or troughs (or any similar point on the wave for that matter) is called the *wavelength* (symbol λ).

The bottom part of Figure 201-4 represents what happens if we look at one position in space and watch as the wave passes by that point over time. An example would be sitting on the edge of a pier and watching the waves pass by. In this figure the amplitude, A, still represents the maximum displacement from equilibrium and the distance between peaks (which is really a separation in time!) is called the *period* (symbol T) of the wave.

Remember that the *periodicity* of the sine function is 2π (Swab math!), which means that if the argument of the sine function (what is inside the parentheses of equation 201-3 above) increases by 2π we return to the same value for the entire function. How do we relate this to the two graphs in Figure 201-4? First, consider 201-4a, which is related to the kx term in our sine function. We know that the term must be related to our periodicity of 2π, and that it returns to the same position on the wave after moving through one wavelength, which suggests

$$kx = 2\pi \frac{x}{\lambda},$$

(201-4)

or

$$k = \frac{2\pi}{\lambda}.$$

(201-5)

k is called the *angular wavenumber* and has standard units of radians/meter. Remember, what this is really telling us is that every time the wave moves through one wavelength (or part thereof), the phase of the sine function increases by 2π (or some part thereof). The angular wavenumber gives us a way to relate changes in position to this change in angle.

A very similar situation arises for the ωt term in our sine function, which is related to Figure 201-4b. Everytime we pass through one period of the wave, the sine function returns to the same value, so

$$\omega t = 2\pi \frac{t}{T},$$

(201-6)

or

$$\omega = \frac{2\pi}{T}.$$

(201-7)

ω is known as the *angular frequency* and has standard units of radians/second. Similar to what we discussed for angular wave number, the angular frequency gives us a way to relate changes in time to changes in angle for the sine function. Remember that the period, T, is the time it takes to move through one cycle. Another very important quantity for waves is the reciprocal of the period, which would represent the cycles per time, known as the *frequency* of the wave,

$$f = \frac{1}{T} = \frac{\omega}{2\pi}.$$

(201-8)

Now that we have the basic definitions of angular wave number, angular frequency, frequency and period down, we'll take a look at another important relationship between these concepts. Consider a point on the wave in Figure 201-4b that is crossing the equilibrium position, which is to say that the value of the wave function at that point is $f(x,t) = 0$. Let's consider what happens if we let time move on. For the value of a sine function to be zero, the argument of the function can be zero (or some multiple of 2π, but we'll consider the simplest solution, zero). Therefore, for

$$f(x,t) = A\,sin(kx - \omega t) = 0,$$

(201-9)

we must have

$$kx - \omega t = 0.$$

(201-10)

To find the speed that the wave moves with, called the *phase speed* since we are looking at the phase of the function, we use the definition of speed, $v = dx/dt$. So, solving for x, we find

$$x = \frac{\omega t}{k},$$

(201-11)

so that

$$v_p = \frac{dx}{dt} = \frac{d}{dt}\left(\frac{\omega t}{k}\right),$$

(201-12)

or

$$v_p = \frac{\omega}{k}.$$

(201-13)

There is a very important point to be made. Note that we started with a minus sign in our phase $(kx - \omega t)$, and we wound up with a positive velocity. If you were to follow through the same derivation for phase speed starting with $(kx + \omega t)$, you would find a negative phase velocity. ***The direction of propagation of a sinusoidal wave is along the axis given in the phase of the sine function and the sign of the velocity is <u>opposite</u> of the sign between the kx and ωt terms***.

You may be starting to notice that one of the problems in learning about waves is that there are quite a few new concepts and each of these concepts is related by a number of small equations - each of which are inter-related with each other. For example, the phase speed equation (equation 201-3) we just found can be written in many ways, employing our other new relationships:

$$v_p = \frac{\omega}{k} = \lambda f = \frac{\lambda}{T}. \tag{201-14}$$

It is hard to sum this all up in a succinct summary box, so below I have the both a summary box for the sinusoidal wavefunction as well as a box relating many of the important concepts and equations we just discussed.

Sinusoidal Waves

$$f(x, t) = A\sin(\omega t - kx) \tag{201-3}$$

Description – This equation gives the displacement from equilibrium of a point on a sinusoidal wave with amplitude A, angular frequency, ω, and angular wave number, k at position x and time t.

Note 1: $\omega = 2\pi f$, $k = \frac{2\pi}{\lambda}$

Note 2: For a transverse wave, this displacement will be measured perpendicularly from equilibrium and for a longitudinal wave the displacement is measured parallel to the wave from equilibrium.

Some Important Relationships
Sinusoidal Waves

Angular Frequency, $\omega = 2\pi f$

Angular Wavenumber, $k = \frac{2\pi}{\lambda}$

Phase Speed, $v_p = \frac{\lambda}{T}$
  (moves one wavelength in one period)

$f = \frac{1}{T}$, frequency is cycles per second, period is time per cycle.

Using these relationships, the phase speed can be determined in a number of ways:

$$v_p = \frac{\omega}{k} = \frac{\lambda}{T} = \lambda f$$

Example 201-2: Wave on a rope

A transverse sinusoidal wave is created on a stretched piece of rope. Imagine that you measure the amplitude of the wave to be 13 cm, the speed of the wave as 2.3 m/s, and that frequency of oscillation creating the wave as 3.4 Hz. What is the transverse position of the wave 1.2 meters down the rope after 3.4 seconds has passed.

Solution:

First, since we know that we have a sinusoidal wave traveling on a string, it should fit the equation

$$f(x, t) = A\sin(\omega t - kx),$$

where $\omega = 2\pi f$ and $k = \frac{2\pi}{\lambda}$. We know the amplitude, A, is 13 cm. Next, let's find the angular frequency

$$\omega = 2\pi f = 2\pi(3.4\,Hz) = 21.3\,rad/s.$$

In order to find the angular wave number, we can use one of our relationships for the phase speed:

$$v_p = \frac{\omega}{k} \quad \rightarrow \quad k = \frac{\omega}{v_p} = \frac{21.3\,rad/s}{2.3\,m/s} = 9.26\,rad/m.$$

We now have all the information needed to find the displacement:

$$f(x, t) = A\sin(\omega t - kx)$$

$$f(x, t) = 13cm \sin\left(21.3\frac{rad}{s}(3.4s) - 9.26\frac{rad}{m}(1.2m)\right)$$

$$f(x, t) = -12.99\,cm \approx -13\,cm.$$

It turns out that this displacement is very close to the amplitude.

Example 201-3: Wave manipulations

The equation of a wave on a stretched string is given by

$$f(x,t) = 2.0 \ m \sin\left(62\frac{rad}{s}t + 26\frac{rad}{m}z\right).$$

(a) What is the direction of propagation of the wave?
(b) What are the amplitude, frequency and wavelength of this wave?
(c) What is the phase speed of this wave?
(d) Find a time when a point on the wave -1.2 meters down the string will be zero?

Solution:

This problem requires us to use many of the small equations for sinusoidal wave propagation.

(a) The direction of the wave is along the $-z$ **axis**. This is always given along the axis in sine or cosine function and is directed opposite to the sign in the trig function.

(b) The amplitude is 2.0 m. This is read directly from the equation of the wave.

The frequency is given by

$$\omega = 2\pi f \quad \rightarrow \quad f = \frac{\omega}{2\pi} = \frac{62 \ rad/s}{2\pi} = 9.9 \ Hz.$$

The wavelength is

$$k = \frac{2\pi}{\lambda} \quad \rightarrow \quad \lambda = \frac{2\pi}{k} = \frac{2\pi}{26 \ rad/m} = 0.24 \ m.$$

(c) The phase speed is

$$v_p = \frac{\omega}{k} = \frac{62 \ rad/s}{26 \ rad/m} = 2.4 \ m/s.$$

(d) In order to find the time that a specific point will be zero, start by writing down the wave equation will all known values

$$0 = 2.0 \ m \sin\left(62\frac{rad}{s}t + 26\frac{rad}{m}(-1.2m)\right).$$

This can be simplified to

$$0 = \sin\left(62\frac{rad}{s}t - 31.2 \ rad\right).$$

The sine function is zero when the argument is some is $0, \pi, \ 2\pi$, etc. Although any of these values can be chosen, I will choose 0 for simplicity. Therefore

$$0 = 62\frac{rad}{s}t - 31.2 \ rad,$$

Which can be solved to yield

$$t = 0.52 \ s \approx 0.5 \ s.$$

Note, the way the question was worded, any multiple of π could be used to find the time. Each multiple of π changes the time by about 0.05 s, so that point on the string is also at zero displacement at 0.55 s, 0.60 s, etc., which makes sense because this is a periodic wave passing the point!

You must be very careful solving problems with sinusoidal waves! Because there are so many small, inter-related equations, wave problems present different conceptual and analytical issues when compared to other areas of our course. I strongly suggest:

- you become familiar with, and maybe even memorize, the equations listed in the *Important Relationships* box above.
- you write down all your known quantities and unknown variables, using the correct notation, symbols and units. (Note: This is a good general tip, but especially important here.)
- you think about the relationship you need *in words*, and then find if you have readily available relationships that will give you the answers you need mathematically. (Note: again, this is a good general tip as well.)

201-4: Fourier's Theorem

Consider: *How can different wave shapes be created using just sinusoidal waves?*

Sinusoidal waves, as we explored in the last section, are very important in their own right since they represent a close approximation to many waves we experience in everyday life (small water waves for example). However, it also turns out that periodic waves of *any* shape can be formed using a **Fourier Series** of sines and cosines. One of the important reasons that this works is that the wave equation is a *linear equation*, meaning that if two functions are a solution to the wave equation, then the sum of those two solutions is also a solution to the wave equation. Forming combinations of waves by linearly adding components is generally known as the **superposition principle**.

Fourier Theorem

$$f(x) = \sum_{n=0}^{\infty} A_n sin(k_n x) + B_n cos(k_n x) \qquad (201\text{-}15)$$

Description – Fourier's Theorem describes how a wave of any shape can be made as the combination of sine and cosine waves of different frequencies.

Note 1: A_n is amplitude of the nth sine term and B_n is the amplitude of the nth cosine term. k_n is the angular wave number of the nth terms.

Note 2: An ω_n term is assumed in each sine and cosine term above but not written. This term would be consistent with our definitions in this chapter, including wave speed.

As an example, consider what is known as a **square wave**. The square wave can be formed by adding up odd harmonics, with each higher harmonic diminishing in amplitude as

$$f(x) = A\frac{\pi}{2}\left(sin(kx) + \frac{sin(3kx)}{3} + \frac{sin(5kx)}{5} + \frac{sin(7kx)}{7} + \cdots\right). \qquad (201\text{-}16)$$

The entire effect can be seen in Figure 201-5. The left side of this Figure gives a pictorial representation of each individual harmonic (n = 1, 3 and 5). On the right, the Figure shows the superposition of each harmonic up to that point. For example, the combined wave on the right side second down is the combination of the lowest two frequencies added together. Even within the three lowest frequencies shown, you can see how the combined wave is approaching a series of squares – thus why it is called a square wave. The final row in the Figure shows how the amplitude of the very high frequencies goes to zero and then the shape of the wave when all frequencies are combined – the full square wave. We will see in the next couple of units why changing the shape of a periodic wave is important.

For completeness, I want to mention a couple other important points about the square wave. If you compare the equation given for the square wave to the general form for Fourier's Theorem, you will notice that all of the cosine terms are zero. This corresponds to all $B_n = 0$. In addition, note that it only contains the odd values of n, meaning that it does not contain all multiples of the lowest frequency, only the odd multiples of that frequency.

Fourier's theorem is an amazingly powerful technique that is used in music to break down sounds into their components and filter them, in optics to break light into its different colors, as well as many, many other fields.

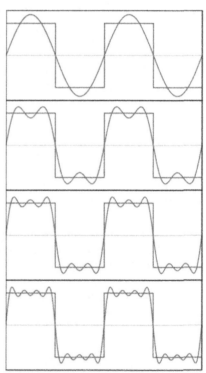

Figure 201-5. Depiction of how a square wave is formed through the superposition of harmonics (Fourier's theorem).

Example 201-4: More shapes

Another 'shape' wave is given by the function

$$f(x) = A\frac{8}{\pi^2}\left(\sin(kx) - \frac{\sin(3kx)}{3^2} + \frac{\sin(5kx)}{5^2} - \frac{\sin(7kx)}{7^2} + \cdots\right).$$

Using analogies to the square wave, determine the shape of the combined wave using Fourier analysis. Note the similarities between this equation and the square wave – we have only changed a couple of signs.

Solution:

The major difference between the function given and the square wave above is that the amplitudes of each term drop of with the square of the denominator and that every other terms is negative.

Making an analogy to the square wave above, we can create a figure with just the first term, the first two terms, the first three terms, etc. to get:

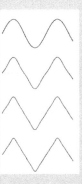

As you can see, even with just four terms (the bottom part of the figure), this looks like a set of triangles, and is, in fact, called a **triangle wave.**

202 – Combining Waves

At the end of unit 201, we discussed how Fourier's theorem allows us to create waves of any shape through the superposition principle. In unit 202, we explore the superposition principle in more detail, and learn how waves can combine to give acoustic beats and standing waves. In the end, we will also find out why a flute sounds different than a trumpet and why you sound different than your brother/sister/roommate etc.

Integration of Ideas

Understanding of sinusoidal waves from unit 201.

The Bare Essentials

- The superposition principle for waves simply states that when waves combine, the result is given by the sum of the individual waves.

- When waves of two frequencies propagate together, they produce beats.

Beat Frequency

$$f_B = |f_1 - f_2|$$

Description – This equation defines the beat frequency of two co-propagating waves with different frequencies as the magnitude of the difference of those frequencies.
Note: The pitch heard is the average of the two frequencies $|f_1 + f_2|/2$.

- Standing waves are the superposition of a wave with its own reflection. Said another way, they are the superposition of waves with the same frequency traveling in opposite directions.

Standing Waves

$$f_{sw} = 2A \sin(kx) \cos(\omega t)$$

Description – This equation describes a standing wave in terms of the amplitude, angular frequency and angular wavenumber of the wave that created it with its own reflection.

- The frequencies supported by standing waves between two endpoints are *quantized*, that is, only certain frequencies are supported and these frequencies are called the *harmonic series* of the system.

Frequencies Supported Between Like Ends

$$f_n = \frac{v}{2L} n \qquad n = 1, 2, 3, \ldots$$

Description – This equation defines the frequencies supported in a standing wave between like ends (two free or two fixed ends) in terms of the phase speed and the length between the ends.
Note: n = 1 is called the fundamental mode *and* first harmonic. n = 2 is known as the second harmonic and so on.

Frequencies Supported Between Unlike Ends

$$f_n = \frac{v}{4L} n \qquad n = 1, 3, 5, \ldots$$

Description – This equation defines the frequencies supported in a standing wave between unlike ends (one free and one fixed) in terms of the phase speed and the length between the ends.
Note: n = 1 is called the fundamental mode *and* first harmonic. n = 3 is known as the third harmonic and so on.

- Various sounds (human voices, musical instruments, industrial sounds) are distinct because their *harmonic series* differ from each other (Fourier's Theorem).

202-1: The Superposition Principle

Consider: *What happens to waves when they run into each other?*

A T THE END OF UNIT 201, we saw that waves with multiple frequencies can add together to give waves of any shape we desire. The process of adding waves, known as the superposition principle, allows us to find many other interesting wave properties as well. However, what do we really mean by adding waves and the superposition principle?

The superposition principle states that a net wavefunction is the simple algebraic sum of the individual wave functions:

$$f(x,t) = f_1(x,t) + f_2(x,t) + \cdots. \tag{202-1}$$

where $f_1(x,t)$ and $f_2(x,t)$ represent individual waveforms and $f(x,t)$ is the combined, or net waveform. Technically, this is only true for *linear* waves; however, most waves that we encounter in everyday life are very nearly linear, so this is a good approximation.

In order to find some of the important properties of linear waves and superposition, consider figure 202-1. In this figure, you can see two pulses, one initially on the left and one initially on the right moving to the right. For the sake of this example, let's say that the two pulses are initially separated by 12 cm (so that the middle is 6 cm from each pulse) and that each pulse is moving at 2 cm/s. Another way of saying this is that pulse 1 (on the left) is initially 6 cm to the left of center and pulse 2 (on the right) is initially 6 cm to the right of center.

The second row in figure 202-1 shows what the pulses look like one second later, the third row shows the pulses two seconds later, etc. At three seconds, the pulses collide and their amplitudes add together. The real question is what happens after this collision? Well, remember, that if linear superposition is to hold the combined waveform is simply the addition to the two individual waveforms. Pulse 1 still represents a pulse moving to the right at 2 cm/s and pulse 2 still represents a pulse moving to the left at 2 cm/s. The results can be seen in the last few lines of figure 202-1 – the pulses continue on their merry way as if they never encountered the other! This is a very important property of linear waves – **linear waves pass through each other undisturbed!**

We will now use the superposition principle to investigate what happens when waves of different frequency travel together and what happens when a periodic wave interacts with its own reflected self. Much of the rest of this chapter is related to the field of **musical acoustics**, and although the details of sound waves are given in the next unit, for now it is only important to know that sound waves are mechanical pressure waves produced in air.

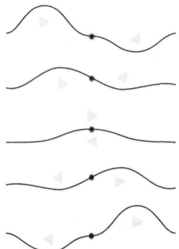

Figure 202-1. Two wave pulses passing through each other undisturbed.

202-2: Beats

Consider: *Why do two sounds, similar in pitch, seem to warble when I listen to them?*

Consider two sinusoidal waves of the same amplitude traveling together at the same speed, but with different frequencies such that

$$f_1(x,t) = A\cos(k_1 x - \omega_1 t), \tag{202-2}$$

$$f_2(x,t) = A\cos(k_2 x - \omega_2 t). \tag{202-3}$$

As a first approximation, the speed of a wave in a given material is given by the properties of the material itself and not the frequencies of the waves. This is not true in **dispersive** materials; however, as long as the frequencies are relatively similar in the two waves, the speed of the waves is usually very similar. So, if we assume that they are moving at the same speed, we know that

$$v = \frac{\omega_1}{k_1} = \frac{\omega_2}{k_2}. \tag{202-4}$$

Since we know there is a distinct relationship between the angular wave number and angular frequency terms, I am going to simplify the notion for the following couple of calculations and write our two waveforms as

$$f_1(x,t) = A\cos(\omega_1 t), \tag{202-5}$$

$$f_1(x,t) = A\cos(\omega_2 t). \tag{202-6}$$

Since we know the waves are traveling at the same speed, no information is lost.

The superposition principal tells us that the net waveform is the sum of the two waves:

$$f(x,t) = f_1(x,t) + f_2(x,t), \tag{202-7}$$

so that

$$f(x,t) = A\cos(\omega_1 t) + A\cos(\omega_2 t), \tag{202-8}$$

$$f(x,t) = A[\cos(\omega_1 t) + \cos(\omega_2 t)]. \tag{202-9}$$

In order to add the two cosine functions, we can employ an angle-addition formula for cosines (Swab math!)

$$\cos\alpha + \cos\beta = 2\cos\left[\frac{1}{2}(\alpha - \beta)\right]\cos\left[\frac{1}{2}(\alpha + \beta)\right], \tag{202-10}$$

leading to

$$f(x,t) = 2A\cos\left[\frac{1}{2}(\omega_1 - \omega_2)t\right]\cos\left[\frac{1}{2}(\omega_1 + \omega_2)t\right]. \tag{202-11}$$

First, keep in mind that all we have done is take the superposition principle included our two wave functions and applied one trigonometric identity. Now, we can analyze this equation – let me rewrite it slightly by using the fact that $\omega = 2\pi f$ regrouping slightly:

$$f(x,t) = \left(2A\cos\left[\frac{2\pi}{2}(f_1 - f_2)t\right]\right)\cos\left[\frac{2\pi}{2}(f_1 + f_2)t\right] \tag{202-12}$$

This equation can be further simplified to

$$f(x,t) = (2A\cos[\pi f_B t])\cos[2\pi f_R t], \tag{202-13}$$

where $f_B = f_1 - f_2$ is called the **beat frequency** and $f_R = (f_1 + f_2)/2$ is the **apparent frequency**. Figure 202-2 shows a graph of this function in time. The term in parentheses in Equation 202-13 represents a time-dependent amplitude (the long-term oscillation in the figure) that oscillates at the beat frequency.

If the interacting waves are sound waves, what we hear is a sound whose loudness increases and decreases at this frequency, which gives rise to a pulsating, or beating, sound. For example, when two instruments are slightly out of tune, you can hear this repetitious *beating*. In fact, many string players (violin and guitar players for example) use these *beats* to tune the strings of their instruments precisely.

It is also important to note that the internal oscillation (as seen in Figure 202-2) oscillates at the average frequency of the two original waves (f_R). Again, if this were a musical sound, this means that the pitch we hear (*apparent pitch*) is the average of the pitches of the two original sound waves.

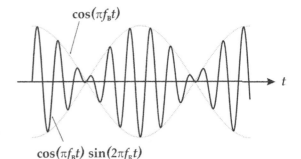

$\cos(\pi f_B t)$

$\cos(\pi f_B t)\sin(2\pi f_R t)$

Figure 202-2. Depiction of how copropagating waves of two frequencies form beats.

Beat Frequency

$$\boldsymbol{f_B = |f_1 - f_2|} \qquad (202\text{-}14)$$

Description – This equation defines the beat frequency of two co-propagating waves with different frequencies as the magnitude of the difference of those frequencies.

Note: The pitch heard (apparent pitch) is the average of the two frequencies $f_R = |f_1 + f_2|/2$.

13

Example 202-1: Beat Frequency

What is the beat frequency and apparent pitch when both middle C (261.6 Hz) and the D-key next to it (293.7 Hz) are played simultaneously on a piano?

Solution:

This is a direct application of the equations for beat frequency and apparent pitch.

For beat frequency:

$$f_b = |f_1 - f_2| = |261.6\ Hz - 293.7\ Hz| = 32.1\ Hz.$$

Note that our ears cannot hear beating at 32.1 *Hz* as individual beats – we perceive this as ***dissonance*** (unpleasant sound).

The apparent pitch is given by the average of the two frequencies being played together:

$$f_R = \frac{|f_1 + f_2|}{2} = \frac{|261.6\ Hz + 293.7\ Hz|}{2} = 277.7\ Hz.$$

As expected, this frequency lies directly between C and D on the piano.

202-3: Reflection and Transmission of Waves at a Boundary

Consider: *What happens when a wave reaches a boundary?*

When a wave reaches a boundary between two media, the wave will both continue into the new medium and reflect back. The effect happens whether we are dealing with waves on a string, sound waves, light waves, or any other type of wave for that matter. ***In almost all cases, both transmission into the new medium and reflection from the boundary occur***. However, there are two ideal situations in which only reflection occurs. Thinking about a wave on a string, if the string reaches a completely fixed point, the wave will reflect and be inverted, as can be seen in figure 202-3. In this case, as the wave reaches the boundary point, if the wave pulls up on the fixed point, the fixed point reacts by pulling down on the string (Newton's third law). Since the reflected wave is started by a downward force, the wave becomes ***inverted***. The other special circumstance is when the wave reaches a completely free end. In this case, the oscillatory motion of the endpoint of the string continues unencumbered, and the reflected wave is ***upright***. This effect can be seen in figure 202-4.

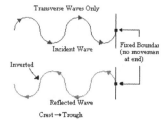

Figure 202-3. Reflection from a fixed point.

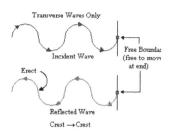

Figure 202-4. Reflection from a free point.

Finally, if the wave encounters a boundary where both transmission and reflection occur, the transmitted pulse will always be upright; however, the reflected pulse can be either upright (if the new medium is less dense than the original medium), or inverted (if the new medium is more dense than the original medium). Think of it this way, if the pulse reaches a new medium that is denser than the old medium, the reflected wave acts as if it came from a fixed point; if the new medium is less dense than the original medium, the new wave acts as if it reflected from a free point. This effect can be seen in figure 202-5, where the medium on the right has a higher density than on the left.

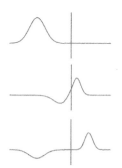

Figure 202-5. Wave reflection and transmission at a boundary.

202-4: Standing Waves

Consider: *What happens when a wave interferes with its own reflection?*

Now, let us return to the idea of the superposition principle and inspect what happens when a sinusoidal wave reflects off of a boundary and interferes with itself. As discussed, a wave that is reflected off of a perfect boundary would look like the same type of wave just moving in the other direction. If you remember, from unit 201, the direction of wave propagation for a sinusoidal wave is given by the sign between the kx and ωt terms, so our two functions are

$$f_1(x,t) = A\sin(kx - \omega t), \tag{202-15}$$

$$f_2(x,t) = A\sin(kx + \omega t). \tag{202-16}$$

Applying the superposition principle, we find

$$f(x,t) = A\sin(kx - \omega t) + A\sin(kx + \omega t). \qquad (202\text{-}17)$$

We now have to find a trig identity that will allow us to add these two terms together:

$$\sin(\alpha \pm \beta) = \sin\alpha\cos\beta \\ \pm \cos\alpha\sin\beta. \qquad (202\text{-}18)$$

If we now use this trig function on our above wave function, we find

$$f(x,t) = A[\sin(kx)\cos(\omega t) - \cos(kx)\sin(\omega t)] \\ + A[\sin(kx)\cos(\omega t) + \cos(kx)\sin(\omega t)]. \qquad (202\text{-}19)$$

Although this wavefunction looks long and complicated, notice that the second term of each of the brackets is the same but opposite sign, so they cancel. Also, the first term in each of the brackets is the same with the same sign, so they add. This leaves us with just:

$$f(x,t) = 2A\sin(kx)\cos(\omega t). \qquad (202\text{-}20)$$

Note that this is a fundamentally different equation than the wave equations we have been dealing with. Our traveling waves all have terms with a $\sin(kx - \omega t)$, whereas equation 202-20 has separate sine and cosine terms. Again, let me rewrite this slightly to bring out the important point:

$$f(x,t) = [2A\sin(kx)]\cos(\omega t). \qquad (202\text{-}21)$$

The first, bracketed term represents an amplitude that varies in space, and the second term tells us there is an oscillation that is not *traveling* in space. This equation represents what is known as a ***standing wave***, a wave that appears to stand still in space, but yet vibrates with an angular frequency given by ω.

Standing Waves

$$\boldsymbol{f_{sw} = 2A\sin(kx)\cos(\omega t)} \qquad (202\text{-}21)$$

Description – This equation describes a standing wave in terms of the amplitude, angular frequency and angular wavenumber of the wave that created it with its own reflection.

Remember, standing waves are caused by two waves of the *same frequency and amplitude* traveling in *opposite directions*. In the next section, we will see that in real situations it is only certain frequencies that lead to well defined standing waves. Figure 202-6 depicts a standing wave between two fixed points (walls), and defines the wavelength, λ of that wave. Please note that this picture also defines an ***antinode*** as a position of maximum oscillation along the wave and a ***node*** as a position of zero oscillation. Figure 202-6 contains four nodes and three antinodes (excluding the fixed positions on the walls).

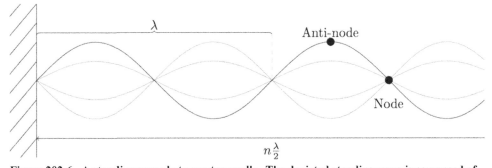

Figure 202-6. A standing wave between two walls. The depicted standing wave is composed of two full wavelengths. The picture also helps define and visualize a node and antinode for the wave.

Example 202-1: Forming a standing wave

A sinusoidal wave on a string interferes with its own reflection. If the frequency of the wave is 215 Hz and it travels at 22 m/s, what is the equation for the standing wave created? The amplitude of the incoming wave is 2.2 cm.

Solution:

At first glance, this appears to be a direct application of the standing wave equation with

$$\omega = 2\pi f = 2\pi(215\ Hz) = 1351\ rad/s,$$

and

$$A = 2.2\ cm.$$

However, we do not know the angular wave number of the wave. Looking over the information given, we have yet to use the speed of the wave, 22 m/s. Remembering back to unit 201, there is a relationship between the phase speed of a wave, its angular frequency and angular wave number:

$$v_p = \frac{\omega}{k} \quad \rightarrow \quad k = \frac{\omega}{v_p}.$$

Therefore, the angular wave number is $k = 61.4\ rad/m$. We can now substitute into the standing wave equation to find the solution:

$$f_{sw} = 4.4\ cm\ \sin(61.4\ rad/m \cdot x)\cos(1351\ rad/s \cdot t)$$

202-5: Resonance

Consider: *How does a violin or flute produce its pitch?*

Consider what happens if a standing wave is confined to a specific region. A guitar string is the classic example – what happens if you pluck the string in exactly the middle of the string? Well, as can be seen in figure 202-7, a standing wave is set up such that the length of the guitar string represents one half-wavelength of the standing wave. Is this the only possible standing wave that could be created on this string? Think about the boundary conditions for this situation – all we really require is that the standing wave have zero amplitude at the ends, since the string is tied down there. In fact, any number of half-wavelengths can fit this description, as shown in figure 202-8. Let's say that the distance between the two ends of the string is L. In the top part of figure 202-8, there is ½ wavelength between the posts, in the second there is one (2/2) wavelengths, in the third there is 3/2 wavelength, etc. If you consider the pattern, we can fit

Figure 202-7. Fundamental standing wave on a string.

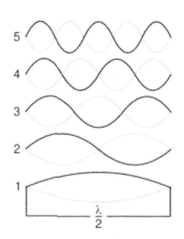

Figure 202-8. First five harmonics of a wave on a string fixed at both ends.

$$L = \frac{n\lambda}{2} \qquad (202\text{-}22)$$

wavelengths into the space, where n is any positive integer. Solving for the possible wavelengths, we find

$$\lambda_n = \frac{2L}{n}. \qquad (202\text{-}23)$$

As mentioned briefly before, the pitch we hear is the frequency of the wave. We know from our earlier discussions in unit 201, that $v_p = \lambda f$, so we can solve this for frequency and plug in the possible wavelength we just found:

$$f_n = \frac{n v_p}{2L}. \qquad (202\text{-}24)$$

This represents all of the possible frequencies that can be supported as standing waves on the string, and therefore the frequencies that the string can produce as sound waves when coupled to air. Note that we have only considered the situation where both ends of the string must have zero amplitude (two fixed ends). The same is also true of any boundaries where both sides of the string would be free to move (two free ends).

Frequencies Supported Between Like Ends

$$f_n = \frac{v_p}{2L} n \qquad n = 1, 2, 3, \ldots \qquad (202\text{-}24)$$

Description – The equation of a standing wave between like ends (two free or two fixed ends) in terms of the phase speed and the length between the ends.
Note: $n = 1$ is called the fundamental mode **and** first harmonic. $n = 2$ is known as the second harmonic, etc.

It is also possible to have a tube that is only open at one end. In this case, the string will be allowed to vibrate at one end, but must be a node at the other end, as can be seen in 202-9. In this figure, the fundamental, first harmonic and second harmonic of our one-end-open tube can be seen. As before, let's think about this in terms of wavelengths. The fundamental in this case is ¼ of a wavelength, the second harmonic is ¾ of a wavelength, which means that we have added ½ a wavelength. Similarly, to go from the second to third harmonic, we must add a half a wavelength. In order to write this mathematically, we can say

$$L = \frac{n\lambda}{4}. \qquad (202\text{-}25)$$

However, n can only be odd integers in this case since we need it to be 1, 3 and 5 for the first couple of harmonics. We can now do the same transformation we did before – so, solving for λ:

$$\lambda_n = \frac{4L}{n}. \qquad (202\text{-}26)$$

Now we use the known relationship between wavelength and frequency ($v_p = \lambda f$) to get

$$f_n = \frac{nv_p}{4L}, \qquad (202\text{-}27)$$

again remembering that n must be an odd integer in this case.

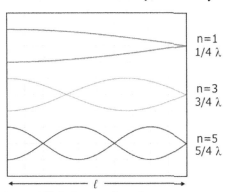

Figure 202-9. Lowest three harmonics of an air column open at one end and closed at the other end. The number of wavelengths present is also shown for each n.

Frequencies Supported Between Unlike Ends

$$f_n = \frac{v_p}{4L} n \qquad n = 1, 3, 5, \ldots \qquad (202\text{-}27)$$

Description – The equation of a standing wave between unlike ends (one free and one fixed) in terms of the phase speed and the length between the ends.
Note: $n = 1$ is called the fundamental mode **and** first harmonic. $n = 3$ is known as the third harmonic, etc.

Example 202-3: Six-string guitar

A guitar is an example of a string system fixed at both ends (like ends). The vibrating length of guitar strings from major manufacturers is 66 cm. If the wave speed on a guitar string is 259 m/s, what are the first three harmonic frequencies supported by this string?

Solution:

This is a direct application of standing waves between two fixed points, with n = 1, 2 and 3.

$$f_1 = \frac{259\ m/s}{2(0.66\ m)} 1 = 196\ Hz,$$

$$f_2 = \frac{259\ m/s}{2(0.66\ m)} 2 = 392\ Hz,$$

$$f_3 = \frac{259\ m/s}{2(0.66\ m)} 3 = 588\ Hz.$$

Note: 196 Hz is a G which characterizes the G-string.

Extension: Harmonics

If the G-string in the example above were freely oscillating at its fundamental mode and you were to place your finger directly in the middle of the guitar, you would extinguish this 1st harmonic since the middle of the string needs to be vibrating to produce that note. However, if you consider Figure 202-7, the 2nd harmonic does not require the middle of the string to be vibrating. So, when you place your finger in the middle, you stop the 1st, 3rd, 5th, etc., harmonics, but allow the 2nd, 4th, 6th, etc. harmonics to continue. In this case, you would hear the pitch jump one octave when you place your finger in the middle.

A similar effect happens if you place your finger 1/3 the way down the string – you extinguish the 1st and 2nd harmonics, but not the 3rd. In this case, the pitch jumps an octave and a fifth.

Example 202-4: Organ pipe length

Typical examples used for a pipe that is closed at one end and open at the other end are organs and muted brass instruments. For now, let us say you want to make a set of organ pipes to play a major C-chord consisting of C-4 (261.6 Hz), E4 (329.6 Hz) and G-4 (392.0 Hz). How long must our three pipes be to create this chord? You may take the speed of sound waves to be 340 m/s.

Solution:

Starting with our equation for frequencies supported between unlike ends, we can solve for the length:

$$f_n = \frac{v_p}{4L}n \quad \rightarrow \quad L = \frac{v_p}{4f_n}n.$$

We can now apply this equation to our three pipe frequencies. Although not used in this problem, you must be very careful with this equation because only odd values of n are used! However, for our lengths, you find:

$$L_C = \frac{340\ m/s}{4(261.6\ Hz)}1 = 0.32\ m,$$

$$L_E = \frac{340\frac{m}{s}}{4(329.6\ Hz)}1 = 0.26\ m,$$

$$L_G = \frac{340\ m/s}{4(392.0\ Hz)}1 = 0.22\ m.$$

202-6: Why Does a Flute Sound different than a Violin?

All musical instruments produce their sounds using standing waves as we've just described. When a note is played on an instrument, the fundamental frequency that is played is the pitch that you hear – A(440Hz) for example. So, why is it, then, that a violin playing the A at 440 Hz sounds different than a flute playing the same 440 Hz note? This once again comes down to Fourier's theorem (unit 201). Each instrument is producing the fundamental and many of the harmonics of that note. However, the geometry of the instruments gives different harmonics different amplitudes, as can be seen in figure 202-10. In this figure, the relative amplitude of the lowest eight harmonics are

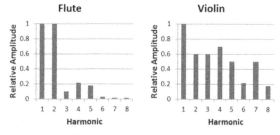

Figure 202-10. Approximate amplitudes of the harmonics of a flute and a violin playing the same frequency.

shown for both the flute and violin. As you can see, the flute has a very strong fundamental *and* second harmonic, whereas in the violin the second harmonic has a much smaller amplitude than the fundamental. Thinking back to the section on Fourier's Theorem from the last unit, these differences in the strength of the higher harmonics change the *shape* of the resulting wave, while maintaining the underlying frequency and period. So, while each are playing the same note – the underlying frequency – we hear very different timbre from each based on the shape of the waves (see figures 202-11 and 202-12). The same can be said of humans' voices – we each produce different amplitudes for the different harmonics and so the shape of the waves you produce are literally different than the shape of your neighbor's waves, leading to each and every person sounding different and unique.

Figure 202-11. Flute waveform based on the harmonics in Figure 202-10

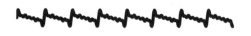

Figure 202-12. Violin waveform based on the harmonics in Figure 202-10

203 – Real Waves are Complicated

As mentioned in the last two units, waves are all around us. In this unit, we apply our knowledge of waves to two important areas – hearing and earthquakes.

Integration of Ideas

The two previous units on waves.

The Bare Essentials

- The speed and intensity of mechanical waves depends on material properties of elasticity and density.

- The speed of a transverse wave on a stretched string has to do with the tension and the linear mass density.

Speed of a Transverse Wave on a Stretched String

$$v = \sqrt{\frac{T}{\mu}}$$

Description – This equation describes the speed of a transverse wave on a stretched string in relation to the tension in the string and its linear mass density, μ.

Note: μ is the mass of the string divided by the length of the strings.

- Hearing is based on detection of acoustic waves

Loudness (dB)

$$L_{dB} = 10 \, log \left(\frac{I}{I_0}\right)$$

Description – This equation defines the loudness of an acoustic wave in decibels (dB)

Notes: The reference intensity, I_0, is generally taken as 10^{-12} W/m^2.

- The 'typical' range of human hearing is 20 Hz to 20,000 Hz, although the upper range decreases rapidly with age.

- Seismic waves are created by earthquakes. P-waves (also known as primary, pressure waves), are longitudinal waves and are generally the fastest of the seismic waves. S-waves (also known as secondary or shear waves).

- Surface waves include Rayleigh waves (ground roll) and Love waves (side to side motion).

- P-waves tend to propagate at 5 – 8 km/s. S-waves generally have a velocity around 60% that of p-waves. Surface waves have speeds around 90% that of s-waves.

Speed of p-waves

$$v_p = \sqrt{\frac{M}{\rho}}$$

Description – This equation defines the speed of a primary seismic wave in the earth's interior in terms of the elastic modulus, M, and density, ρ, of the material the wave is traveling in.

Notes: p-waves can travel through both solid and fluid areas of the earth.

- The intensity of earthquakes is measured using the Moment Magnitude Scale (successor to the Richter Scale).

Moment Magnitude Scale

$$M_w = \frac{2}{3} log(M_0) - 6.0$$

Description – This equation defines the size of an earthquake, M_w, in terms of the seismic moment M_0.

Notes: The seismic moment is given by $M_0 = GAD$, where G is the shear modulus of the material, A is the surface area of the rupture and D is the displacement of the rupture.

- The ratio of energy released between two earthquakes can be found by manipulating the equation for the moment magnitude scale.

Ratio of Earthquake Energy

$$f_E = 10^{1.5(M_2 - M_1)}$$

Description – This equation gives us the ratio of the energy released between two earthquakes with magnitudes of M_2 and M_1.

203-1: Mechanical Wave Speed

Consider: *What determines the speed at which waves move?*

THE SPEED OF A MECHANICAL WAVE depends on the material through which the wave travels (i.e., the ***medium*** in which it travels), and more specifically always has the form

$$v_{wave} = \sqrt{\frac{Some\ measure\ of\ elasticity}{Some\ measure\ of\ density}}. \tag{203-1}$$

The exact form of the equation depends on the type of wave and medium through which the wave is traveling. Take for example a wave traveling on a stretched string. For this case, the measure of elasticity is the tension in the string and the measure of density is the linear mass density ($\mu = M/L$), so that

$$v_{string} = \sqrt{\frac{T}{\mu}} = \sqrt{\frac{TL}{M}}. \tag{203-2}$$

To prove the point, consider this quick derivation for the speed of a wave on a string: Figure 203-1 shows a small section of a wave on a string under a tension T. Since the section at the top is curved, we can find a small enough piece of the string to treat it as if it is part of a circle, meaning that the net force on that piece should be the centripetal force and point straight downward. So, consider that there is tension on both sides of the pulse and that the tension is not horizontal, but down at an angle of θ on each side. If we consider the vertical component of this tension, for each it would be $T \sin \theta$, meaning that there is a combined downward force on this piece of $2T \sin \theta$. Also, if the piece of string is very small, the angle, θ, will be small and we can approximate $\sin \theta \approx \theta$. However, since we know this must be a centripetal force, we can write (assuming the tension is the only force acting on this piece of string)

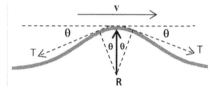

Figure 203-1. Forces on a wave pulse of a stretched string.

$$2T\theta = \frac{mv^2}{R} = \frac{2R\theta\mu v^2}{R}, \tag{203-3}$$

where I have used the fact that the mass of the piece of string can be written as its mass density (μ) times its length (arc length of $2R\theta$). Solving for v in this equation gives us

$$v = \sqrt{\frac{T}{\mu}}, \tag{203-4}$$

which is exactly what we hoped to find.

The specific example for the speed of a wave on a string is important, but the general relationship between wave speed, elasticity and density will come back a few times throughout this unit.

Speed of a Transverse Wave on a Stretched String

$$v = \sqrt{\frac{T}{\mu}} \tag{203-4}$$

Description – This equation describes the speed of a transverse wave on a stretched string in relation to the tension in the string and its linear mass density, μ.

Note: μ is the mass of the string divided by the length of the strings.

Example 203-1: Guitar string

A guitar string has a linear mass density of $2.4x10^{-4}kg/m$.
    (a) What is the speed of a transverse wave on this string when it is under 67 N of tension?
    (b) What is the fundamental frequency of this string if it is 1.07 m long?

Solution:

(a) This is a direct application of the wave on the string equation (203-4), since we know both the tension and the mass density:

$$v = \sqrt{\frac{T}{\mu}} = \sqrt{\frac{67\ N}{2.4x10^{-4}kg/m}} = 528\ m/s.$$

(b) This is a review question related to the fundamental frequency of a standing wave between two like (fixed) ends – equation 205-24:

$$f_n = \frac{v_p}{2L}n.$$

We just found the speed of the wave on this string and we were given the length in the problem. $n = 1$ for the fundamental frequency. Therefore,

$$f_1 = \frac{v_p}{2L}(1) = \frac{528\ m/s}{2(1.07\ m)}(1) = 247\ Hz.$$

This corresponds to a tuned B-string on a guitar.

203-2: Sound Waves

Consider: In what ways are sound waves different and in what ways are they the same as the waves we've already studied?

In unit 201, I mentioned that sound waves are longitudinal waves, meaning that the displacement of the particles in the wave is along the same axis that the wave propagates. First, it is important to mention that this is true for sound waves in fluids. Sound waves which propagate in solids can have both transverse and longitudinal components as we'll see in the upcoming section on seismic waves. For now, though, let's consider just the longitudinal part. Figure 203-2 shows a snapshot of a sound wave along with the representation of the pressure along the wave and the average displacement of the particles along the wave. Sound waves are a type of pressure wave because the oscillation of the particles causes areas of high pressure (called condensed areas) and areas of low

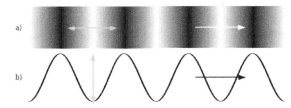

Figure 203-2. Properties of a longitudinal wave. (a) White areas show high pressure and dark areas show low pressure. (b) The displacement of particles from their equilibrium position as the wave passes.

pressure (called rarefaction areas). The key is that the areas of highest and lowest pressure are where the particles have moved the least - that is to say that particles have moved to the point of highest pressure (which the particles already there didn't have to move at all) and away from the points of lowest pressure (where the particles already there also didn't have to move at all). So, although both a plot of particle motion and pressure look like sine or cosine waves, and have the same frequency and wavelength, they are out of phase by a quarter of a wave. This is only important conceptually, because sometimes it makes sense to discuss the pressure and sometimes it makes sense to discuss the average particle position – they follow the same pattern, but are slightly different as to where they are on the wave.

Speed of Sound Waves

$$v_{liquid} = \sqrt{\frac{K}{\rho}} \qquad v_{gas} = \sqrt{\frac{\gamma P_0}{\rho}} \qquad (203\text{-}5)$$

Description – This equation describes the speed of longitudinal sound waves in a fluid in terms of the density of the fluid and either the bulk modulus, K, or the ambient pressure, P_0, of the fluid..

Note: γ is called the **adiabatic constant**. $\gamma = 1.4$ for diatomic molecules.

What is the speed of sound in air at standard temperature (293.15 K) and pressure (101.325 kPa)? Note: the density of air at STP is 1.2057 kg/m^3

Solution:

This is a direct application of the equation for the speed of sound waves

$$v = \sqrt{\frac{\gamma P_0}{\rho}}.$$

Substituting our known values, we find

$$v = \sqrt{\frac{(1.4)(101.325\ kPa)}{1.2057\ kg/m^3}} = 343\ m/s.$$

Note that this standard value for the speed of sound in air depends on the ambient pressure of air – which changes regularly with height, weather, etc.

203-3: Hearing

Consider: *How does human hearing work?*

Hearing is, of course, one of the five senses that a typical human possesses. Hearing is almost entirely based on our ability to detect and interpret sound waves as they approach the body. So, how do you do it? Figure 203-3 depicts the basic anatomy of the human ear. The *pinna* is the external part of each of your ears that acts to focus sound waves down the **ear canal** (acoustic meatus) towards the **tympanic membrane** (ear drum). The shape of the pinna acts to collect many sound waves and focus them down the ear canal. You can see how this works and even improve the amplification by cupping your hand over an ear and notice that sounds you are facing get amplified.

The sound wave then exerts pressure on the tympanic membrane. This is why we just discussed how sound waves are pressure waves – they can exert force on the membrane just as any other pressure would. The vibrations of the membrane that come from the sound waves are then transmitted to the three smallest bones in your body, known as the ossicles. The ossicles are designed to amplify the

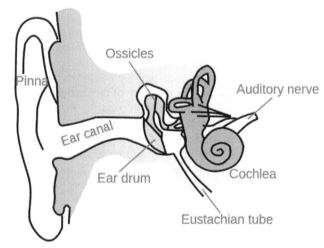

Figure 203-3. Anatomy of the human ear.

vibrations and tap on the opening to the inner ear called the oval windows. On the far side of the oval window is the snail-shaped *cochlea*, the organ that does the actual magic of hearing. I should also note that the inner ear also contains the *vestibular system*, which is extremely important for balance. Therefore, issues with your inner ear not only affect hearing, but can also affect your balance.

The Cochlea

Once vibrations from the ossicles reach the oval window, they are transferred into the fluid filled ducts of the cochlea. A basic diagram of the cochlea is shown in figure 203-4. Down the middle of the cochlea is a stiff line of cells called the basilar membrane. The stiffness of the basilar membrane changes along the length of the cochlea, and this change in addition to a small change in the inner diameter of the cochlear chamber causes the resonant frequency of the cochlea to change along its length. The inside of the cochlea is lined with about 15,000 cells called **hair cells** (because they resemble hair), which are connected to the basilar membrane. The hair cells respond to the incoming vibrations if they are at the position of the cochlea where the resonant frequency matches the

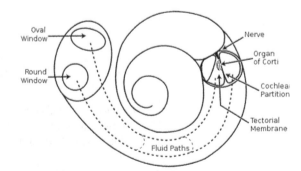

Figure 203-4. The human cochlea.

frequency of the vibrations. The resonant frequency of the cochlea starts at about 20,000 Hz and decreases to about 20 Hz at the center.

When hair cells vibrate, they create an electric signal that is then sent to the brain, telling the brain that you heard that particular frequency. The sounds that we hear everyday are generally made up of a wide range of frequencies and so it is actually quite a feat for the brain to decode all of the information coming to it from the ear. Putting all this together is *hearing*.

A few observations

- The range of human hearing (as a baby) is 20 – 20,000 Hz because that is the range to which the cochlea can respond.
- Damage to human hearing tends to start at the upper range, because sound waves have the most energy at the entrance to the cochlea where the highest pitches are detected.
- Damage to the hair cells at a particular location in the cochlea cause you to lose sensation to that particular frequency. Said another way – if you have ever left a loud concert with a high pitched ringing in your ear - the pitch you hear is the swan song of that hair cell as it dies – *you will never hear that particular frequency again.*
- The upper-range of hearing decreases rapidly for most people with age. Although babies can hear 20,000 Hz, the upper range typically decreases to ~18,000 Hz by the time you are 20, and ~15,000 Hz by the time you are 30. It's all downhill from there, and, on average, at 70 you can only hear up to about 6,000 Hz. So, yes, those iPhone apps that claim to play a tone that you can hear but not professors are absolutely telling the truth.

Based on these human characteristics, we break sound waves up into three distinct categories each with their own uses as noted in Table 207-1.

Table 203-1. Ranges of Acoustic Waves

Category	Frequency Range	Perception and/or Use
Infrasonic	< 20 Hz	• Humans do not have detectors for this range. • At high intensity, infrasonic waves can be felt by the human body (as in the guttural shaking from a subwoofer).
Sonic	20 Hz – 20,000 Hz	• Frequencies that can be detected by the human auditory system.
Ultrasonic	> 20,000 Hz	• Humans do not have detectors for this range, although frequencies close to 20,000 Hz can be perceived as audible by bone conduction in the skull. • Therapeutic Ultrasound (800 kHz – 2 MHz) – sound waves are used to break up scar tissue and aid in drug delivery. • Medical Imaging (>2 MHz) – sound waves used to image structure in the body (babies, ulcers, etc.).

203-4: Sound Intensity and Loudness

Consider: *How can we describe the loudness of sound?*

As we discussed earlier in the unit, sound waves are characterized by a pressure disturbance that propagates through space. As the pressure of the wave increases, so does the energy that the wave carries. Unfortunately, it is hard to describe the total energy of a sound wave since it is spread out over a large area, so we tend to discuss the *intensity* of the wave, which is the power of the wave per unit area (W/m^2).

In terms of the pressure the wave creates, the intensity can be written

$$I = \frac{P^2}{2v\rho},$$
(203-6)

where P is the pressure of the wave (in Pascals), v is the speed of the wave (in m/s) and ρ is the density of the material (in kg/m^3) in which the wave propagates.

We won't, however, often hear people talk about the W/m^2 for sound. Part of the reason is that it covers a tremendous range. Humans can start to hear sounds with intensities as low as 10^{-12} W/m^2, and the threshold for pain is around 1 W/m^2. That upper limit is a *trillion* (1,000,000,000,000) times the lower limit!!

In order to combat this, scientists came up with the decibel (dB) scale, a logarithmic scale that compresses this huge range into a smaller, more easily understood range.

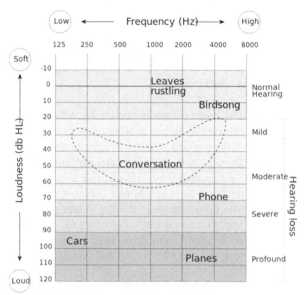

Figure 203-5. Frequency and loudness of common sounds.

Loudness (dB)

$$L_{dB} = 10\ log\left(\frac{I}{I_0}\right) \qquad (203\text{-}7)$$

Description – This equation defines the loudness of an acoustic wave in decibels (dB)

Notes: The reference intensity, I_0, is generally taken as the lower limit of human hearing: 10^{-12} W/m^2.

The decibel scale is structured so that sounds at the lower limit of human hearing (10^{-12} W/m^2) give a zero decibel value. Continued exposure to sounds above 90 dB tends to cause hearing damage and the threshold for pain in humans is around 120 dB. Again, I should note that these values are for fresh, newborn ears and that our ability to hear sounds at the lower level tends to decrease with age. Figure 203-5 shows a number of common sounds and where the sit in terms of both frequency (pitch) and loudness. Again, the range of 0 to 120 dB for normal human hearing is much more tractable than the factor of a trillion for pressure levels.

Example 203-3: Examples

What is the sound-level threshold for pain in dB?

Solution:

This is a direct application of the dB equation. The threshold for pain was given as 1 W/m^2, so

or

$$L_{dB} = 10\ log\left(\frac{I}{I_0}\right) = 10\ log\left(\frac{1\ W/m^2}{10^{-12}\ W/m^2}\right),$$

$$L_{dB} = 120\ dB.$$

So, the threshold for pain is 120 dB, well above the level for damage to your hearing (90 dB).

203-5: Seismic Waves and Earthquakes

Consider: *How do earthquakes relate to waves?*

Another set of important mechanical waves are those caused by earthquakes, known as *seismic waves*. An earthquake is a sudden release of energy from the earth's surface due to the motion of large sections of earth relative to each other. The U.S. Geological Survey (USGS) estimates that there are *several million earthquakes per year* around the world. A large majority of these go undetected because they are in areas of the world not covered by seismic detectors or are in areas where the earthquakes are too small for the equipment available to measure. The USGS's National Earthquake Information Center actually registers about 20,000 earthquakes per year.

The crust of the earth is divided into a number of *tectonic plates*, large land masses that are literally floating on top of the earth's mantle. As the large masses try to move relative to each other, friction between surfaces of the plates acts to stop their relative motion. In many cases, this causes a great deal of potential energy to be stored as parts of the plate move and the friction prevents other parts from moving – essentially bending the entire plate. Just as with a spring, when you compress or stretch the plates, *elastic potential energy* is stored. Once the potential energy reaches a certain point (or rather the force related to the potential energy overcomes the force of friction), the surfaces give way and move catastrophically relative to each other. The energy that was stored must be released and it is often radiated away in the form of seismic waves.

To give an idea of the size and scale of these motions, Figure 203-6 shows the major tectonic plates superimposed over the visible continental features of the earth. Most of the large earthquakes happen at the boundaries of these tectonic plates along boundaries called *faults*. An example of this is the ***San Andreas Fault***, the famous area in Southern California where the North American Plate is rubbing up against the Pacific Plate. If you look closely at figure 203-6, you will see the arrows along the plates in Southern California that show how the plates are sliding by each other. If friction were not present, the Pacific

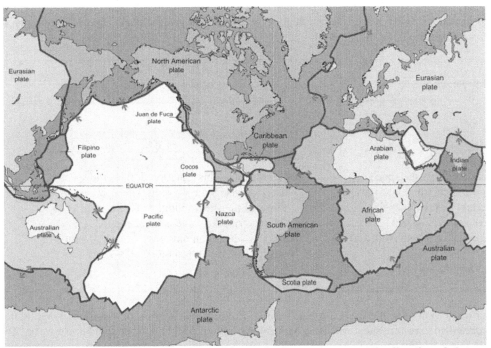

Figure 203-6. The major tectonic plates of the earth. Arrows represent the direction of plate movement along the ring of fire.

plate and the North American plate would slide by each other at 1.3 to 1.5 inches per year. The Pacific and North American Plates are the largest of the tectonic plates and together comprise a whopping 35% of the earth's surface area!

The outer boundary of the Pacific Plate, the Filipino Plate (in the west) and the Nazca plate (in the east) is called the ***Ring of Fire***. Along these boundaries, there is tremendous earthquake and volcanic activity. In fact, most of the recent devastating earthquakes occurred along the Ring of Fire, including the 2004 Indian Ocean Quake (which created a tsunami that devastated Indonesia), the 2010 Chile Earthquake and the 2011 Tōhoku earthquake (which created a tsunami that destroyed the Fukushima nuclear power plant).

Although the vast majority of earthquakes happen along or near plate boundaries, earthquakes do happen in the middle of plates as well – these are known as ***intraplate earthquakes***. Intraplate earthquakes tend to be much smaller than earthquakes that occur along plate boundaries, although there are a few exceptions, such as the devastating 1811-1812 New Madrid Earthquakes in what is now Missouri and the 2011 earthquake in Virginia, which caused cracks in the Washington Monument in DC. Any earthquakes here in Connecticut would be of the intraplate variety and we would therefore expect them to be relatively small.

Seismic Waves

Most earthquakes do not happen at the surface of the earth, but rather deep underground (many kilometers). When the fault slips, much of the energy released is radiated away the point of slippage, known as the *focus*. These seismic waves that carry the energy away are generally radiated spherically in all directions. Seismic waves come in both longitudinal and transverse forms. Solids respond differently to deforming forces that try to compress them when compared with shearing forces (forces that try to slide one part relative to another). Another way to say this is the elasticity of solids is different for the two types of deformations.

Now, remembering from earlier in the unit that the speed of mechanical waves is dependent on the elasticity of a material, we can now say that longitudinal seismic waves travel at different speeds than transverse seismic waves. Longitudinal seismic waves are called ***primary waves, or p-waves***, because they

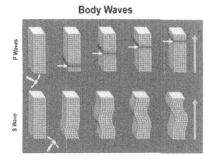

Figure 203-7. Representations of p-waves (top) and s-waves (bottom). Hammers depict how the direction of each wave is created.

travel faster than transverse seismic waves, which are called ***secondary waves, or s-waves***. Note: Some authors say the p in p-wave stands for pressure and the s in s-waves stands for shear, although these are less common. Figure 203-7 gives a representation of p-waves and s-waves in the bulk of the earth.

The speed of p-waves depends on the material properties of the rock they are traveling through (as expected). Since they are pressure waves, elasticity measurement is given by a parameter known as the elastic p-waves moduli, M.

Secondary waves typically travel at only 60% the speed of primary waves, although this varies depending on the type of material the waves are traveling in. Since s-waves are transverse, their speed depends on the elasticity of the material to shearing forces, known as the **shear modulus** (G) of the material.

<table>
<tr><td>

Speed of p-waves

$$v_p = \sqrt{\frac{M}{\rho}} \qquad (203\text{-}8)$$

Description – This equation defines the speed of a primary seismic wave in the earth's interior in terms of the elastic modulus, M, and density, ρ, of the material the wave is traveling in.
Note 1: p-waves can travel through both solid and fluid areas of the earth.
Note 2: p-waves tend to move at 5-8 km/s.

</td><td>

Speed of s-waves

$$v_s = \sqrt{\frac{G}{\rho}} \qquad (203\text{-}9)$$

Description – This equation defines the speed of a secondary seismic wave in the earth's interior in terms of the shear modulus, G, and density, ρ, of the material the wave is traveling in.
Note 1: s-waves can only travel through solids.
Note 2: s-waves tend to move at 3-4 km/s.

</td></tr>
</table>

If you remember back to our discussion of sound waves, we said that in fluids, sound waves could only be longitudinal. This is because fluids have no mechanism to restore shear forces – so transverse motion could not be coupled into a wave. The same is true of seismic waves. In fact, seismic waves in bulk material resemble sound waves exactly. So, both p-waves and s-waves can travel through solids, but only p-waves can travel through fluids. This is actually very important because of the structure of the earth. You see, since seismic waves radiate in all directions from the focus, multiple locations all around the earth can be used to detect the waves and even locate the earthquake with quite good precision – but not all waves reach all locations.

The earth is made up of a number of layers as can be seen in figure 203-8. On the surface is the **crust** – the layer we live on. The crust is only, on average, about 35 km deep (compare this to the radius of the earth – 6380 km). Below the crust is the **mantle**, a region up to about 2900 km thick made of dense silicate rock. Next is the **outer core**, a layer of *liquid iron and nickel* 2300 km thick. Finally, in the center of the earth is the **inner core**, a region of solid iron-nickel crystals about 1200 km thick.

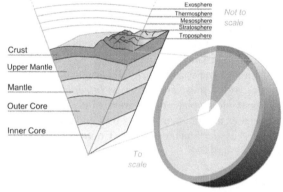

Figure 203-8. Schematic view of the interior of Earth with a cutout showing the earth's layers.

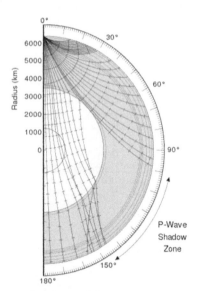

Figure 203-9. Propogation of p-waves and s-waves through the bulk of the earth.

We know from our studies of fluids that the pressure in a fluid increases with depth because of the weight of the fluid above it. The same is also true of solids in earth. For example, the pressure in the mantle of the earth continues to increase as you move towards the center of the earth. Even though we think of rock as incompressible, its density actually will increase as pressures become huge – such as in the interior of the earth (the pressure at the bottom of the mantle is approximately 140 GPa!). Because the density of rock changes, this causes seismic waves to curve as they move through the rock of the mantle as can be seen in Figure 203-9. To make this more complicated, the elasticity of materials actually increases with depth as well. We will see more as to why the waves curve in the units on optics later in the course.

The other important feature here is the liquid outer core. Since it is liquid, s-waves cannot travel through it! This creates a region around the earth where s-waves from an earthquake are never detected, known as the **s-wave shadow zone**. In addition, since the p-waves curve due to the changing density of the earth, there is also a **p-wave shadow zone**, even though there are regions on the other side of the earth that will receive p-waves from the earthquake. All of these features can be seen in figure 203-9.

Surface Waves

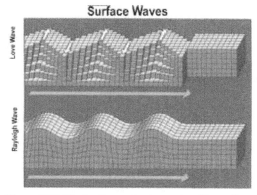

P-waves and s-waves are ***body waves***, meaning that they are the waves that travel though the bulk of the earth. When seismic waves reach the surface of the earth, they do propagate along the surface, but their waveforms are more complicated. ***Rayleigh waves*** are responsible for ***ground roll*** in an earthquake. These waves are a combination of longitudinal and transverse forms and the overall motion is slightly elliptical. The other main type of surface waves are ***Love waves***, which are transverse and responsible for ***side to side motion***. The motion of these two waves are described in figure 203-10.

Both Rayleigh waves and Love waves are ***dispersive***, which means that their speed depends on their frequency. Since this is a much more complicated situation, we will not give an equation for their speed, but will note that they tend to move at about 90% the speed of s-waves in the same material.

Figure 203-10. Surface seismic waves. Top – the side to side motion of Love waves. Bottom – the roll of Rayleigh waves.

203-6: The Moment Magnitude Scale

Consider: *How do we measure the strength of an earthquake?*

When an earthquake occurs, one of the important features that is reported is the ***magnitude of the earthquake***, which gives an indication of the amount of energy released, and is therefore related to the amount of damage that the earthquake can produce. Before going into this, it is important to note that the damage associated with earthquakes has to do with very many features, including the type of rock or soil, the depth of the earthquake, and the distance from the ***epicenter*** (the place on the surface of the earth directly above the focus).

When we hear earthquake reports, they are always accompanied by a number – 4.5, 6.2, 7.8, etc. Many people think these magnitudes are from the Richter Scale – however, the Richter Scale has not been regularly used by seismologists for decades. Today, they use the ***Moment Magnitude Scale (MMS)***. These two scales are related; however, MMS was designed take better account of the surface waves (Love waves and Raleigh waves), especially in earthquakes that are distant from the detectors.

Like the loudness of sound, the energy released from the largest known earthquakes is about a *trillion* times that of the smallest earthquakes we measure. So, any scale (including both MMS and the Richter scale) that is designed to give a reasonable number for earthquake magnitudes must use a logarithmic scale. What are the important quantities? We know that we are starting with elastic energy stored in the rock. So, we need to know how much rock was moved and by how much, which is done by taking the surface area of the rock that moved, A, and multiplying it by the distance over which it traversed during the earthquake, D. Then, we also have to worry about the type of rock and how much it can bend, which is again quantified with the shear modulus (G) described above.

This leaves us with a quantity called the ***seismic moment***, M_0, which includes all of these values: $M_0 = GAD$. Then, all that is left is to take the log of the seismic moment and add some constants to make it match up with the older Richter scale as much as possible, and we have the *MMS*.

Moment Magnitude Scale

$$M_w = \frac{2}{3}log(M_0) - 6.0 \qquad (203\text{-}10)$$

Description – This equation defines the size of an
   earthquake, M_w, in terms of the seismic moment M_0.
Notes: The seismic moment is given by $M_0 = GAD$,
   where G is the shear modulus of the material, A is the
   surface area of the rupture and D is the displacement of
   the rupture. The units of M_0 are Nm or J.

27

I should note that M_0 gives the total energy released in an earthquake. Much of that energy goes into rock deformation, temperature changes, sound, etc., so the value of M_0 cannot be used directly to measure the seismic energy alone.

Earlier in this section, I mentioned four recent important earthquakes and one historical earthquake swarm. Table 203-2 gives some information about each of these earthquakes, including their magnitudes on the MMS. Of particular note, I wanted to mention more about the 2004 Indian Ocean earthquake and tsunami. At 9.1 (M_w) it is the third highest recorded earthquake ever recorded on a seismograph. The fault slipped for somewhere between 8.3 and 10 *minutes*, making it the longest recorded earthquake as well. The earthquake was so powerful that Rayleigh waves of at least 1 cm were recorded *everywhere on earth*; another way to say this is that the Indian Ocean earthquake vibrated the entire planet with an amplitude of at least 1 cm. The total energy released is estimated at *1.1 x 10^17 J*, which is about the same energy as 26 modern thermonuclear bombs or the equivalent energy usage for the United States *for 370 years*! Truly remarkable.

Table 203-2. Major earthquakes noted in the unit with magnitude, death toll and estimated costs.

Earthquake	Magnitude (M_w)	Deaths	Cost *
2004 - Indian Ocean	9.1	250,000 (est.)	$ 15 billion
2010 - Chile	8.8	525	$ 4-7 billion
2011 - Tōhoku	9.0	15,885	$ 300 billion
1811 - New Madrid	7.0 – 8.1 (est.)	small	unknown
2011 - Virginia	5.8	0	$ 200 million

* Note that the cost of a disaster is dependent upon the economics of the area in which it occurred.

How does the magnitude of an earthquake relate to what you feel? Since the magnitude of an earthquake on the MMS is based on energy released, there is no perfect correlation between this scale and what is felt. However, the ***Modified Mercalli Intensity scale*** attempts to bridge this gap. Information about this scale can be found in table 203-3. Please note that the Mercali scale depends on position such that a reading of VII near the epicenter of an earthquake may be a II 100 miles from the center. In addition, there is a great deal of subjectivity that goes into comparing earthquake magnitude and intensity – the idea of intensity in this case is based on personal observations and reports.

Table 203-3: The Modified Mercalli Intensity Scale

Scale	Intensity	Description of Effects	Aprrox. M_w
I	Instrumental	Detected only on seismographs	
II	Feeble	Very few people feel it	< 4.2
III	Slight	Felt by people resting; feels like a rumbling truck	
IV	Moderate	Felt by people walking	
V	Slightly Strong	Sleepers awake; church bells ring	< 4.8
VI	Strong	Trees sway; objects fall off shelves; chandeliers swing	< 5.4
VII	Very Strong	Walls crack; plaster falls; Feeling of alarm in most people	< 6.1
VIII	Destructive	Moving cars uncontrollable; masonry fractures	
IX	Ruinous	Some houses collapse; ground cracks; pipes break open	< 6.9
X	Disastrous	Many buildings destroyed, Liquefaction* and landslides occur	< 7.3
XI	Very Disastrous	Most buildings and bridges collapse; infrastructure destroyed	< 8.1
XII	Catastrophic	Total destruction; trees fall; triggering of other hazards	> 8.1

* Liquefaction is a process where soil loses its strength and stiffness when shaken. In essence, the soil acts like liquid.

Finally, a little bit on comparing earthquakes. As noted, logarithmic scales give us a great tool for reporting huge ranges with reasonable numbers; however, they also make it harder to compare two values conceptually. As an example, even though we just discussed it a few pages ago, does it seem like 120 dB sound should be a *trillion* times as loud as 1 dB sound?

Let's try comparing two earthquakes, and call their magnitudes M_1 and M_2, with M_2 larger than M_1. First, I'm going to rearrange our equation for the M_w and solve for M_0 since it is directly related to energy:

$$M_0 = 10^{1.5M_w+9}.$$

$$\text{(203-11)}$$

Then, we can find the ratio of our two M_0 values to gives us the ratio of the energy released by two earthquakes. I will call this ratio f_E:

$$f_E = \frac{10^{1.5M_2+9}}{10^{1.5M_1+9}} = 10^{1.5M_2+9-(1.5M_1+9)}.$$

$$\text{(203-12)}$$

This equation can be simplified to give us our final answer for the ratio of two energies:

Ratio of Earthquake Energy

$$f_E = 10^{1.5(M_2 - M_1)} \qquad (203\text{-}13)$$

Description – This equation gives us the ratio of the energy released between two earthquakes with magnitudes of M_2 and M_1.

Therefore, a difference in magnitude of one, say comparing an earthquake with magnitude 8 to one with magnitude 7, is $10^{1.5}$, which is 32. This means that a magnitude 8 earthquake releases *32 times* the energy of a magnitude 7 earthquake. A magnitude 8 earthquake releases *1,000 times* the energy of a magnitude 6 earthquake. You can see that the energy differences increase very fast as we increase the difference in magnitudes.

To again try and put this in perspective, the largest earthquake magnitude every recorded was an earthquake off the coast of Chile on May 22, 1960, with a magnitude of 9.5. If you compare this to the 2004 Indian Ocean earthquake, you will find that the 1960 Chile earthquake released 2.5 times as much energy!! We already discussed the effects of the 2004 earthquake. Take that and consider it two and a half times over, and you have the effects of the 1960 Chile quake.

Example 203-1: Devastating or not?

An earthquake is measured in continental crust ($G = 3x10^{10} N/m^2$) to have an average fault slip of 6.2 m over an area of 1200 km^2.
  (a) What is the seismic moment of this earthquake?
  (b) What is the magnitude of this earthquake, M_w?
  (c) Compare the energy released by this earthquake to one with a magnitude of $M_w = 6.0$.
  (d) Describe the type of damage that might be expected from the earthquake if it were near land.

Solution:

(a) The seismic moment of an earthquake is given by

$$M_0 = GAD.$$

Using information from the problem, we find

$$M_0 = (3x10^{10} N/m^2)(1200\ km^2)(6.2\ \text{m}),$$

where we must be careful to convert the km^2 to m^2. Simplifying, we find

$$M_0 = 2.23x10^{20}\ N \cdot m.$$

(b) In order to find the magnitude of this earthquake, we use the seismic moment we just found in the equation for M_w

$$M_w = \frac{2}{3} log(M_0) - 6.0:$$

$$M_w = \frac{2}{3} log(2.23x10^{20}) - 6.0 = 7.56.$$

Therefore, this is a 7.6 on the moment magnitude scale.

(c) Since the magnitude we just found is larger than 6.0, we know that it releases more energy than the 6.0 this part asks us to compare to. Using the equation comparing energy release by earthquakes, we find

$$f_E = 10^{1.5(M_2 - M_1)} = 10^{1.5(7.56 - 6.0)} = 219.$$

Therefore, the earthquake described in this problem releases 219 times as much energy as a 6.0 earthquake.

(d) Although the Mercalli intensity scale and the Moment Magnitude scale do not correlate exactly, Table 203-3 suggests that a 7.6 earthquake near land would be disastrous – causing many buildings to be destroyed, landslides to occur, and any material built on loose dirt to be in danger of sinking (liquefaction).

29

203-7: Real Waves are Complicated

Honestly, we only touched the surface of the science behind hearing and seismic waves, with both being active areas of research for physicists, engineers and other professionals. I hope that this unit gave you a nice introduction to how some of the basic ideas of waves we learned in Units 204 and 205 can be applied to cutting-edge research as well as areas of societal interest. I also hope that this unit highlights some of the more interesting and complex wave topics, giving you the vocabulary and tools to explore them further on your own if you choose.

Here are a couple of resources:

ASHA Hearing Loss: http://www.asha.org/public/hearing/Hearing-Loss/
> ASHA is the American Speech-Language-Hearing Association – the professional society for those working in the fields. This website gives a tremendous amount of information on the science and impacts of hearing loss

How Bad are iPods for Your Hearing?: http://content.time.com/time/health/article/0,8599,1827159,00.html
> This is a nice article from Time Magazine on how iPods (and all smart phones and mp3 players) are leading to increased hearing loss in young people.

USGS Earthquake Hazards Program: http://earthquake.usgs.gov/
> This website gives real-time information on earthquakes around the world as well as detailed information on an earthquake as it is analyzed. You'd be surprised at how many earthquakes there are and where the happen. At the time I'm writing this, the website listed 27 earthquakes with magnitude 2.5 or greater around the world in the last 24 hours – 11 of which were on the Ring of Fire and two of which were >5.0 and in Chile!

Earthquakes for Kids: http://earthquake.usgs.gov/learn/kids/
> I feel a little weird sending you to a site for kids, but man, there is so much cool information - pictures, animations, games, etc. - on this site, that if you are interested in earthquakes, you can't miss it!

204 – Electromagnetic Waves

Although our initial introduction to waves covered all waves, the last couple of units were focused on mechanical waves - waves that travel by the mechanical oscillation of some type of medium. In this unit we extend our discussion to electromagnetic waves, waves that do not require a medium to travel in and that describe microwaves, radio waves, light, ultraviolet rays and x-rays all at once.

The Bare Essentials

- Electromagnetic waves are a combination of electric and magnetic fields

Electromagnetic Radiation

$$\vec{E} = E_0 sin(\omega t - kx)\hat{e}$$

$$\vec{B} = B_0 sin(\omega t - kx)\hat{b}$$

Description – These equations define the propagation of an electromagnetic wave as made of perpendicular oscillating electric and magnetic fields.
Note1: $E_0 = cB_0$
Note2: The crux of this equation is the EM radiation acts as a sinusoidal wave like we discussed in Unit 201.

- Electromagnetic waves travel at the speed of light in a vacuum and slower in material given by the index of refraction.

Speed of EM Radiation in Matter

$$v = \frac{c}{n}$$

Description – This equation relates the speed of EM radiation in matter to the speed of light in a vacuum and the index of refraction, n.
Note1: n is greater than one for all ordinary matter.
Note2: $c = 299,792,458 \, m/s$

- The electromagnetic spectrum accounts for many types of radiation

Regions of the EM Spectrum

Region	Wavelength	Frequency (Hz)
Gamma ray	< 0.02 nm	> 15 EHz
X-ray	0.01 nm – 10 nm	30 EHz – 30 PHz
Ultraviolet	10 nm – 400 nm	30 PHz – 750 THz
Visible	390 nm – 750 nm	770 THz – 400 THz
Infrared	750 nm – 1 mm	400 THz – 300 GHz
Microwave	1 mm – 1 meter	300 GHz – 300 MHz
Radio	1 m – 100,000 km	300 MHz – 3 Hz

- Electromagnetic waves are produced by objects at any temperature as well as by accelerated charged particles.

Stefan's Law

$$P = \sigma A e T^4$$

Description – Stefan's law relates the power emitted via EM radiation of an object with surface area A, emissivity e and absolute surface temperature T.
Note 1: σ is the Stefan-Boltzmann constant and is equal to 5.6703 x 10^{-8} W/m²K⁴.
Note 2: Stefan's law is for blackbody radiation and not accelerated charged particles.

Wien's Law

$$\lambda_{max}T = 2.898 \times 10^{-3} \, m \cdot K$$

Description – Wien's Law relates the peak wavelength emitted by an object to the object's absolute temperature T.
Note: The temperature *must* be in Kelvin to use Wien's Law

Lamor Formula

$$P = \frac{2}{3}\frac{kq^2a^2}{c^3}$$

Description – The Lamor formula relates the power radiated by an accelerated particle to the charge of the particle, q, the acceleration of the particle, a, and the speed of light, c.
Note: k is the Coulomb constant and is given by 8.99 x 10^9 Nm²/C².

- The effects of EM radiation depend on the frequency and intensity of the waves.

204-1: Electric Charges and Electric and Magnetic Fields

Consider: *I've heard of electric charges and magnetic fields, but what are they?*

THIS UNIT IS A BRIDGE between our study of waves and our study of electric and magnetic fields. Historically, one of the largest problems for students learning to deal with these 'fields,' is that they lacked context and connection to reality. So, here, I'm going to give a very cursory introduction to the idea of electric charges and the two fields of interest and then we are going to talk about one of the most important constructs of these fields – light. The light you use every day to see the world is a wave. Not only is it a wave as we've discussed in the last couple of units, light is very special – it can travel through a vacuum. Remember that in the last couple of units, we always needed a medium – a material through which the wave could move; some material to oscillate as the wave passes. Electromagnetic waves (light) do not need a medium at all – they can travel through completely empty space. If you think about it, this is a very important property because most of the light we use during daylight comes from the sun, and those waves have traveled through mostly empty space on their path from the sun to the earth.

Electromagnetic waves (EM waves) are created by accelerating electric charges. We will go into electric charges in depth in the next unit, but for now, it is enough to know that all subatomic particles have a property known as **charge**. For example, the protons in the nucleus of an atom have a positive charge, and the electrons that surround the nucleus have a negative charge. Many atoms have the same number of protons as electrons (studied later in the course), so the net electric charge is zero, or very small at least. However, we can manipulate charged particles so that they appear on their own and this leads to things such as static electricity, electric currents, magnets, and, yes, light.

Before we get into the details of light, it is also important to talk about **fields**, specifically electric and magnetic fields. Just as all matter with mass produce a gravitational field such that two masses will attract each other (gravitational force), charged particles produce an **electric field** such that they will either attract or repel other charge particles (unit 205). Even more, *moving* charged particles produce a **magnetic field** such that they will attract or repel other moving charges (unit 209). Finally, if a charged particle is *accelerating*, it produces a combination of electric and magnetic fields that radiate away as a wave, know as electromagnetic radiation.

To summarize:

Charged particles produce **electric fields,**
Moving charged particles produce **magnetic fields,**
Accelerating charged particles produce **electromagnetic waves.**

The effect is cumulative as well, so a moving charge particle produces both an electric field (because it *is* a charge particle), and a magnetic field (because it *is* a moving charged particle).

A note on units and charged particles: As we will see over the next couple of days, charge is measured in Coulombs (C), electric fields are measured in Newtons per Coulomb (N/C) and magnetic fields are measured in Tesla (T).

204-2: Electromagnetic Waves

Consider: *How do we describe waves such as light, ultraviolet rays, infrared light, microwaves, etc.?*

So, let's say we have an accelerating charged particle that produces an EM wave. What does it look like? Let's say that the wave is sinusoidal and is traveling in the z-direction. From our recent study of waves, we know that a sinusoidal wave traveling in the z-direction must include $\sin(kz - \omega t)$. Electromagnetic waves, as their name implies, are actually made up of two parts, the part containing the electric field and the part containing the magnetic field. These two parts are intimately related; in fact, they have to both have the same sine function, so that

$$\vec{E} = E_0 sin(kz - \omega t)\hat{e}, \tag{204-1}$$

$$\vec{B} = B_0 sin(kz - \omega t)\hat{b}. \tag{204-2}$$

Many of the symbols just introduced need explanation. First, a capital E is often used for electric field. So, $\vec{E}$ represents the electric field at some time and place and E_0 represents the amplitude of the electric field (the largest value of the electric field). Similarly, a capital B is often used for magnetic field such that $\vec{B}$ represents the magnetic field at some time and place and B_0 represents the amplitude of the magnetic field.

The unit vectors $\hat{e}$ and $\hat{b}$ tell us the direction of the electric and magnetic field, respectively. It turns out that the directions of the electric and magnetic fields are also connected to each other and to the direction of propagation (the z-direction in our current case) – they are all perpendicular to each other. In fact

$$direction\ of\ propagation = \vec{E}\ x\ \vec{B}, \tag{204-3}$$

where $\vec{E}$ and $\vec{B}$ are also perpendicular to each other. The cross product in this equation is the same cross product we dealt with for torque and angular momentum, and you will use the right-hand rule to determine the relationship between the fields and the direction of propagation. This relationship can be seen in figure 204-1, which shows how the electric field, magnetic field and direction of propagation can all be perpendicular to each other.

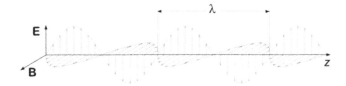

Figure 204-1. Depiction of an electromagnetic wave.

Another connection comes between the magnitudes of the electric and magnetic fields:

$$E_0 = cB_0, \tag{204-4}$$

where c is the speed of light. There is a fundamental reason that the relationship between the magnitudes of our two fields is the speed of light and we'll see that in a couple of units.

Electromagnetic Radiation

$$\vec{E} = E_0 sin(\omega t - kx)\hat{e}$$

$$\vec{B} = B_0 sin(\omega t - kx)\hat{b}$$

$$\tag{204-5}$$

Description – These equations define the propagation of an electromagnetic wave as made of perpendicular oscillating electric and magnetic fields.

Note1: $E_0 = cB_0$

Note2: The crux of this equation is the EM radiation acts as a sinusoidal wave we discussed in Unit 201.

One more property of EM waves is so important that I will set it off and cannot emphasize it enough:

All electromagnetic waves in a vacuum move at the speed of light, c.

In fact, the speed of light in a vacuum is now defined as exactly $c = 299,792,458\ m/s$. We will usually abbreviate this to either $3.00\ x\ 10^8$ m/s or even $3\ x\ 10^8$ m/s, but you should know that the value is defined to this level.

Once EM waves enter any sort of matter, however, the speed of the wave does change. There is a simple factor, known as the index of refraction, $n = c/v$, that helps us defined the speed of EM waves in matter. The index of refraction is always greater than one (meaning that EM waves move slower in material than in a vacuum). However, there is no theoretical upper limit on the index of refraction, so it is possible to create materials where these waves moves relatively slow.

The current world record for the index of refraction is n = 38.6. An index of refraction this high does not exist naturally, and must be created in a laboratory. Ordinary transparent materials, such as plastic, glass and water have indices of refraction between one and two. A list of such indices can be found in table 204-1. We will return to the index of refraction in unit 219 when we start to discuss optics.

Table 204-1. Indices of Refraction

Material	Index of Refraction
Vacuum	1.00000
Air (STP)	1.00029
Ice	1.31
Water (20°)	1.33
Eye (lens)	1.41
Eye (cornea)	1.38
Plexiglas	1.49
Glass (typical)	1.52
Sugar Water	1.49
Diamond	2.42

Speed of EM Radiation in Matter

$$v = \frac{c}{n} \qquad (204\text{-}6)$$

Description – This equation relates the speed of EM radiation in matter to the speed of light in a vacuum and the *index of refraction*, n.
Note1: n is greater than one for all ordinary matter.
Note2: $c = 299{,}792{,}458 \, m/s$

Example 204-1: What makes up a wave?

The electric field part of an electromagnetic wave is given by

$$\vec{E} = 22\frac{N}{C}\sin\left[\left(3x10^{14}\frac{rad}{s}\right)t - \left(1.0x10^6\frac{rad}{m}\right)z\right]\hat{x}$$

(a) What is the frequency of this wave?
(b) What is the wavelength of this wave?
(c) In what direction in the wave propagating?
(d) What is the amplitude of the magnetic field associated with this wave?
(e) What is the equation for the magnetic field associated with this wave?

Solution:

Each part of this problem can be solved using a direct application of sinusoidal waves, or information about em waves specifically.

(a) Comparing the equation for the electric field to the standard form for sinusoidal wave, we know

$$\omega = 3x10^{14}\frac{rad}{s} = 2\pi f.$$

Therefore, we can solve for f, to find

$$f = \frac{3x10^{14} \, rad/s}{2\pi} = 4.77x10^{13} Hz.$$

(b) Similarly, we can find the wavelength, by noting

$$k = 1.0x10^6\frac{rad}{m} = \frac{2\pi}{\lambda}.$$

Solving for the wavelength, we find

$$\lambda = \frac{2\pi}{1.0x10^6 \, rad/m} = 6.28x10^{-6} \, m.$$

(c) Since the argument of the sine function contains a negative in the z dimension, the wave is propagating along the *positive z-axis*.

(d) The amplitude of the magnetic field is found from

$$B_0 = \frac{E_0}{c} = \frac{22 \, N/C}{3x10^8 m/s} = 7.33x10^{-8} \, T.$$

(e) We know the amplitude of the magnetic field from part d, and we also know that the argument of the sine function is the same for both the electric and magnetic field parts of the wave. Also, the right hand rule tells us that if the electric field is along the +x-axis and the wave propagates on the +z-axis, then the magnetic field must be along the +y-axis (equation 204-3). Therefore,

$$\vec{B} = 7.33x10^{-8} \, T \sin\left[\left(3x10^{14}\frac{rad}{s}\right)t - \left(1.0x10^6\frac{rad}{m}\right)z\right]\hat{y}.$$

204-3: The Electromagnetic Spectrum

Consider: *So, what is the difference between visible light, x-rays, radio waves, etc.?*

The light that we use to see is only a small part of all possible electromagnetic waves. Like all other waves, EM waves are well-characterized by their frequency, wavelength and intensity. We now know that (in a vacuum), EM waves move at the speed of light, which allows us to write the following relationship.

$$f\lambda = c. \qquad (204\text{-}7)$$

That is, the frequency of the EM waves multiplied by its wavelength gives the speed of light. Since the speed (again in a vacuum) is known, this reduces one of our unknowns for the EM waves.

It turns out that the difference between types of EM radiation are well characterized by their frequency, and that many things that you are used to hearing about, ultraviolet light, infrared light, x-rays, etc., are all members of the electromagnetic spectrum with different frequency ranges. The situation is summarized in figure 204-2.

Figure 204-2 gives us a wealth of information. First, notice that both the wavelength and frequency of each section of the spectrum is given. In

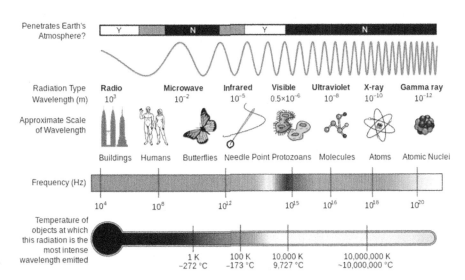

Figure 204-2. Desciption of the electromagnetic spectrum and the types of radiation that make it up.

producing this figure, it was assumed that the EM radiation is propagating in a vacuum. Next to each of the types of radiation is a visual picture of something that is on the order of size for that part of the spectrum. For example, looking at the figure, we can see that humans are sized on the order of the wavelength of microwaves or radio waves (the distinction between microwaves and radio waves is blurred.). Below the pictures in the figure, the frequency range for each of the areas of the spectrum are listed.

Each of the regions of the EM spectrum has a different story in terms of how it was discovered and its effect on the environment. Here are the important regions of the EM spectrum (in increasing frequency / decreasing wavelength) and some information about EM waves in each of the regions:

Radio Waves – Radio waves are the largest wavelength and therefore the lower frequency of the EM spectrum. Their wavelengths generally range from hundreds of meters to about one meter. Radio waves are often created by the oscillations of electrons in antennae. The radio in your car receives radio waves in the kHz range (AM radio) and MHz range (FM). Over-the-air TV transmission is also carried on radio waves. Information is carried by varying the amplitude, frequency and phase of the waves. Radio waves readily travel through many types of material.

Microwaves – Microwaves have wavelengths in the millimeter range. The corresponding microwave frequency of 2.45 GHz turns out to be the rotational resonant frequency for liquid water. In a microwave oven, standing waves of microwaves cause water molecules to rotate and the friction of these water molecules with their surrounding cause the material to warm up.

Terahertz Radiation – Traditionally left off of the electromagnetic spectrum because they were not studied extensively until recently, terahertz radiation (named for the frequency) is becoming very important for military applications such as imaging (TSA uses some terahertz imagers), wireless communications (the data rate record is currently in the terahertz range) and interdiction (terahertz waves can be used to disrupt enemy electronics).

Infrared Radiation (IR) – This region contains wavelength approximately 750 nm to 1 mm. Infrared radiation is also close in frequency to the resonant frequency of atoms in many materials – meaning that infrared radiation will tend to either cause materials to vibrate or rotate, leading to friction and heating. The warmth you feel standing in the sun is the IR increasing the energy of molecules in your skin.

Visible Light – (400 nm to 750 nm). This is the area of the spectrum we use to see. There is more about this area of the spectrum below and in future units.

Ultraviolet (UV) Radiation – (10 nm – 400 nm). Waves in the UV region and the regions below are known as *ionizing radiation*. The waves have enough energy to disrupt electrons in many materials, including proteins and DNA in your body. When excessive UV radiation hits your skin, it causes damage in two different ways. Part of the UV spectrum known as UVA (320 – 400 nm wavelengths) causes oxygen ions to form in the cells of your skin, which then interact with and damage your DNA. UVB waves (280 – 320 nm wavelengths) causes direct damage to

DNA and other proteins in your skin cells by ionization. ***Tanning, sunburn and melanoma (cancer) are all caused by these damage mechanisms.*** UV rays on the short side of the spectrum (10 – 100 nm) can be devastating to life; however, the upper parts of our atmosphere filters out most of these rays.

X-Rays – (0.01 nm – 10 nm). We all know x-rays from their imaging properties. X-rays tend to pass through material that is relatively not dense (soft biological material) and absorbed by dense materials (bone and harder cartilage). An x-ray image is actually a 'negative' because the areas that are light (bones for example) are where the x-rays did *not* make it through the material. Like UV, x-rays are ionizing radiation and therefore the long-term exposure must be limited.

Gamma Rays – (less than 0.02 nm). Usually created by energy-level transitions inside of atomic nuclei, gamma rays are the highest energy EM waves with no lower limit to their wavelengths. Exposure to high levels of gamma rays is devastating to biological materials, including humans: a single gamma ray can destroy many strands of DNA. At very low doses, gamma rays are used for medical imaging and cancer treatment although even these have stark medical consequences.

Table 204-2 contains a summary of the wavelengths and frequencies of the regions of the spectrum described above.

So much of our modern technology is based on electromagnetic waves that they are very important to understand, both qualitatively and quantitatively. Consider a typical conversation you might have with a friend from home on a cell phone while standing outside:

Table 204-2. Wavelength and Frequencies of the electromagnetic spectrum.

Name	Wavelength	Frequency (Hz)	Photon energy (eV)
Gamma ray	< 0.02 nm	> 15 EHz	> 62.1 keV
X-ray	0.01 nm – 10 nm	30 EHz – 30 PHz	124 keV – 124 eV
Ultraviolet	10 nm – 400 nm	30 PHz – 750 THz	124 eV – 3 eV
Visible	390 nm – 750 nm	770 THz – 400 THz	3.2 eV – 1.7 eV
Infrared	750 nm – 1 mm	400 THz – 300 GHz	1.7 eV – 1.24 meV
Microwave	1 mm – 1 meter	300 GHz – 300 MHz	1.24 meV – 1.24 µeV
Radio	1 m – 100,000 km	300 MHz – 3 Hz	1.24 µeV – 12.4 feV

Your cell phone is using radio/micro waves to communicate between your phone and the nearest cell-tower. At the cell tower, the signal is converted to infrared light and then travels down on optical fiber to a cell phone tower near your friend, where it is converted back to a radio/micro wave and on to his or her phone. While you are standing there, you can look around because of the visible light all around you and the warmth you feel comes from the infrared light reaching earth from the sun. Hopefully, not much of the UV radiation from the sun interacting with your skin, but at least a little probably is causing damage.

You can see from this quick demonstration that electromagnetic waves are all around us in various forms and intensities. As modern technology continues to advance, we will likely see more use of the EM spectrum as well. This is a very important modern topic to take away from your physics class.

I do want to spend just a couple of sentences on visible light. Above, I noted that visible light is EM radiation in wavelength range of 400 nm – 750 nm. All the colors of the rainbow fit into that range, and the range of each color is given in Table 204-3. What we actually *see* is complicated. Materials that are emitting light tend to give off many colors and what we see is a combination of those colors. For most materials (like the clothes you are wearing now), some colors of light are reflected off the surface and those colors mix to give us what we actually see. This is also why different light sources make things look different – the initial light hitting the surface (of your clothes for example) is different for different sources, so the reflected light is also different.

Table 204-3. The visible spectrum

Color	Frequency (Thz)	Wavelength (nm)
Violet	668-789	400-450
Blue	606-668	450-495
Green	526-606	495-570
Yellow	508-526	570-590
Orange	484-508	590-620
Red	400-484	620-750

204-4: Production of Electromagnetic Waves

Consider: How are electromagnetic waves produced?

EM waves are generally produced in two ways – the acceleration of charged particles such as electrons, and ***blackbody radiation*** - radiation emitted by all matter that is dependent on the absolute temperature of the material. Have you ever felt the warmth of another body near you on a cold day? That is actually infrared radiation that the person is emitting. Humans, it turns out, are the right temperature to give off IR. What about the sun? Why is it yellow? It turns out that the temperature of the sun is just right to give off radiation that peaks in the green portion of the spectrum (near the peak of human vision!), but includes all types of radiation from UV rays to infrared. The reason the sun looks yellow is complicated, but has to do with filtering and scattering effects of the atmosphere as well as the way our eyes work.

Another good example is a heating element in a toaster oven. As the tines start to get 'warm' they are giving off infrared radiation that we feel as an increase in temperature. As they get even hotter though, the tines start to glow red. This is the peak of the blackbody spectrum of the tines entering into the visible region of the spectrum as opposed to the infrared region. There is still plenty of infrared radiation coming off of the tines because the blackbody spectrum is broad, meaning that it gives off many wavelengths at the same time. The approximate spectrum for a number of temperatures can be seen in figure 204-3.

The surface temperature of the sun is approximately 5800 K, which you can see in the figure has the peak of its blackbody spectrum in the green part of the spectrum. However, also notice that even though the spectrum peaks in the visible part of the spectrum, there is both UV and IR being emitted at this temperature (as noted earlier in this section).

There is a relationship, known as Wien's Law that lets us determine the peak wavelength emitted by an object based on its surface temperature as shown below.

Connection: Cell Phones and Cancer

At various times of the last couple of decades, reports have come out trying to link cell phone use with cancer. From physics and biophysics standpoints, it is hard to find a mechanism by which this would occur. In terms of its effects on biological materials, em radiation can be broken into two categories: ionizing radiation and non-ionizing radiation. Ionizing radiation is considered to be radiation with energy greater than that of visible light – therefore, UV rays, X rays and gamma rays are all ionizing radiation. They can cause damage to the cells in your body by ionizing DNA or other proteins causing cells to mutate or die (more about this in unit 225 on nuclear radiation). This is why there is a notable danger from these three types of em radiation – each are capable of causing cell damage, including the mutations that lead to cancer.

Non-ionizing radiation, on the other hand, **_does not_** have the energy to ionize biological material. Cell phone conversations using low-energy and low-intensity radio/microwaves really just don't have enough energy to cause cell damage.

Wien's Law

$$\lambda_{max}T = 2.898 \times 10^{-3} \; m \cdot K \qquad (204\text{-}8)$$

Description – Wien's Law relates the peak wavelength emitted by an object to the object's absolute temperature T.

Note: The temperature *must* be in Kelvin to use Wien's Law

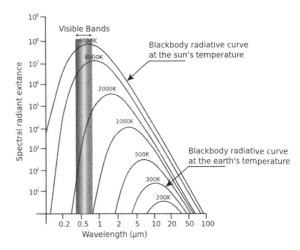

Figure 204-3. Blackbody radiation curves for a number of temperatures.

Example 204-2: Human Body Emission

A typical human has a temperature around 98 degrees Fahrenheit (310 K). What is the peak wavelength emitted by a human at that temperature assuming they are a blackbody?

Solution:

This is a direct application of Wien's Law.

$$\lambda_{max}T = 2.898 \times 10^{-3} \; m \cdot K,$$

so,

$$\lambda_{max} = \frac{2.898 \times 10^{-3} \; m \cdot K}{T} = \frac{2.898 \times 10^{-3} \; m \cdot K}{310 \; K}$$

$$\lambda_{max} = 9.35 \times 10^{-6} \; m = 9350 \; nm.$$

This wavelength is firmly in the infrared part of the spectrum and agrees with Figure 204-3.

Infrared cameras use detectors that are sensitive to light in the infrared part of the spectrum and then use electronics to convert the IR signals into visible light so that we can see it.

Since humans (and many mammals) have a blackbody spectrum that peaks in around 10 microns, cameras set to detect wavelengths in this range work very well for low-light or nighttime imaging.

The fact that we emit more radiation than we absorb creates a state of thermal non-equilibrium, and is one of the reasons we need continuously fuel our bodies with food!

The other means of production of electromagnetic radiation is the acceleration of charged particles. This can come about in an antenna, where we use some sort of power supply to cause electrons to oscillate (we'll discuss this in future units), or it can arise from electrons moving from one energy level to another in an atom, or even protons and neutrons transiting energy levels inside of atomic nuclei (production of gamma rays as noted before).

Whether EM radiation is produced by blackbody radiation or acceleration of charged particles, it is important to know how much power is being emitted by the system. Each of our two mechanisms has their own law that governs how much power is released.

For blackbody radiation, we would expect the power emitted to be dependent on how large the object is – that is its surface area, as well as its temperature. There is one more important factor that comes into place, known as the emissivity. *Emissivity* is a relative measure of a body's ability to emit radiation – you see, a true blackbody radiator must be in thermal equilibrium with its environments. This is usually not true for many of the objects we discussed. Are humans in thermal equilibrium with their environment? Usually not, since we are constantly producing heat through our biological processes. Regardless, for real objects, the emissivity is a value less than one and we will give it to you if you need it.

Stefan's Law

$$P = \sigma A e T^4 \qquad (204\text{-}9)$$

Description – Stefan's law relates the power emitted via EM radiation of an object with surface area A, emissivity e and absolute surface temperature T.

Note 1: σ is the Stefan-Boltzmann constant and is equal to $5.6703 \times 10^{-8} \, W/m^2 K^4$.

Note 2: Stefan's law is for blackbody radiation and not accelerated charged particles.

A perfect Blackbody has an emissivity of 1 ($e = 1$), and for real objects varies between 0 and 1.

For accelerated charged particles, the power radiated away is given by the Lamor formula, which relates the charge of the particle and its acceleration to the power emitted. We will discuss the production and detection of electromagnetic waves via accelerated electrons more in unit 218.

Lamor Formula

$$P = \frac{2}{3} \frac{kq^2 a^2}{c^3} \qquad (204\text{-}10)$$

Description – The Lamor formula relates the power radiated by an accelerated particle to the charge of the particle, q, the acceleration of the particle, a, and the speed of light, c.

Note 1: k is the Coulomb constant and is given by 8.99 x 10^9 Nm^2/C^2

Note 2: the charge of an electron (our usual culprit) is 1.6 x 10^-19 C.

Note 3: c is the speed of light in a vacuum noted earlier in the unit.

Example 204-3: Earth emission

How much electromagnetic energy per second does the earth radiate into space due its temperature? You may assume the average temperature of the earth to be 270 K and the emissivity to be 0.8.

Solution:

Noting that energy per second is power, this problem is a direct application of Stefan's Law. We have all of the important information, except the surface area of the earth. This can be found as

$$A = 4\pi r^2 = 4\pi(6380 \ km)^2 = 5.12x10^{14} \ m^2,$$

We can then start with Stefan's law,

$$P = \sigma A e T^4,$$

to find

$$P = 5.67x10^{-8} \ \frac{W}{m^2 K^2} (1.28x10^{14} \ m^2)(0.8)(270K)^4.$$

Simplifying, we find

$$P = 1.23x10^{17} \ W.$$

Although this is a very large power, it is less than one-billionth the power emitted by the sun.

Example 204-4: Lamor Formula Problem.

An electron oscillating in a wire undergoes a maximum net force of $3.52x10^{-16} \ N$. How much energy per second does this electron emit while under this net force?

Solution:

In order to find the power emitted by a charged particle, we must use the Lamor formula. However, before we are able to do this, we must find the acceleration of the electron.

Since we know the net force on the electron, we can use Newton's Second Law to find the magnitude of its acceleration:

$$a = \frac{F_{net}}{m} = \frac{3.52x10^{-1} \ N}{9.11x10^{-31} kg} = 3.86x10^{14} \ m/s^2.$$

First, note how large an acceleration this electron undergoes even with such a small force!

We can now use our known values in the Lamor formula to find the power emitted:

$$P = \frac{2}{3} \frac{kq^2 a^2}{c^3}.$$

Therefore,

$$P = \frac{2}{3} \frac{\left(8.99 \ x \ 10^9 \frac{Nm^2}{C^2}\right)(1.6x10^{-19}C)^2 \left(3.86x10^{14} \frac{m}{s^2}\right)^2}{\left(3x10^8 \frac{m}{s}\right)^3},$$

which gives us

$$P = 8.47x10^{-2} \ W.$$

Although this is a very small power, keep in mind it is for just one electron.

Extension: How many electrons.

The values in example 204-1 are very realistic for an electron in a radio tower transmitter. How many electrons must be undergoing this oscillation for a 25-kW station?

This is relatively easy to figure out, since the total power emitted is the power of each electron multiplied by the number of electrons:

$$P_{total} = P_{one} n_{electrons},$$

therefore

$$n_{electrons} = \frac{P_{total}}{P_{one}} = \frac{25x10^3 \ W}{8.47x10^{-25} \ W}.$$

This gives us

$$n_{electrons} = 2.95x10^{28} \ electrons.$$

Although this seems like a huge number of electrons, keep in mind that the free electron density of copper is around 10^{29} electrons per cubic meter – there are a lot of electrons available.

39

205 – Charge and Electric Force

As noted in the previous unit, electromagnetic waves are composed of time varying electric and magnetic fields. A very qualitative view of these fields was given in Unit 204, we will now begin to learn about these fields. We will start by investigating electric charges and the force between them...

Integration of Ideas

- Review the fundamental electromagnetic interaction.
- Review vector computations in 2-D and 3-D
- Review basic force problems in 2-D and 3-D.

The Bare Essentials

- **Electric charge**, a fundamental, intrinsic particle property, comes in two flavors, positive and negative. Like charges repel and unlike charges attract.

- The fundamental unit of electric charge is the **Coulomb (C).**

- All particles known to exist on their own in nature have charges that are either zero or multiples of the charge on the proton: $e = 1.60 \times 10^{-19}$ **C.** (The symbol is named for the electron, which has the same magnitude of charge, but the opposite sign.). This charge is known as the **fundamental charge**.

- Electric charge is **conserved** meaning that it is neither created or destroyed, only transferred.

- Coulomb's Law gives the force between two charged particles.

Coulomb's Law

$$\vec{F}_{12} = \frac{k|q_1 q_2|}{r^2}\hat{r} = \frac{k|q_1 q_2|}{r^3}\vec{r}$$

Description – Coulomb's Law describes the force, $\vec{F}$, between two charges, q_1 and q_2, separated by a distance r.

Note 1: $k = 8.99 \times 10^9 \, Nm^2/C^2$

Note 2: $\hat{r}$ is the direction of the force on the charge in question. The force is always along a line connecting the two charges.

205-1: Electric Charge

Consider: *What is electric charge and how do charges interact?*

WHAT HAPPENS WHEN YOU WALK across a rug in socks on a cold winter day and then try to touch a metal door handle? You have probably experienced the shock that causes you to recoil, and may have even seen a flash and heard an audible sound. What is happening here? Why does this happen in the middle of winter but not during the summer? Why does it happen in socks, but not in shoes?

We can answer these questions with an understanding of *electric charge* or just charge. Charge is one of the most fundamental properties of matter. In fact, each subatomic particle is imbued with this property we call charge and the macroscopic effects we see are due to how billions of charged particles move and interact. However, in order to start to understand electric charge, we must think about the small particles that carry charge. As discussed earlier, the three main subatomic particles are the proton, the neutron and the electron. By definition, the electron is said to have negative charge and the proton is said to have positive charge. The neutron, on the other hand, has zero electric charge. Although most macroscopic materials contain tremendous numbers of both protons and electrons, the number of each is very nearly equal - which is to say that an electrically neutral object has approximately the same number of protons as electrons.

You might remember that the electron is a considerably smaller particle than either the proton or the neutron. In addition, in an atom, protons and neutrons form the small, tightly bound nucleus of the atom and electrons 'orbit' the nucleus. Some electrons are held onto tightly by their atoms or molecules and some are held with less strength – and this depends upon the molecular structure of the material. This is where the sock and the rug come back into play. When you walk across a wool carpet in cotton socks, the atoms in the wool have a lower affinity for electrons than the cotton atoms; that is to say that when the two materials are rubbed together, some of the electrons moved from the wool to the cotton – or rather from the rug to the socks. Since the socks now have an excess of electrons, they have become *negatively charged* and since the rug has lost electrons, it now has more protons than electrons and it is *positively changed*.

Another important property of charge particles is how they interact. Like charges are repelled from each other and unlike charges are attracted to each other. That is to say that two positively charged objects will repel each other, as will two negatively charged objects, and a positively charged object and a negatively charged object will attract each other. We will discuss this more in the next section.

> In most real-world situations, it is electrons (negative charge) that are transferred when charging or discharging an object.

For now, let's return to your walk across the carpet. We now know that you are positively charged because of the interaction of the cotton sock with the wool carpet. When your hand approaches the doorknob two important things happen. First, the excess positive charge on your skin, which is actually a lack of electrons, tries to attract negatively charged electrons in the doorknob. Since the metal doorknob is a *conductor* charge is free to move and those attracted negative charges (electrons) move to the surface of the doorknob closest to your hand. These excess charges tend to stay on the surface of the material where they are; however, when positively charged hand and the negatively charged doorknob get close enough together, the attraction between the charges becomes very large and the excess electrons can jump across the air gap between the objects – ZAP!

> Like charges repel and unlike charges attract.
>
> **Connection**: **All red blood cells in your body have a slight negative charge so they don't stick together!**

The little scenario we just discussed introduced a number of important concepts related to electric charge that I would like to summarize. First, charge comes in two flavors, positive (carried by protons) and negative (carried by electrons) such that

> - negatively charged objects have an excess of electrons;
> - positively charged objects have too few electrons.

Charge can move in all materials; however, some materials allow charge to move more freely than others, a property known as conductivity. Therefore, in a good conductor, charges are relatively free to move and in a good insulator, charges do not move easily and therefore excess charge tends to stay where it is applied. We can take this idea to the extreme and define

> Ideal Conductor – an object through which charge moves freely
> Ideal Insulator – an object through which charge does not move

Although we tend to think of only metals as conductors, there are other materials that are good conductors. Much of the way your body works is based on the flow of charge into and out of cells, and the near perfect conductor, in this case, is water. Although water is electrically neutral, its shape causes it to act as if one side has a slight positive charge and the other side has a slight negative charge. This "partial charge" reacts with the ions in your body and allow water to act as a near perfect conductor.

Charge is quantized, meaning electric charge comes in discrete little chunks, where the smallest free-floating chunk is the charge on the electron or the proton. The SI unit of charge is the Coulomb (C) and is defined such that one Coulomb of charge is made of $6.242 \, x \, 10^{18}$ protons. The magnitude of charge on an electron or proton can then be written

$$e = |q_e| = 1.60217657 \, x \, 10^{-19}C. \tag{205-1}$$

Again, please note that the charge on an electron is negative and the charge on a proton is positive. We discussed earlier in Physics I how the quarks that make up protons and neutrons have fractional charges relative to the electron; however, isolated quarks are not found in nature and come in chunks that do have integer multiple of the electron charge.

So, what *is* electric charge? It turns out that charge is very hard to define. Electric charge is one of the fundamental quantities associated with subatomic particles; just like mass. Charge, again like rest mass, is a relativistic invariant, meaning that charge does not change with the particle's motion; it is intrinsic to the particle itself.

Electric charge is ***conserved*** – it is not created or destroyed.

Finally, electric charge is a ***conserved quantity***, meaning that the total charge in the universe does not change (as far as we know). So, even though charge can be transferred between materials, as we will see in the next couple of sections, the net charge in a closed system must remain constant.

205-2: Charging Objects - Triboelectricity

Consider: *What determines how objects get excess electric charge?*

We now have at least a bit of understanding of what charge is and know that electrons are one of the primary carriers of charge. So, how does this relate to getting that shock when we reach for the door handle? It turns out that some atoms hold on to their electrons tighter than others. For example, when you are wearing cotton socks and walk across a wool carpet, the cotton in the socks is more likely to lose electrons and the wool in the carpet is more likely to accept electrons, meaning that your socks (and therefore you) become more positive. This type of qualitative description of how materials tend to exchange electrons when in contact with each other is known as ***triboelectricity***. Table 205-1 presents a ***triboelectric series***, which is a list of materials in order of how they *tend* to transfer electrons.

Above I said that the cotton socks become positive when rubbed against the wool carpet. You can see this in the table because wool is above cotton on the table. When two objects are rubbed together, the one that is higher on the triboelectric series *tends* to have a positive charge and the one that is lower tends to be left with a negative charge. Another example would be a rubber balloon rubbed against human hair. Human hair is considerably above the rubber balloon in the series and therefore the hair is left with excess positive charge, while the balloon would be left with excess negative charge.

The triboelectric series is by no means perfect. It is experimentally derived and there are many factors that can lead to different results than the series would suggest. One example of this is humidity in the air – depending on how humid the air is at a given time can change the order of materials on the triboelectric series. However, for simplicity, we will assume that the series shown in table 205-1 holds for the sake of this course.

Table 205-1 Triboelectric Series
Electropositive (lose electrons)
Air
Human Skin
Adhesive on transparent tape
Rabbit fur
Glass
Mica
Human Hair
Wool
Fur
Silk
Paper
Cotton
Wood
Amber
Rubber Balloon
Hard Rubber
Nickel
Copper
Acetate
Polyester
PVC (Polyvinyl chloride)
Electronegative (gain electrons)

Example 205-1: **The Tape Did What?**

We can explore a number of the properties just discussed by doing a simple experiment with cheap transparent tape. I would suggest you grab a couple pieces of transparent tape and follow along, but you can also visualize what happens if you don't have tape on hand. It also helps to do this with a friend, but you can do it yourself if necessary.

The Setup:

First, you need four pieces of tape, each approximately 4-5 inches long. As you remove each piece from the spool, fold over just a small piece (~1 cm) at the top to make a **handle** (as shown). Then, stick each piece of tape to the side of a table while you prepare the others.

Now, using a pen or pencil, mark two of the pieces of tape with a "T" (for top) and two with a "B" (for bottom), and place them back against the table.

The Experiment:

Grab the handle of a "T" in your right hand and a "B" in you left hand, both with the tacky side facing away from you. Place the two piece of tape together so that the tacky side of "T" (in your right hand) is up against the non-tacky side of "B" (in your left hand). Use one to make sure that the two pieces of tape are completely stuck to each other.

Then, grab each handle and quickly pull the two pieces apart. You can now place the two pieces back on the side of the table and follow the same direction for the other T and B.

Now, in turn, bring two T's together, two B's together and then a T close to a B and observe what happens. You can do this by either holding two pieces by the handles you created, or you can hold one piece while the other dangles from the table.

What did you find?

If everything went well, the two T's and two B's should have repelled each other and the T and B should have attracted each other and probably stuck together. Kind of cool, eh?

What you did was transfer charge just like what happens with the cotton socks and the wool carpet. The difference here is that we were able to control the situation and make observations as opposed to being shocked!!

Modern transparent tape is made of acetate. If you compare "Acetate" and "Adhesive on transparent tape" in the triboelectric series above, you can see that the T tape became positively charged (lost electrons) and the B tape became negatively charged. Personally, this always seems backwards to me, but it has been verified by many experiments and since they are so far apart on the triboelectric series, it is almost always this way.

205-3: Coulomb's Law

Consider: *How can we quantify the force on charged particles?*

We saw in the last example that charged objects tend to either attract each other or repel each other. Put another way, they exert a force on each other. Careful measurements have revealed that this force is proportional to how much charge is on each object as well as how far separated they are. For point charges, this relationship is described by Coulomb's Law:

Coulomb's Law

$$\vec{F}_{12} = \frac{k|q_1 q_2|}{r^2}\hat{r} = \frac{k|q_1 q_2|}{r^3}\vec{r} \qquad (205\text{-}2)$$

Description – Coulomb's Law describes the force, $\vec{F}$, between two point charges, q_1 and q_2, separated by a distance r.

Note 1: $k = 8.99 \times 10^9 \, Nm^2/C^2$

Note 2: $\hat{r}$ is the direction of the force on the charge in question. The force is always along a line connecting the two charges.

The constant of proportionality in Coulomb's Law, k, is known as Coulomb's Constant and has a value of $k = 8.99 \times 10^9 \, Nm^2/C^2$. The Coulomb Constant is related to one of the fundamental constants of physics, known as the **permittivity of free space**, $\epsilon_0 = 8.85 \times 10^{-12} C^2 N^{-1} m^{-2}$. Today, we tend to use k in Coulomb's Law because it is easier to write and still contains all of the important information about the force. However, we will see the permittivity of free space a number of times, so I want to introduce it now. Written with this fundamental constant, Coulomb's Law is

44

$$\vec{F}_{12} = \frac{1}{4\pi\epsilon_0}\frac{|q_1 q_2|}{r^2}\hat{r} = \frac{1}{4\pi\epsilon_0}\frac{|q_1 q_2|}{r^3}\vec{r}. \tag{205-2}$$

I also want to take just a moment to point out the similarities between Coulomb's Law and Newton's Universal Law of Gravity:

$$\vec{F}_{12} = \frac{Gm_1 m_2}{r^2}\hat{r}. \tag{205-3}$$

The two equations have the same form with constants of proportionality (G and k), "charges" (electric charge, q, and mass charge, m) and the square of the separation between the charges. It's interesting that two of the fundamental interactions have such similar macroscopic relationships.

Example 205-2: How big is a Coulomb?

What is the magnitude of the force between two one Coulomb charges separated by one meter?

Solution:

This is a direct application of Coulomb's Law

$$F_{12} = \frac{k|q_1 q_2|}{r^2} = \frac{\left(8.99 \; x \; 10^9 \frac{Nm^2}{C^2}\right)|(1C)(1C)|}{(1\;m)^2}$$

$$F_{12} = 8.99 \; x \; 10^9 \; N$$

This is a huge force and helps reiterate that 1 C of charge is an immense amount of charge. The net charge on objects is more along the line of mC, μC and nC.

Example 205-3: One Dimensional Example

Imagine that a 2 μC point charge is placed at the origin of a coordinate system. A -4 μC point charge is then placed at 2 cm along the positive x-axis. What is the force (magnitude and direction) of the force on the 2 μC charge due to the 4 μC charge?

Solution:

Although this is still a direct application of Coulomb's Law, we must now be careful to include the vector nature of the law. First, drawing a quick sketch, the situation looks like this:

There are two slightly different ways we can approach this: First, we could qualitatively say that since one of the charged particles is positive and one is negative, we would expect them to attract each other. Therefore, the 2 μC charge should feel a force to the right (+x-direction). Then, we can find the magnitude of the force

$$F_{12} = \frac{k|q_1 q_2|}{r^2} = \frac{\left(8.99 \; x \; 10^9 \frac{Nm^2}{C^2}\right)|(2 \; x \; 10^{-6}C)(-4 \; x \; 10^{-6}C)|}{(2 \; x \; 10^{-2}\;m)^2}$$

$$F_{12} = 180 \; N$$

So, we would describe the force as 180 N to the right.

Alternatively, we could write this as a full vector equation. I will use the second version of Coulomb's Law since we know the vector separating the two charged particles is

$$\vec{r} = \begin{bmatrix} 2 \; x \; 10^{-2} \; m \\ 0 \\ 0 \end{bmatrix}.$$

Therefore

$$\vec{F}_{12} = \frac{k|q_1 q_2|}{r^3}\vec{r}$$

$$= \frac{\left(8.99 \; x \; 10^9 \frac{Nm^2}{C^2}\right)|(2 \; x \; 10^{-6}C)(-4 \; x \; 10^{-6}C)|}{(2 \; x \; 10^{-2}\;m)^3}\begin{bmatrix} 2 \; x \; 10^{-2} \; m \\ 0 \\ 0 \end{bmatrix}.$$

Which reduces to

$$\vec{F}_{12} = \begin{bmatrix} 180 \; N \\ 0 \\ 0 \end{bmatrix}.$$

This form gives us the exact same information as before; however, it is good practice for when we will need to use the vector forms below.

Newton's 3rd Law

It is important to note that everything we learned about forces in Physics I holds now, including Newton's 3rd Law. Since we found the force on the 2 μC due to the 4 μC charge, we can immediately write the force on the 4 μC charge due to the 2 μC charge as

$$\vec{F}_{21} = \begin{bmatrix} -180\,N \\ 0 \\ 0 \end{bmatrix},$$

That is, it has the same magnitude and opposite direction.

Example 205-4: Two-dimensional Fun

Consider the square of charges as shown below. Each side of the square is 21 μm long.

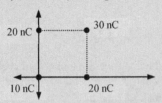

Find the force on the 30 nC charge due to the other three charges.

Solution:

First, I will label each charge. Call the 10 nC charge at the origin q_1, the 20 nC charge on the x-axis as q_2, the 20 nC charge on the y-axis q_3 and the 30 nC charge q_4.

As with all forces, the net force on q_4 is found by first determining the force on q_4 due to each of the other charges individually and then adding them using the superposition principle.

First, think about this qualitatively. Since all of the charges are positive, they should all be repulsive. Therefore, q_1 will push q_4 up and to the right, q_2 will push q_4 up and q_3 will push q_4 to the right. Therefore, I would expect the net force to be up and to the right. Also, given the symmetry of the situation, I would expect that the force will be at 45 degrees north of east.

Let's first find the force on q_4 due to q_2.

$$\vec{F}_{42} = \frac{k|q_1 q_2|}{r^3}\vec{r}$$

$$= \frac{\left(8.99\ x\ 10^9 \frac{Nm^2}{C^2}\right)|(30\ x\ 10^{-9}C)(20\ x\ 10^{-9}C)|}{(21\ x\ 10^{-6}\ m)^3}\begin{bmatrix} 0 \\ 21\ x\ 10^{-6}\ m \\ 0 \end{bmatrix}$$

$$\vec{F}_{42} = \begin{bmatrix} 0 \\ 1.22\ x\ 10^4 N \\ 0 \end{bmatrix}$$

Next, I will find the force on q_4 due to q_3:

$$\vec{F}_{43} = \frac{k|q_1 q_2|}{r^3}\vec{r}$$

$$= \frac{\left(8.99\ x\ 10^9 \frac{Nm^2}{C^2}\right)|(30\ x\ 10^{-9}C)(20\ x\ 10^{-9}C)|}{(21\ x\ 10^{-6}\ m)^3}\begin{bmatrix} 21\ x\ 10^{-6}\ m \\ 0 \\ 0 \end{bmatrix}$$

$$\vec{F}_{43} = \begin{bmatrix} 1.22\ x\ 10^4 N \\ 0 \\ 0 \end{bmatrix}$$

Note that this force has the same magnitude as the force from q_2, which we could have expected considering they have the same charges. Using this, we could have written the force from q_3 down right away, but I chose to write the whole thing out.

For the force on q_4 due to q_1, we must first find the distance between the two charged particles. To do this, we use the Pythagorean theorem:

$$r_{14} = \sqrt{(21\ x\ 10^{-6}m)^2 + (21\ x\ 10^{-6}m)^2}$$
$$= 2.97\ x\ 10^{-5}\ m$$

Now, since we know the angle of the force should be at 45 degrees (along the line connecting them), I will use the first version of Coulomb's Law

$$\vec{F}_{41} = \frac{k|q_1 q_2|}{r^2}\hat{r}$$

$$= \frac{\left(8.99\ x\ 10^9 \frac{Nm^2}{C^2}\right)|(30\ x\ 10^{-9}C)(10\ x\ 10^{-9}C)|}{(2.97\ x\ 10^{-5}\ m)^2}\begin{bmatrix} \cos 45 \\ \sin 45 \\ 0 \end{bmatrix}$$

$$\vec{F}_{41} = \begin{bmatrix} 2.16\ x\ 10^3 N \\ 2.16\ x\ 10^3 N \\ 0 \end{bmatrix}.$$

Finally, the net force is given by the sum of the three forces we have found:

$$\vec{F}_4 = \vec{F}_{42} + \vec{F}_{43} + \vec{F}_{41}$$

$$\vec{F}_4 = \begin{bmatrix} 0 \\ 1.22\ x\ 10^4 N \\ 0 \end{bmatrix} + \begin{bmatrix} 1.22\ x\ 10^4 N \\ 0 \\ 0 \end{bmatrix} + \begin{bmatrix} 2.16\ x\ 10^3 N \\ 2.16\ x\ 10^3 N \\ 0 \end{bmatrix}$$

$$\vec{F}_4 = \begin{bmatrix} 1.44\ x\ 10^4 N \\ 1.44\ x\ 10^4 N \\ 0 \end{bmatrix}$$

Discussion:

This solution contains many of the features we were expecting. First, the net force is up and to the right (both x- and y-components are positive) and, in fact, it is at 45 degrees since both the x- and y-components are equal.

This example shows the important features for solving Coulomb's Law problems with multiple charges. First, find the force from each *pair*, and then sum them to find the net force.

206 – Introduction to Electric Fields

We begin our discussion of individual electric and magnetic fields. The basic tenet we will follow is that charges cause electric fields, moving charges create magnetic fields and accelerating charges create electromagnetic radiation. In this unit, we will explore the basic idea of the electric field – a field created by all charged particles that pervades all of space.

Integration of Ideas

Review the idea of a field from Physics 1.

Review the idea of forces in 2-D and 3-D.

Review the idea of charges from unit 205.

The Bare Essentials

- The electric field is analogous to the gravitational field: the gravitational force on an object is the mass of the object multiplied by the gravitational field at the point of the object, and the electric force on a charged particle is the charge of the particle multiplied by the electric field at the point of the charge.

- The electric field in a region of space is defined by the force on a *test particle* placed in the electric field.

Definition of Electric Field

$$\vec{F} = q\vec{E} \quad \text{or} \quad \vec{E} = \frac{\vec{F}}{q}$$

Description – This equation defines the electric field in terms of the force $\vec{F}$ on a charge q in a field caused by other charges, $\vec{E}$.

Notes: This is a definition.

- Electric field lines give us a way to visualize the electric field in a region of space.
  - Electric field lines point towards a negative point charge and away from a positive point charge,
  - Electric field lines show the direction of the force on a positive particle placed on that line,
  - The density of electric field lines is proportional to the strength of the electric field in that region.

- The electric field due to a point charge is a vector field that is proportional to the charge of the particle and inversely proportional to the square of distance from the charged particle to the point in space we want to find the field.

Electric Field due to a Point Charge

$$\vec{E} = \frac{kq}{r^2}\hat{r}$$

Description – This equation describes the electric field (vector) produced by a charge q, in a region of space a distance r from the charged particle. $\hat{r}$ is the directional for the field (directed radially away from the point charge).

Note 1: $k = 8.99 \times 10^9 \; Nm^2/C^2$

Note 2: If the point charge is negative, the electric field vector will point towards the point charge, however, $\hat{r}$ still points away. The negative in the charge will change the direction of the field.

- The electric field in a region of space due to a number of point charges is found by summing the electric field of each individual point charge at that point.

206-1: The Notion of a Field

Consider: *What is a field in physics? Have we studied some already?*

We have dealt with fields through our study of Physics I without much thought. Although many authors call $\vec{g} = 9.8 \, m/s^2$ down, the acceleration due to gravity, we chose to use the far more accurate term *gravitational field strength*. The mass within the earth creates a *gravitational field* around the entire planet, such that if we place a massive object at any place near the surface of the earth, the object will feel a force directed towards the center of the earth with a force given by $\vec{F} = m\vec{g}$. Let's deconstruct this equation:

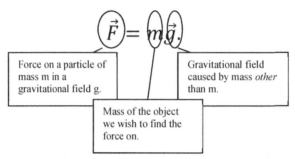

We can even make this discussion more precise by considering Newton's Universal Law of Gravitation,

$$\vec{F} = \frac{Gm_1m_2}{r^2}\hat{r}. \tag{206-1}$$

In this law, G is the Universal Gravitational Constant, m_1 and m_2 are the masses that are interacting and r is the distance between the two masses. The $\hat{r}$ represents the fact that the two masses are attracting each other along the line that connects them. It turns out that our two force law equations are exactly the same. We can see this by rearranging the Universal Law of Gravitation just a slight bit,

$$\vec{F} = m_1\left(\frac{Gm_2}{r^2}\right)\hat{r}. \tag{206-2}$$

If you plug in values for the gravitational constant, the mass of the earth and the radius of the earth for the term in parentheses above, you get 9.81 m/s², or the gravitational field strength at the surface of the earth. The important point about the Universal Gravitational law is that it is, well, more universal – it can be used anywhere, the surface of the earth or the moon, or even 10,000 km above the surface of the sun. To see this, let's break down this equation, the same way we did for out weight equation above:

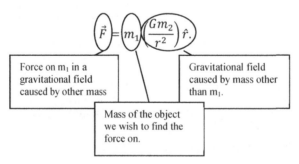

So, now you can see, if we define little-g a bit more generally as

$$\vec{g} = \left(\frac{Gm_2}{r^2}\right)\hat{r}, \tag{206-3}$$

we have a *gravitational field* a distance r away from a massive object with mass m_2. Then, we know that if we place an object of mass m_1 in that field, we can find the gravitational force on m_1. This is the general idea of a ***vector field*** in physics and it turns out that each of the fundamental interactions has at least one (and often more) of these fields associated with it.

The rest of this unit will be devoted to developing your sense for the field created by *electrically* charged particles – called the electric field.

206-2: The Electric Field

Consider: *Why is a field so important when talking about electric charges?*

Just as massive particles create a gravitational field given by $\vec{g} = \left(\frac{Gm}{r^2}\right)\hat{r}$, any charged particle creates an ***electric field***,

$$\vec{E} = \frac{kq}{r^2}\hat{r}. \tag{206-4}$$

The gravitational interaction is always attractive (as far as we know!), which is because we only know one type of gravitational "charge," which is mass. Since electric charge comes in two flavors, positive and negative, we have to be careful with the equation for the electric field. Consider a positive point charge, q_1, placed at the origin of a coordinate system as shown in the Figure 206-1. If we now place another positive point charge, q_2, somewhere along the positive *x-axis*, what is the direction of the force on q_2 due to q_1? Since two positive charges repel each other, q_1 will feel a force directed to the right. Similar to our discussion of the gravitational force above, this is because a charge placed in an electric field will feel an electric force given by the relationship

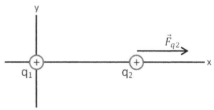

Figure 206-1. Setup describing the force between two charged particles

$$\boxed{\vec{F}_E} = \boxed{q}\boxed{\vec{E}.}$$

Force on the charge, q, due to the electric field created by other charges.

Charge of the object we wish to find the force on.

Electric field produced by other charges

This electric field pervades all of space, and is such that if a charge particle is placed into this electric field, it will feel a force given by

$$\vec{F}_1 = q_1\vec{E} \tag{206-5}$$

This equation would be read "The force on charged particle q_1 is given by the product of its charge with the electric field produced by other charges at the point of charge q_1." This may seem like a long winded way to describe the equation, but we have to be very careful – a charge particle does *not* respond to its own electric field, only to the electric field of the *other* particles.

An electric charge does not respond to its own electric field, only fields created by other charges.

The relationship between electric field and force shown above is actually used as the definition of the electric field – that is, an electric field exists in all of space such that if we placed a charged particle at that point, it would feel an electric force.

Definition of Electric Field

$$\vec{F} = q\vec{E} \quad \text{or} \quad \vec{E} = \frac{\vec{F}}{q} \tag{206-5}$$

Description – This equation defines the electric field in terms of the force $\vec{F}$ on a charge q in a field caused by other charges, $\vec{E}$.
Notes: This is a definition.

The electric field is real! You cannot see it directly, and it may feel very new and artificial to you, but just as there is gravity all around since the earth is under your feet, there are electric fields around you due to rogue charged particles.

Field from Force

If an electron is undergoing an instantaneous acceleration of 123 m/s² directed 30 degrees north of east, what is the electric field at the position of the electron?

Solution:

This problem has two parts to it. First, we must find the net force on the electron causing its acceleration and then we can use our relationship between force and field to find the electric field at the position of the electron.

Force:

We know that the net force on the electron is related to the acceleration by

$$F_{net} = ma.$$

Therefore

$$F_{net} = (9.11x10^{-31}kg)\left(123\frac{m}{s^2}\right) = 1.12x10^{-28}N.$$

The direction of this force must be the same direction as the acceleration.

Electric Field:

Now we have to make the assumption that the electric force is the only force acting on the electron at that moment, so that

$$E = \frac{F}{q}.$$

Therefore,

$$E = \frac{1.12x10^{-2}\ N}{-1.60x10^{-19}C} = -7.00x10^{-10}N/C.$$

The negative sign tells us that the electric field is in the opposite direction of the force, and is therefore directed 30 degrees south of west.

This problem exposes an important point. For electrons, and other negatively charged particle or particles, the electric force and electric field will be in opposite directions to each other. This can be seen from the vector form of the equation

$$\vec{F} = q\vec{E}.$$

Remember that a negative in a vector equation means the opposite direction!!

Electric Kinematics

A proton, initially moving at $6\ x\ 10^5\ m/s$ enters a region of electric field, with the field pointed in the same direction as the velocity of the proton. If the proton is measured to move 80 cm in 1 ms, what is the magnitude of the electric field in that region, assuming the electric field is constant?

Solution:

Knowing that the electric field is constant, means that the electric force on the proton will also be constant (F = qE). Therefore, we can use our kinematic relationships from Physics I. We know initial velocity, change in position and time, so I choose

$$x_f = x_i + v_i t + \frac{1}{2}at^2.$$

The only thing we do not know in this equation is the acceleration; however

$$a = \frac{F}{m} = \frac{qE}{m}.$$

Therefore,

$$\Delta x = v_i t + \frac{1}{2}\frac{qE}{m}t^2,$$

or

$$E = \frac{2m(\Delta x - v_i t)}{qt^2}$$
$$= \frac{2(1.67x10^{-2}\ kg)\left(0.8m - 6x10^5\ m/s\ (1x10^{-3}s)\right)}{(1.6x10^{-19}C)(1x10^{-3}s)^2}.$$

$$E = 4.18x10^{-3}N/C$$

Example 206-3: Electric Kinematics 2

Let's say the same proton in example 206-2 instead entered the region of the electric field perpendicular to the field as shown in the figure. What would be the proton's velocity after 1 ms in the field?

Solution:

From example 206-2, we know the proton's initial velocity is 6×10^5 m/s, which we will now say is in the +x-direction (to the right in the figure). The electric field was found to be $4.18 \times 10^{-3} N/C$, which we now take to be in the $-z$- direction (down in the figure above).

We now have a two-dimensional kinematics problem, very similar to a projectile motion problem from physics I. Since the electric field direction and initial velocity of the proton are perpendicular to each other, the field has no effect on the x-velocity of the proton (that is, the x-velocity of the proton is constant).

However, in the z-direction, the proton had no initial velocity, but will have an acceleration due to the electric field

$$v_{f,z} = v_{i,z} + at = 0 + \frac{qE}{m}t.$$

Therefore

$$v_{f,z} = \frac{(1.6x10^{-1}\ C)(-4.18x10^{-3}N/C)}{1.67x10^{-27}kg}(1x10^{-3}s),$$

where I took the electric field as negative since it is down in the picture. Finishing the calculation gives

$$v_{f,z} = -401\ m/s.$$

Combining the two directions, the final velocity is

$$\vec{v}_f = \begin{bmatrix} 6\ x10^5 m/s \\ 0 \\ -401\ m/s \end{bmatrix}.$$

206-2: The Electric Field of a Point Charge

Consider: *How do we determine the electric field due to a very small bit of excess charge?*

There are direct ways to calculate electric fields due to charged particles and distribution of charges. In the next unit and later in a unit on Gauss's Law, we will look at how to calculate the electric field for distributed charge, but for now, let's stay with our ideas of simple charges particles. Let's combine our equation for electric field with that of Coulomb's Law (force between two charged particles) from last unit:

$$\vec{E} = \frac{\vec{F}_{12}}{q_2} = \frac{1}{q_2}\frac{k|q_1 q_2|}{r^2}\hat{r} = \frac{kq_1}{r^2}\hat{r}. \qquad (206\text{-}6)$$

Because it is derived from Coulomb's Law, this equation is sometimes called Coulomb's Law for the electric field; however, the name is technically reserved for the force law.

Electric Field due to a Point Charge

$$\vec{E} = \frac{kq}{r^2}\hat{r} \qquad (206\text{-}6)$$

Description – This equation describes the electric field (vector) produced by a charge q, in a region of space a distance r from the charged particle. $\hat{r}$ is the directional for the field directed radially away from the point charge

Note 1: $k = 8.99\ x\ 10^9\ Nm^2/C^2$

Note 2: If the point charge is negative, the electric field vector will point towards the point charge, however, $\hat{r}$ still points away. The negative in the charge will change the direction of the field.

Just as with Coulomb's Law, the electric field equation can be used to find the field due to a single charged particle, or it can be used multiple times to find the net field due to a collection of point charges (see section 206-4 below).

Example 206-4: Electric Field Magnitude

What is the magnitude of the electric field 2.7 meters from a $3.2x10^{-16}$ C point charge?

$$|E| = \frac{kq}{r^2}.$$

Solution:

This problem is a direct application of our electric field equation and since we are asked for the magnitude, we do not even need to worry about the directional part of the vector equation:

Plugging in our values, we find

or

$$|E| = \frac{(8.99 \times 10^9 \, Nm^2/C^2)(3.2x10^{-16} \, C)}{(2.7 \, m)^2},$$

$$|E| = 3.95x10^{-7} \, N/C.$$

Example 206-5: Electric Field Vector

An electron is placed at the origin of coordinate system. What is the electric field (magnitude and direction) at a point 3 meters east and 4 meters north of the electron?

Solution:

Like example 206-4, this is a direct application of our electric field equation, except that we now need to include the direction.

First, we start with the equation for electric field from a point charge:

$$\vec{E} = \frac{kq}{r^2}\hat{r}.$$

Remember that the directional, $\hat{r}$, gives the direction of the vector, but has a magnitude of one. The best way to accomplish this is to use trigonometric functions. Since we want to find the electric field at a position 3 meters east and three meters north, this is equivalent to (3,4) on a standard graph as shown to the right.

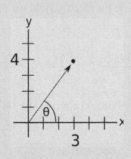

Therefore

$$\hat{r} = \begin{bmatrix} \cos\theta \\ \sin\theta \\ 0 \end{bmatrix} = \begin{bmatrix} 3/5 \\ 4/5 \\ 0 \end{bmatrix},$$

where I have used the fact that the hypotenuse of a 3-4-5 right triangle is, in fact, 5.

The electric field can now be calculated directly using the equation above:

$$\vec{E} = \frac{kq}{r^2}\hat{r} = \frac{\left(8.99 \times 10^9 \frac{Nm^2}{C^2}\right)(-1.6x10^{-19} \, C)}{(5 \, m)^2}\begin{bmatrix} 3/5 \\ 4/5 \\ 0 \end{bmatrix}.$$

$$\vec{E} = \begin{bmatrix} -3.45x10^{-11} \\ -4.60x10^{-1} \\ 0 \end{bmatrix}\frac{N}{C}.$$

Although this is a perfectly good way to report a vector, the problem specifically asks for magnitude and direction:

$$mag(\vec{E}) = \sqrt{E_x^2 + E_y^2 + E_z^2} = 5.75x10^{-1} \, N/C,$$

$$\theta = \tan^{-1}\frac{E_y}{E_x} = 53°.$$

So, the magnitude of the electric field at (3,4) m is $5.75x10^{-11} \, N/C$, and the direction is 53° south of west.

As we will see in the next section, the electric field should point towards a negative charge – if you think about the geometry of this problem, that is exactly true in this case!

206-3: Electric Field Lines

Consider: *How do I visualize a field – specifically the electric field?*

The concept of the electric field is tough! In fact, the whole idea of fields - be it gravitational, electric, magnetic, nuclear – are thought to be some of the hardest ideas to grasp in all of physics. Fortunately, we do have a visual tool that we can use to help us gain an understanding of how particles react to fields. These visualizations are called **field lines**. Let's start off by considering gravity again. As you sit there, you are being attracted to the center of the earth by the gravity. As I described earlier in the chapter, this is because you have mass and are situated in the earth's gravitational field. Now, no matter where you sat on earth, you would still be pulled towards earth's center. So, if I were to try to make a visual of this situation, it might look something like figure 206-2. A couple of observations about the figure: First, each of the lines, known as field lines, points towards the center of the planet. Second, the field lines are closer near the surface of earth than they are farther away. This is actually a representation that the gravitational field is stronger near the surface than farther away. Making a connection again to what we discussed earlier, this fits with our equation for gravitational field

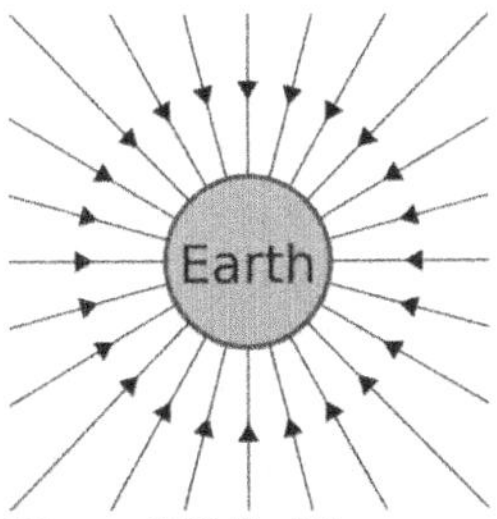

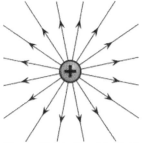

Figure 206-2. Diagram of the earth's gravitational field.

$$\vec{g} = \left(\frac{Gm_2}{r^2}\right)\hat{r}. \qquad (206\text{-}7)$$

In this equation, the r^2 in the denominator makes the field weaker the farther we are from the center of earth.

There are direct analogies between electric field lines and gravitational field lines, just as I tried to make connections between the analytical versions of the fields earlier in the unit. Although it may seem odd to start with, I'd like to consider a negatively charged particle first. Just as mass is attracted to the center of the earth's gravitational field, a positive point charge is attracted directly to the center of a negatively charged particle. Therefore, I would say that its not a huge stretch to write electric field lines for a negative particle like those seen in figure 206-3. Note that the qualitative features are the same as figure 206-2 for the gravitational field, including the fact that the field is stronger near the charge than it is farther away.

Figure 206-3. Electric field of a negative point charge.

Now, consider what would happen if we try to place two positively charged particles near each other. Well, we know that they will repel and, in fact, will repel each other along the line that connects them. There is no known analogy to gravity in this case since gravity is always attractive; however, it is not too much of a stretch so say that the fields lines are qualitatively the same, just that they point away from the positive charge as can be seen in figure 206-4.

When more charges are present, field lines are allowed to curve and bend. There are a few rules and conventions to follow when drawing electric field lines

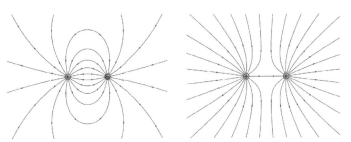

Figure 206-4. Electric field lines of a positive point charge.

1) Electric field lines only begin or end on charges or at infinity.
2) Field lines point away from positive charged particles and towards negative charged particles
3) Electric field lines never cross.
4) The number of field lines a charged particle has is proportional to its charge (i.e. a charged particle with twice the charge will have twice as many field lines).
5) The density of field lines is related to the field strength (field is stronger where field lines are closer together).
6) The actual electric field at a point is tangent to the electric field line passing through that point.

Using these few rules and conventions, it is relatively easy to come up with pictorial representations of more complicated fields. To gain a fuller view of field lines we'll consider the two fields depicted in figure 206-5. The picture on the left in this figure shows what is

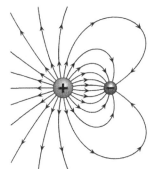

Figure 206-6. Electric field between positive and negative point charges of different magnitudes.

Figure 206-5. Electric field lines between a positive and a negative point charge (left) and between two positive point charges (right).

known as an *electric dipole*, that two equal and opposite charges separated by a small distance. You can see in this figure how the field lines emerging from the positive charge curl towards the negative charge except for the field line directed exactly away from the negative charge. The two charges have the same number of field lines since they have the same magnitude of charge, and going by the picture, you might guess that the field is strongest directly in between the two charges (which it is).

In the figure on the right, you can see how the field lines curl away from each other. Since field lines can never cross, they are 'repelled,' which forms a vertical line directly in between the charged particles where the field is zero (no field lines). You could also guess that the field is strongest near each of the point charges (which it is), although the field lines do not give us enough information to know exactly where.

Finally, consider figure 206-6, which shows a situation where the positively charged particle has three times the magnitude of the negatively charged particle. Note how the positive particle has twice the number of field lines as the negative particle and how the four field lines closest to the negative particle were curved around to end on that particle whereas the remaining eight field lines go to infinity. All of the other characteristics can be read from this diagram the same way – close field lines means stronger field, direction of lines gives direction of field, etc.

Again, field lines as we've just developed them are a tool that we can use to visualize the characteristics of an electric field. The field lines themselves do not exist although the actual fields do!

206-4: Superposition of Electric Fields

Consider: *What happens if there is more than one field? Can they be combined?*

If we want to find the electric field due to multiple point charges, we use the **superposition principle** for fields. As with all other superposition principles, the superposition principle for fields states that the net field at a point in space is the vector sum of all the fields due to an individual particle. Mathematically, this is written

$$\vec{E}_{total} = \vec{E}_1 + \vec{E}_2 + \vec{E}_3 + \cdots \qquad (206\text{-}8)$$

The next two examples will show how to apply this principle to multiple charges.

Example 206-6: Two point charges

A point mass with a charge of $1.23 \times 10^{-5}\ C$ is placed at $x = 5\ cm$, and a point mass with a charge of $2.34 \times 10^{-5}\ C$ is placed at $x = -5\ cm$. What is the net electric field at the origin of the coordinate system ($x = 0$).

Solution:

Following equation 206-8, this key to this problem is to find the electric field at $x = 0$ due to each point mass individually and then add the electric fields together.

Although this appears to be a one dimensional problem, let's use column vectors for consistency with higher dimensional problems. Since the electric field points away from positive charges, we would expect the field to look like the diagram below. First, let's consider the positive charge at $x = 5\ cm$. The direction from x=5 to $x = 0$ is along the negative x-axis, so the directional will be

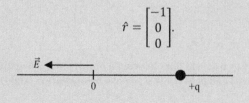

$$\hat{r} = \begin{bmatrix} -1 \\ 0 \\ 0 \end{bmatrix}.$$

Therefore, the electric field at the origin due to the positive point charge at x = 5 cm is

$$\vec{E}_1 = \frac{kq}{r^2} \begin{bmatrix} -1 \\ 0 \\ 0 \end{bmatrix} = \frac{(8.99 \times 10^9 \frac{Nm^2}{C^2})(1.23 \times 10^{-5}\ C)}{(0.05\ m)^2} \begin{bmatrix} -1 \\ 0 \\ 0 \end{bmatrix},$$

$$\vec{E}_1 = -4.42 \times 10^7 \frac{N}{C} \begin{bmatrix} 1 \\ 0 \\ 0 \end{bmatrix},$$

which is along the –x-axis as expected. Similarly, we can find the electric field at the origin due to the negative charge at $x = -5\ cm$. The direction from $x = -5$ to $x = 0$ is directly to the right ($+x$), so we find

$$\vec{E}_2 = \frac{kq}{r^2} \begin{bmatrix} 1 \\ 0 \\ 0 \end{bmatrix} = \frac{(8.99 \times 10^9 \frac{Nm^2}{C^2})(-2.34 \times 10^{-5}\ C)}{(0.05\ m)^2} \begin{bmatrix} 1 \\ 0 \\ 0 \end{bmatrix},$$

$$\vec{E}_2 = -8.41 \times 10^7 \frac{N}{C} \begin{bmatrix} 1 \\ 0 \\ 0 \end{bmatrix}.$$

So combined:

$$\vec{E}_T = \vec{E}_1 + \vec{E}_1 = -1.28 \times 10^8 \frac{N}{C} \begin{bmatrix} 1 \\ 0 \\ 0 \end{bmatrix}.$$

Example 206-6: A square

Charged particles are made to form a square as shown to the figure at the right. The distance between each charge is a. What is the electric field at the center of the square in terms of q and a?

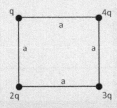

Solution:

Just as in example 206-5, we must find the electric field at the center of the square due to each individual charge and then add them together.

First, for each charge, the distance to the center of the square is

$$r = \sqrt{(a/2)^2 + (a/2)^2} = a/\sqrt{2},$$

since the distance along both the x- and y- directions to the center of the square is $a/2$.

We must also find the directional of the electric field at the center due to each charge. For q (upper left hand corner), we expect the electric field to point away from the charge along the line connecting the corner and the center of the square – an angle of 45°. Therefore, the directional is

$$\hat{r}_q = \begin{bmatrix} \cos 45° \\ -\sin 45° \\ 0 \end{bmatrix},$$

Since it is to the right (positive in the x-component) and down (negative in the y-component).

Similarly, we can write the other directionals as

$$\hat{r}_{2q} = \begin{bmatrix} \cos 45° \\ \sin 45° \\ 0 \end{bmatrix}, \quad \hat{r}_{3q} = \begin{bmatrix} -\cos 45° \\ \sin 45° \\ 0 \end{bmatrix}, \quad \hat{r}_{3q} = \begin{bmatrix} -\cos 45° \\ -\sin 45° \\ 0 \end{bmatrix}.$$

We can now write each of the individual electric fields by substituting in the charge, distance and directional for each charge:

$$\vec{E}_q = \frac{kq_q}{\left(a/\sqrt{2}\right)^2} \begin{bmatrix} \cos 45° \\ -\sin 45° \\ 0 \end{bmatrix} = \frac{\sqrt{2}kq}{a^2} \begin{bmatrix} 1 \\ -1 \\ 0 \end{bmatrix},$$

$$\vec{E}_{2q} = \frac{kq_{2q}}{\left(a/\sqrt{2}\right)^2} \begin{bmatrix} \cos 45° \\ \sin 45° \\ 0 \end{bmatrix} = \frac{\sqrt{2}kq}{a^2} \begin{bmatrix} 2 \\ 2 \\ 0 \end{bmatrix},$$

$$\vec{E}_{3q} = \frac{kq_{3q}}{\left(a/\sqrt{2}\right)^2} \begin{bmatrix} -\cos 45° \\ -\sin 45° \\ 0 \end{bmatrix} = \frac{\sqrt{2}kq}{a^2} \begin{bmatrix} -3 \\ 3 \\ 0 \end{bmatrix},$$

$$\vec{E}_{4q} = \frac{kq_{4q}}{\left(a/\sqrt{2}\right)^2} \begin{bmatrix} -\cos 45° \\ \sin 45° \\ 0 \end{bmatrix} = \frac{\sqrt{2}kq}{a^2} \begin{bmatrix} -4 \\ -4 \\ 0 \end{bmatrix}.$$

To find the total electric field, we add the vectors of each of the individual fields

$$\vec{E}_{total} = \vec{E}_q + \vec{E}_{2q} + \vec{E}_{3q} + \vec{E}_{4q}$$

$$\vec{E}_{total} = \frac{\sqrt{2}kq}{a^2} \begin{bmatrix} -4 \\ 0 \\ 0 \end{bmatrix}.$$

This net electric field is pointed towards the left, which makes sense because there is up/down symmetry in terms of where the charges are, but not left/right symmetry – there is more charge on the right side of the square.

Example 206-6: Between electrons - symmetry

What is the electric field at a point directly between two electrons?

Solution:

The electric field lines between two electrons is similar to Figure 206-4. Note that in this figure, there are no electric field lines directly between the positive charges, suggesting that there is no electric field.

Conceptually, we can argue the same conclusion. Let's say the electrons are placed on the x-axis of a coordinate system. The electron on the right produces an electric field at the origin pointed to the right. The electric field on the

Left produces an electric field to the left (at the origin). Both fields would have the same magnitude (same charge and distance), but opposite directions – therefore they cancel.

Mathematically, we would write

$$\vec{E}_{total} = \vec{E}_1 + \vec{E}_2 = \frac{ke}{r^2} \begin{bmatrix} -1 \\ 0 \\ 0 \end{bmatrix} + \frac{ke}{r^2} \begin{bmatrix} 1 \\ 0 \\ 0 \end{bmatrix} = 0.$$

Anyway we look at it, the electric field between two electrons should be zero. Conceptual reasoning of this type can be a very powerful aid to problem solving and is called using *symmetry arguments*.

207 – Calculating Electric Fields

In unit 206, we explored the basics of electric fields, including the field due to individual point charges and simple distributions. In this unit, we continue that discussion and extend it to continuous charge distributions and dipoles. Although the mathematics is more complex in unit 204, the results of this unit more closely match realistic situations that those from unit 203.

Integration of Ideas

Review the basics of electric field from unit 206.
Review the relationship between fields and forces.

Review Newton's Second Law in 2-D and 3-D.

The Bare Essentials

- Charge may be distributed over a line, a surface or a volume.

    Linear Charge Density: $\lambda = Q/L$
    Surface Charge Density: $\sigma = Q/A$
    Volume Charge Density: $\rho = Q/V$

- The electric field due to a small piece of a continuous charge distribution resembles that of a point charge.

Electric Field due to an Infinitesimal Point Charge

$$d\vec{E} = \frac{k}{r^2} dq \, \hat{r}$$

Description – This equation describes the electric field (vector) produced by a very small piece of a continuous charge distribution, dq. $\hat{r}$ is the directional for the field (directed radially away for a positive point charge.
Notes: 1) $k = 8.99 \times 10^9 \ Nm^2/C^2$
2) $\hat{r}$ is the direction of the electric field in a region

---

- To calculate the total electric field due to distributed charges, you must add all of the small pieces of charge together (integrate).

Electric Field due to an Infinitesimal Point Charge

$$\vec{E} = \int \frac{k}{r^2} dq \, \hat{r}$$

Description – This equation describes how the total electric field of a charge distribution is found by integrating over the entire distribution.

- The electric force on a charged particle is just one possible force on the particle and its reaction is given by Newton's Second Law

- An electric dipole is created by the separation of charged particles with the same magnitude and opposite signs. The dipole is described by the electric dipole moment.

Electric Dipole Moment

$$\vec{p_e} = q\vec{d}$$

Description – The dipole moment describes an electric dipole in terms of the charge of one of the particles, q, and the distance of separation between the particles, d.
Note 1: The dipole moment always points from the negatively charged particle to the positively charged particle.
Note 2: The units of $\vec{p_e}$ are the Coulomb-meter ($C \cdot m$)

- An electric dipole in a uniform electric field feels no net force, however, there is a torque on the dipole due to the electric field (the torque may be zero).

Torque due to a Dipole

$$\vec{\tau} = \vec{p} \, x \, \vec{E}$$

Description – This equation defines the torque, $\vec{\tau}$, on a dipole with dipole moment, $\vec{p}$, in an electric field, $\vec{E}$.

207-1: Charge Distributions

Consider: *How do we describe charges that are not at a point, but distributed in some way?*

lthough unit 206 gave an introduction to electric fields and how to calculate the electric field due to a few point charges, these calculations can quickly become intractable as the number of point charges increases. When you consider that an average rain drop has on the order of 10^{21} water molecules, you can realize pretty quickly that situations with a *large* number of charges are very possible. In order to deal with this, we often talk of **charge distributions** – that is areas over which electric charge is spread out. There are three types of charge distribution that we will be concerned with in Physics II – and each relates to the number of dimensions the charge is distributed over:

$$\text{Linear Charge Density: } \lambda = Q/_L \qquad (207\text{-}1)$$

$$\text{Surface Charge Density: } \sigma = Q/_A \qquad (207\text{-}2)$$

$$\text{Volume Charge Density: } \rho = Q/_V \qquad (207\text{-}3)$$

In each of these equations, Q represents the total charge in the system, L is a length, A is a surface area, and V is a volume over which the charge is distributed, respectively. Also note that the Greek letters used for these charge densities are λ *(lambda)*, σ *(sigma)* and ρ *(rho)*, respectively

The volume charge density is probably closest to the density you are used to – mass density. Almost always when dealing with density in fluids or back in Chemistry, you would have dealt with mass density – that is, mass per unit volume.

Example 207-1: Charge Densities

Consider the cylinder to the right with radius 10 cm and length 25 cm. If there is a total charge of 5.1 μC distributed evenly throughout the cylinder, what is its linear and volume charge density?

Solution:

First, you might be asking how it can have both a linear *and* volume charge density? This is a very good question to ask and it has a straightforward answer – if we do not care about the width of the cylinder and want to just treat it like a line of charge (like a wire), the linear charge density may be all we need. However, we may also care about how the charge is distributed throughout, in which case we need the volume charge density.

Either way, the calculation is quite straightforward and just uses our definitions above.

$$\lambda = \frac{Q}{L} = \frac{5.1 \times 10^{-6}C}{25 \times 10^{-2}m} = 2.04 \times 10^{-5}C/m$$

$$\rho = \frac{Q}{V} = \frac{Q}{\pi r^2 L} = \frac{5.1 \times 10^{-6}C}{\pi(10 \times 10^{-2}m)^2(25 \times 10^{-2}m)}$$

$$\rho = 6.49 \times 10^{-4}C/m^3.$$

Extension: Surfaces

Instead, let's say all of the charge on the cylinder above lies only on the surface of the cylinder. What is the surface charge density in this case?

Solution:

Since the charge is now distributed on just a surface, we care about the surface charge density.

Keep in mind that the total surface area of this cylinder is comprised of both the curved surface and both endcaps:

$$\sigma = \frac{Q}{A} = \frac{Q}{2\pi r^2 + 2\pi r L}.$$

$$\sigma = \frac{5.1 \times 10^{-6}C}{2\pi(10 \times 10^{-2}m)^2 + 2\pi(10 \times 10^{-2}m)(25 \times 10^{-2}m)}$$

$$\sigma = 6.49 \times 10^{-1}C/m^2$$

207-2: Electric Field of Charge Distributions

Consider: *How are the electric fields due to charge distributions calculated?*

In unit 206, we found that the electric field created by a charged point particle is given by

$$\vec{E} = \frac{kq}{r^2}\hat{r},$$

(207-4)

and we also discussed how it could become very hard to use this equation if the number of point particles creating the electric field became large. In this section, we are going to learn how to deal with this, all-too-common a situation. First, since we know we are going to deal with a large number of point particles, I will say that the electric field due to just one of those particles is a very small part, $d\vec{E}$, of the total field that we will find, and that this field is caused by a small part, dq, of the total number of charges available. Now, we have

$$d\vec{E} = \frac{k}{r^2}dq\,\hat{r}.$$

(207-5)

The total electric field at a point in space can then be found by adding up all of the little pieces of charge. If the pieces of charge are spread out continuously, this means integrating $d\vec{E}$ to find the complete field.

The problem with this integration is that I don't know how to integrate over dq! We must first use what we know about how the charge is distributed to transform dq into a variable that we can deal with, such as an x or a θ. For example, if charge is distributed evenly over a straight line, I would expect half of the line to have half of the charge, or even ¼ of the line to have ¼ of the charge. Generalizing this, I can write

$$\frac{dx}{L} = \frac{dq}{Q},$$

(207-6)

That is, the ratio of any small piece dx relative to the entire length, L, should be equal to a corresponding small piece of the charge, dq, relative to the total charge, Q. This relationship can then be solved for dq:

$$dq = \frac{Q}{L}dx = \lambda dx.$$

(207-7)

Note that I wrote this equation both with Q/L if those are the variables we have, but also with the linear charge density, λ, since that is defined as Q/L.

Once we have written dq in a form related to the geometry, we can then integrate over the entire area of space containing charge, which will give us the total electric field at a point in space.

Electric Field due to an Infinitesimal Point Charge

$$\vec{E} = \int \frac{k}{r^2}dq\,\hat{r}$$

(207-8)

Description – This equation describes how the total electric field of a charge distribution is found by integrating over the entire distribution.

I will now use this in an example, which I hope will illustrate why these transformations are important.

Example 207-2: Electric field of a line of charge

Consider a length of wire, L, with a total charge Q distributed evenly along it. What is the electric field directly above the center of the wire as a function of distance from the wire?

Solution:

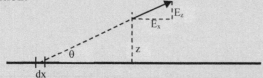

The figure shows a line of charge with length L and total charge Q. I've also shown how one little section of the line (dx) with charge dq would create an electric field at a point a distance z above the center of the line. In addition, the figure shows how the electric field can be broken into components E_x and E_z.

First, we start with our general equation for a small piece of the electric field

$$d\vec{E} = \frac{k}{r^2} dq \, \hat{r}.$$

As noted above, I don't know how to integrate over dq, so I must transform dq into a variable I can handle. Since we have a line of charge, I will use the substitution noted just before this example:

$$dq = \frac{Q}{L} dx.$$

Also, the directional $\hat{r}$ gives us a way to state the direction of the electric field in terms of sines and cosines. Using the figure, I can write

$$\hat{r} = \begin{bmatrix} \cos\theta \\ 0 \\ \sin\theta \end{bmatrix}.$$

Putting this together, we have

$$d\vec{E} = \frac{k}{r^2} \frac{Q}{L} dx \begin{bmatrix} \cos\theta \\ 0 \\ \sin\theta \end{bmatrix}.$$

Inspecting this equation, I want to integrate over x, but the distance r of the little piece dx from where I care about depends on x (Pythagorean theorem), as do the trig functions in the vector. Using standard definitions and our figure, I can then write:

$$r = \sqrt{x^2 + z^2}$$

$$\cos\theta = \frac{adj}{hyp} = \frac{x}{r} = \frac{x}{\sqrt{x^2 + z^2}},$$

$$\sin\theta = \frac{opp}{hyp} = \frac{z}{r} = \frac{z}{\sqrt{x^2 + z^2}}$$

$$r = \sqrt{x^2 + z^2},$$

$$\cos\theta = \frac{adj}{hyp} = \frac{x}{r} = \frac{x}{\sqrt{x^2 + z^2}},$$

$$\sin\theta = \frac{opp}{hyp} = \frac{z}{r} = \frac{z}{\sqrt{x^2 + z^2}}.$$

Now, I can substitute all of these values back into our equation for the electric field

$$d\vec{E} = \frac{k}{x^2 + z^2} \frac{Q}{L} dx \begin{bmatrix} \frac{x}{\sqrt{x^2 + z^2}} \\ 0 \\ \frac{z}{\sqrt{x^2 + z^2}} \end{bmatrix}.$$

Now, this may look like a mess, but what we have done is get everything in terms of constants (k, Q, L) as well as possible variables x and z. We can also treat z as a constant since we are trying to find the electric field *at that point* due to all the little pieces, dx.

All that is left is to simplify the expression a little bit and then take the integral over the entire region of dx that contains charges ($-\frac{1}{2}L$ to $\frac{1}{2}L$).

$$d\vec{E} = \frac{kQ}{L} \begin{bmatrix} \frac{x\,dx}{(x^2+z^2)^{3/2}} \\ 0 \\ \frac{z\,dx}{(x^2+z^2)^{3/2}} \end{bmatrix} \quad \text{or} \quad \begin{aligned} dE_x &= \frac{kQ}{L} \frac{x\,dx}{(x^2+z^2)^{3/2}} \\ dE_z &= \frac{kQz}{L} \frac{dx}{(x^2+z^2)^{3/2}} \end{aligned}$$

These integrals are not easy to do by hand; however, they can be looked up in an integral table:

$$\int_{-\frac{1}{2}L}^{\frac{1}{2}L} \frac{x\,dx}{(x^2+z^2)^{3/2}} = \frac{1}{(x^2+z^2)^{1/2}} \Bigg|_{-\frac{1}{2}L}^{\frac{1}{2}L} = 0$$

$$\int_{-\frac{1}{2}L}^{\frac{1}{2}L} \frac{dx}{(x^2+z^2)^{3/2}} = \frac{x}{z^2(x^2+z^2)^{1/2}} \Bigg|_{-\frac{1}{2}L}^{\frac{1}{2}L} = \frac{L}{z^2\left(\frac{1}{4}L^2+z^2\right)^{1/2}}.$$

These results can now be placed back into our equation for the electric field

$$\vec{E} = \frac{kQz}{L} \frac{L}{z^2\left(\frac{1}{4}L^2+z^2\right)^{1/2}} \begin{bmatrix} 0 \\ 0 \\ 1 \end{bmatrix}$$

$$\vec{E} = \frac{kQ}{z\left(\frac{1}{4}L^2+z^2\right)^{1/2}} \begin{bmatrix} 0 \\ 0 \\ 1 \end{bmatrix}.$$

Extensions:

1) There are some important features to this electric field. First, notice that it is *only* in the z-direction. This is a result of symmetry, and is something that we could have predicted. Take a look at the following figure with just a few more electric field lines drawn in

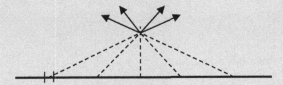

Think about just the x-components of the fields in this figure. It turns out that for every piece of charge to the left of the midpoint of the line, you can find a piece of charge to the right of the midpoint of the line such that their electric fields cancel out. However, every piece of charge creates a vertical component to the electric field that will add together. This is why the x-component canceled out, but the z-component did not.

What I just explained is known as a **symmetry argument**, and these can be used to greatly reduce the amount of calculation that must be done for a field – if we had reasoned out that the x-component of the electric field would be zero through symmetry, we could have not done that part of the calculation!

2) Also, let's look at what happens to the field if we are close to the wire in comparison to the length of the wire, such that $z \ll L$. In this case, we can approximate

$$\left(\frac{1}{4}L^2 + z^2\right)^{1/2} \approx \left(\frac{1}{4}L^2\right)^{1/2} = \frac{1}{2}L.$$

Then, we could write the z-component of the electric field as

$$E_z = \frac{kQ}{z\left(\frac{1}{2}L\right)} = \frac{2kQ}{zL} = \frac{2k\lambda}{z}.$$

Why would I care about this simplification? Well, let's pretend the line of charge is very, very long (infinitely long), then our simplification would hold for almost any distance from the wire, z. What we now see is that our symmetry arguments have made is so that the electric field drops off as $1/r$ from the wire as opposed to the $1/r^2$ that we have for a point charge. This result is actually important enough that I will write it down again on its own.

The electric field a distance r, from an infinitely long line of charge with linear charge density λ, is given by

$$E_z = \frac{2k\lambda}{r} = \frac{\lambda}{2\pi\epsilon_0 r},$$

Where I also included the version with ϵ_0 in addition to k.

I did that last example in extreme detail, to show all of the steps required to solve an electric field problem using integration. As noted in the extension, don't forget to use **symmetry arguments** to try and reduce the number of steps you need to take in your calculations.

What if charge is not distributed along a straight line? We can use the same reasoning in terms of ratios – always. As another example, think about charge distributed over a ring – basically a line of charge that is wrapped around itself to form a circle. A ring contains circular symmetry, meaning that if we rotate it about an angle along the central axis, it does not change. That leads me to believe that the variable of interest might be the angle, θ. So, I'm going to say that if charge is uniformly distributed around a ring that the charge in a small portion of the ring relative to the entire ring should be held in a small angle of the ring relative to the entire ring:

$$\frac{d\theta}{2\pi} = \frac{dq}{Q}, \tag{207-9}$$

where I used the fact that once around a ring is an angle of 2π. Like we did before, this can be solved for dq:

$$dq = \frac{Q}{2\pi}d\theta. \tag{207-10}$$

Example 207-3: Electric field on the axis of a charged ring.

Calculate the electric field on the axis of a ring of charge with radius R and total charge Q.

Solution:

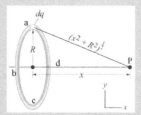

The figure above shows the ring of charge with radius R that holds a total charge Q. We will consider the axis of the ring to be along the x-axis and call the distance from the center to point P (where we want to know the field) x. The figure also shows a little piece of charge, dq, at the top of the ring. As noted before, I do not know how to integrate over dq, and since I have rotational symmetry, I'll use the transformation introduced just before this example:

$$\frac{d\theta}{2\pi} = \frac{dq}{Q} \quad \rightarrow \quad dq = \frac{Q}{2\pi} d\theta.$$

Substituting this into the equation for electric field, we find

$$d\vec{E} = \frac{k}{r^2} dq\, \hat{r} = \frac{k}{r^2} \frac{Q}{2\pi} d\theta\, \hat{r}.$$

This is a problem where symmetry arguments are huge because the electric field at point P is inherently three-dimensional. The electric field from point a is down and to the right at point P. The electric field from point b is to the right and into the page at point P. We are already using all three dimensions (right, down, into the page).

Luckily, point c produces a field that is to the right and up at point P (canceling the downward component from point a) and point d produces a field that is to the right and out of the page (canceling the into the page component of point b).

We could extend these arguments to every point around the ring: Each produces an electric field component to the right, but all other components will be canceled by the point on the opposite side of the ring!

This argument allows us to say the only component of the electric field we need to calculate is the x-component.

Another important thing to keep in mind is that θ is an angle *around the ring*. The component of the electric field in the x-direction has to do with the angle between the x-axis and radius vector, which I'll call ϕ:

$$\cos\phi = \frac{x}{\sqrt{x^2 + R^2}}.$$

Putting this all together

$$dE_x = \frac{k}{r^2} \frac{Q}{2\pi} d\theta \cos\phi = \frac{k}{x^2 + R^2} \frac{Q}{2\pi} d\theta \left(\frac{x}{\sqrt{x^2 + R^2}}\right),$$

which simplifies to

$$dE_x = \frac{kQ}{2\pi} \frac{x}{(x^2 + R^2)^{\frac{3}{2}}} d\theta.$$

In order to get the total electric field, we now integrate over $d\theta$ around the entire right. Inspecting dE_x you'll notice that all of the quantities are constant except for θ, so

$$E_x = \frac{kQ}{2\pi} \frac{x}{(x^2 + R^2)^{\frac{3}{2}}} \int_0^{2\pi} d\theta = \frac{kQ}{2\pi} \frac{x}{(x^2 + R^2)^{\frac{3}{2}}} 2\pi.$$

Again simplifying, we find

$$\boldsymbol{E_x = \frac{kQx}{(x^2 + R^2)^{\frac{3}{2}}}.}$$

There are two important limits. First, at the very center of the ring ($x = 0$), the electric field is zero. This makes sense since every point on the ring creates an electric field vector pointing directly in, so they all cancel!

The other limit is very far away from the ring ($x \gg R$). In this limit, the R in the denominator becomes inconsequential and the expression reduces to

$$E_x = \frac{kQ}{x^2},$$

which is our equation for the electric field due to a point charge! This makes sense because as we get far from the ring, the ring appears to get smaller and smaller and look more and more like a point!

    The examples in this unit so far use a lot of words, and this is ***very important***! It is almost impossible to plug-and-chug your way through electric field calculations. You must use what you know about the situation to set up the equations for dq and then use symmetry arguments to determine how to go about the integrations. Slowing down and using these techniques will help you to get through the problems with less frustration.

    Here are the steps I've used to help guide me through these problems. It's very helpful when learning a new technique to put it into a framework, and so I strongly suggest you try to follow these steps:

Tips for Solving Electric Field Problems with Distributed Charge:

1) Choose a variable that will help you describe the geometry of your distribution and that you know how to integrate (x, θ, r for example). Determine the limits of integration based on your variable.
2) Use ratios to describe how a small piece of the charge, dq, relates to your chosen variable.
3) Express the distance, r, in the electric field equation in terms of your chosen variable and as few other variables as possible.
4) Use symmetry arguments to determine if any of the electric field components will go to zero.
5) Use trig to write the components you do have to calculate in terms of your chosen variable.
6) Integrate. Hopefully following the above steps will simplify the integration that is something easily tractable or that you can look up in an integral table.

We're going to do one more example and then move on to another topic. We've done two situations where charge was distributed over a line (straight line and a line curved into a circle). What if the charge is distributed over a surface, or over a volume? We're going to attack the surface area question and I'll leave the volume for you to try and work out.

We're going to determine the electric field due to a uniformly charged circular disk along an axis through the center of the disk. If the charge is distributed over the surface area, the transformation that I want to use for dq will be

$$\frac{dA}{A} = \frac{dq}{Q},$$
(207-11)

where A is the surface area of the object. This relationship keeps with everything we've been doing up until now. All it does is assume that every small piece of area (dA) will have the same small amount of charge (dq) relative to the total area and charge. This is again the definition of uniformly distributed.

What is new is that we don't have an automatic way to write down dA. But this is ok, because we can figure it out. We know that the area of a circle is given by

$$A = \pi r^2.$$
(207-12)

If we take the derivative of this equation with respect to r, we find

$$\frac{dA}{dr} = 2\pi r.$$
(207-13)

This can immediately be solved for dA:

$$dA = 2\pi r dr.$$
(207-14)

And now we are there. We can substitute this into our above equation for dq to find

$$\frac{2\pi r dr}{\pi R^2} = \frac{dq}{Q},$$
(207-15)

Where πR^2 is the total area of the ring. Now, all that is left is to solve for dq:

$$dq = \frac{2\pi Q}{\pi R^2} r dr = \frac{2Q r dr}{R^2}.$$
(207-16)

I also want to point out that this could be written as $dq = 2\pi \sigma r dr$ if you knew the surface charge density.

Example 207-4: Electric field on the axis of a charged circular disk.

Calculate the electric field on the axis of a circular disk of charge with radius R and total charge Q.

Solution:

In the last example, we found that a single ring of charge gave an electric field along its axis of

$$E_x = \frac{kQx}{(x^2 + R^2)^{\frac{3}{2}}}.$$

We can now think of this ring as one of many that make up the disk. What happens now is that the electric field from a single ring becomes a small part of the entire electric field and the charge of that single ring a small part of the total charge of the disk. This allows us to write

$$dE_x = \frac{kx}{(x^2 + r^2)^{\frac{3}{2}}} dQ$$

for that single disk, acknowledging, also, that the radius of each disk could be different (which is why it is r and not R). Substituting what we know for dQ for an entire disk (described just before the example) gives us

$$dE_x = \frac{kx}{(x^2 + r^2)^{\frac{3}{2}}} \frac{2Qrdr}{R^2}.$$

Finally, to form the full disk, we need only integrate this equation for each little disk for disks of radius zero up to disks of radius R:

$$E_x = \int_0^R \frac{kx}{(x^2 + r^2)^{\frac{3}{2}}} \frac{2Qrdr}{R^2}.$$

Taking out the constants gives us

$$E_x = \frac{2kxQ}{R^2} \int_0^R \frac{rdr}{(x^2 + r^2)^{\frac{3}{2}}},$$

which can then be integrated to find

$$E_x = \frac{-2kxQ}{R^2} \frac{1}{\sqrt{x^2 + r^2}} \Big|_0^R = \frac{-2kxQ}{R^2} \left(\frac{1}{\sqrt{x^2 + R^2}} - \frac{1}{x} \right).$$

By pulling an x out of the denominator and switching the terms in the parentheses to get rid of the negative sign, the solution can be simplified just a bit

$$E_x = \frac{2kQ}{R^2} \left(1 - \frac{1}{\sqrt{1 + (R/x)^2}} \right).$$

Extension:

If we are very close to the center of the disk, or if the disk is very big (x << R), then the second term in the parentheses is quite small, and the electric field no longer depends on x:

$$E_x = \frac{2kQ}{R^2} = \frac{\sigma}{2\epsilon_0}.$$

This result is quite important as it says that near a charged plate that is very large compared to how close you are to the plate (think infinitely sized plate), the electric field is constant and simply points away from the plate.

207-3: Motion of Charged Particles in Electric Fields

> **Consider:** *Since electric fields create forces on electrically charged particles, how do we use Newton's Second Law to account for motion?*

Although we have already dealt with this along the way, it is important to reiterate that a charged particle in an electric field feels a force related to that field:

$$\vec{F_e} = q\vec{E}. \tag{207-17}$$

This force may be only one of many forces that act on the particle and determine its trajectory according to Newton's Second Law

$$\vec{F}_{net} = m\vec{a}. \tag{207-18}$$

Example 207-5: Electron near a charged plate.

An electron is initially moving at a speed of 2.61×10^5 m/s parallel to a large charged plate sitting on top of a table. The charged plate has a radius of 2 meters and the electron is initially 2 mm above the plate. What must be the charge density of the plate in order for the electron to move in a straight line? (Include the force of gravity).

Solution:

Near the surface of the charged plate, the electron will experience two forces, gravity and the electric force. Since gravity is directed downward, we know the charge on the charged plate must be positive to create an electric field directed upward at the point of the electron:

$$\vec{F}_{net} = \vec{F}_e + \vec{F}_g.$$

In order for the electron to travel without deflection (straight line), we need the net force on the electron to be zero. Also, since both forces must be along the z-axis, we do not need to continue with the vector symbols, so

$$0 = F_e - F_g \quad \rightarrow \quad F_e = F_g.$$

We can now substitute what we know about the electric and gravitational forces.

$$qE = mg.$$

Now is where our new knowledge of electric fields comes into play. In the extension to example 207-4, we found that the electric field very close to a charged plate is given by

$$E_x = \frac{\sigma}{2\epsilon_0},$$

Which not only can we now use in our equation for the force on the electron, but also introduces the charge density, σ, that we are solving for:

$$q\frac{\sigma}{2\epsilon_0} = mg.$$

This gives us

$$\sigma = \frac{2\epsilon_0 mg}{q}.$$

We can now plug in the known values to find

$$\sigma = \frac{2(8.85x10^{-1}\ C^2m^{-2}N^{-1})(9.11x10^{-31}kg)\left(9.81\frac{m}{s^2}\right)}{(-1.6\ x\ 10^{-19}\ C)}$$

$$\sigma = -9.89\ x\ 10^{-22} C/m^2.$$

This is an impossibly small charge density (1 extra electron every 161 m²) and shows how weak gravity is in relation to other forces!

Example 207-6: Electron near a charged plate.

If the charge density of the plate in example 207-5 is instead given by $\sigma = 1.2 \times 10^{-5} C/m^2$, how far will the electron move before crashing into the plate?

Solution:

This problem is now a kinematics problem. In this case, both the gravitational force and electric force will be directed down (the positively charged plate will attract the negatively charged electron). So, in this case, I will take down as my positive direction and write

$$F_{net} = F_e + F_g = |q|E + mg.$$

Using what we know about the electric field near a large charged plate, we find

$$F_{net} = |q|\frac{\sigma}{2\epsilon_0} + mg$$

$$F_{net} = (1.6x10^{-1}\ C)\frac{1.2x10^{-5}C/m^2}{2(8.85x10^{-12}C^2m^{-2}N^{-1})} + 9.11x10^{-3}\ kg(9.81\ m/s^2)$$

$$F_{net} = 1.08\ x\ 10^{-1}\ N.$$

Note that in the force calculation, the gravitational force is of the order 10^{-30} N and plays very little role when compared to the electric force.

We now have a kinematics problem (projectile motion problem actually), except the acceleration of the electron is

$$a = \frac{F_{net}}{m} = \frac{1.08\ x\ 10^{-1}\ N}{9.11x10^{-3}\ kg} = \frac{1.2x10^{17}m}{s^2}.$$

First, we find the amount of time it takes the electron to reach the plate, remembering that it had zero vertical velocity to start

$$z_f = z_i + v_{0z}t + \frac{1}{2}a_z t^2 \quad \rightarrow \quad t = \sqrt{\frac{2\Delta z}{a_z}}.$$

This gives us

$$t = \sqrt{\frac{2(2x10^{-3}m)}{1.2x10^{17}\ m/s^2}} = 1.8x10^{-1}\ s.$$

Then we can find how far it moved in that time, remembering that there is no acceleration in the x-direction:

$$x_f = x_i + v_{0x}t + \frac{1}{2}a_x t^2 \quad \rightarrow \quad \Delta x = v_{0x}t.$$

Solving, we find

$$\Delta x = (2.61x10^5 m/s)(1.8x10^{-10}s)$$

$$\Delta x = 4.7x10^{-5}m = 47\ \mu m.$$

207-4: Electric Dipoles

Consider: *Are the positively charged nucleus and negatively charged electron cloud of atoms separated in an electric field?*

So, what happens when we place an atom, at rest, in an electric field? From our recent discussions, we know that the proton will want to move in the direction of the field. However, the electron, being negatively charged, will want to move opposite to the field. But the electron and proton also attract each other because one is positive and one is negative. In reality (for reasonable electric fields), all of these things play a part: the proton tried to move with the field, the electron tried to move against the field and their attraction keep them from separating too far. We've created what's called an electric dipole – a positive and negative charge with the same magnitude, separated by some distance.

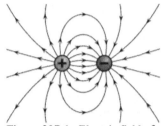

Figure 207-1. Electric field of an electric dipole.

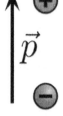

Figure 207-2. Dipole Moment

Without even realizing it, we drew the electric field of a dipole in unit 206 in figure 209-4, which I'm recreating as figure 207-1. Whenever a dipole has formed, we define the dipole moment as the charge of one of particles multiplied by the distance the particles are separated as shown in figure 207-2. The dipole moment is a vector and always points from the negative charge towards the positive charge.

Electric Dipole Moment

$$\vec{p}_e = q\vec{d}$$

Description – The dipole moment describes an electric dipole in terms of the charge of one of the particles, q, and the distance of separation between the particles, d.

Note 1: The dipole moment always points from the negatively charged particle to the positively charged particle.

Note 2: The units of $\vec{p}_e$ are the Coulomb-meter ($C \cdot m$)

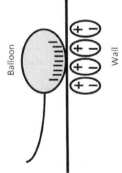

Figure 207-3. A balloon stuck to a wall b induced electric dipoles.

Electric dipoles are the reason you can stick a balloon to a wall after rubbing it on your head! The effect can be seen in figure 207-3. When you rub the rubber balloon against your hair, the balloon tends to become negatively charged close to the wall (see the triboelectric series in unit 206). When you then bring the balloon the balloons electric field creates electric dipoles in the molecules in the wall. But now, the negative charges in the balloon are closer to the positive charges in the wall (attraction) than they are to the negative charges (repulsion). We know that electric force is dependent on the distance between the charged particles, so there it is! The balloon sticks to the wall. This almost seems like circular reasoning – the balloon creates dipoles and then uses dipoles to stick to the wall – but that really *is* the way it works!

Some molecules, such as water, have a **permanent dipole moment**. This comes about because of the shape of the molecule. Consider the picture of water shown in figure 207-4. The oxygen atom in water is sharing electrons with the hydrogen atoms. This leaves the oxygen side of the molecule with a slight negative charge and the hydrogen side with a slight positive charge. The net effect is that all water molecules have this permanent dipole moment. In class, you will see what happens if a charged object is placed near a stream of water.

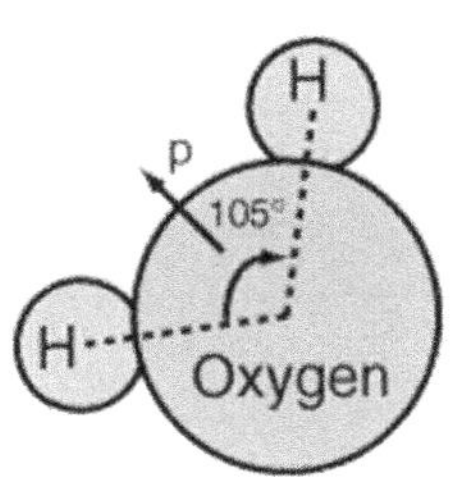

Figure 207-4. The permanent dipole moment of water.

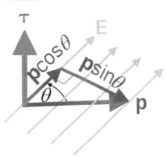

Figure 207-5. Torque of a dipole in an electric field.

Let's think about what happens when we put a dipole in a uniform electric field – it doesn't matter whether we are talking about an induced dipole like the dipoles in the wall, or a permanent dipole like a water. The situation is pictured in figure 207-5. Just like our discussion before, the positive charge will want to move with the field – to the right in the case of figure 207-5. The negative charge, on the other hand, will want to move against the field – to the left. Since the two particles in the dipole are destined to stay together, there is no net force on the system in a uniform electric field. However, they can rotate so that the dipole moment lines up with the external electric field. Since we now have a force causing a rotation, this is a torque! The equation for this torque is relatively easy to obtain from the basic definition of torque:

$$\vec{\tau} = \vec{r}x\vec{F} = 2\left(\frac{1}{2}\,\vec{d}xq\vec{E}\right) = q\vec{d}x\vec{E} = \vec{p_e}x\vec{E}. \qquad (207\text{-}19)$$

Torque due to a Dipole

$$\vec{\tau} = \vec{p_e} \; x \; \vec{E} \qquad (207\text{-}19)$$

Description – This equation defines the torque, $\vec{\tau}$, on a dipole with dipole moment, $\vec{p_e}$, in an electric field, $\vec{E}$.

Example 207-7: Water Rotation

The electric dipole moment of water is $p = 6.2x10^{-30}C \cdot m$. Imagine that an initially stationary water molecule is in region of space where a 120 N/C electric field is instantaneously turned on. If the water's dipole moment is oriented at an angle of 27 degrees to the direction of the electric field, what is the torque on the water molecule at this time?

Solution:

This is a direct application of the torque on a dipole equation.

$$\vec{\tau} = \vec{p_e}x\vec{E}.$$

As a coordinate system, I will say that the electric field is along the x-axis and the initial direction of the dipole moment of the water molecule is 27 degrees from the x-axis towards the y-axis. Therefore

$$\vec{\tau} = 6.2x10^{-30}C \begin{bmatrix} \cos 27 \\ \sin 27 \\ 0 \end{bmatrix} x \begin{bmatrix} 120\,N/C \\ 0 \\ 0 \end{bmatrix}.$$

Expanding this out using the definition of the cross product, we are left with only one term

$$\vec{\tau} = \begin{bmatrix} 0 \\ 0 \\ -(6.2x10^{-30}\,C\,)(\sin 27)(120\,N/C) \end{bmatrix},$$

which gives us

$$\vec{\tau} = \begin{bmatrix} 0 \\ 0 \\ -3.38\ x\ 10^{-28}Nm \end{bmatrix}.$$

The torque is in the negative z-direction as expected since the result of a cross product is perpendicular to the both of the vectors that comprise it.

Note: This problem could also have been done using $|\vec{\tau}| = |\vec{p}||\vec{E}| \sin \theta$, and then finding the direction with the right hand rule.

208 – Creating Magnetic Fields

In this unit, we explore how moving charges create magnetic fields and how different materials create and/or respond to magnetic fields.

Integration of Ideas

Review magnetic fields from the previous unit.
Review cross products.
Review magnetic dipole moments from the previous unit.

The Bare Essentials

- The Biot-Savart Law describes how magnetic fields are created by moving charged particles.

Biot-Savart Law for a Moving Charged Particle

$$\vec{B} = \frac{\mu_0}{4\pi}\frac{q\vec{v} \times \hat{r}}{r^2}$$

Description – This equation describes how a charged particle with charge q, moving with velocity $\vec{v}$ creates a magnetic field, $\vec{B}$, a distance r from the source particle.
Note 1: μ_0 is the permeability of free space and is given by
$\mu_0 = 4\pi \times 10^{-7}\, T \cdot m/A$
Note 2: $\hat{r}$ is directed from the source particle to the point of interest in space

---

Biot-Savart Law for a Piece of Current

$$d\vec{B} = \frac{\mu_0}{4\pi}\frac{I d\vec{l} \times \hat{r}}{r^2}$$

Description – This equation describes how a small piece of wire with length $d\vec{l}$ carrying a current, I, creates a magnetic field, $d\vec{B}$, a distance r from the wire segment.
Note 1: μ_0 is the permeability of free space and is given by
$\mu_0 = 4\pi \times 10^{-7}\, T \cdot m/A$
Note 2: $\hat{r}$ is directed from the source current to the point of interest in space.

- Electrons 'orbiting' the nucleus of an atom create magnetic moments similar to that of a current loop.

- All materials exhibit the property of ***diamagnetism*** where the material itself creates a magnetic field in opposition to a magnetic field applied to it. Diamagnetic materials are repelled by a magnetic field. Diamagnetism is temporary and only exists while the external field is present.

- Some materials exhibit the property of ***paramagnetism*** Paramagnetism generally exists in materials with unpaired electrons and these unpaired electrons cause the material to be attracted by a magnetic field. Paramagnetism is temporary and only exists while the external field is present.

- A few materials, such as iron, nickel, cobalt and a few rare-earth elements, exhibit a cooperative phenomenon known as ***ferromagnetism***, a property whereby the magnetic moments of a large percentage of atoms in a material align and create a *very strong* magnetic field. Ferromagnetism may be very long lasting and exist after the external field is turned off (hysteresis in permanent magnets).

- The magnetic field of the earth resembles that of a bar magnetic. The north geographic pole of the earth is near the south magnetic pole.

208-1: Just a little history

Consider: *How long have we known about magnetic fields?*

WE ALL HAVE EXPERIENCE WITH MAGNETIC FIELDS. Who hasn't used a magnet to stick something to a refrigerator? Haven't we all played with a pair of small magnets and wondered about the mysterious invisible force that causes them to attract or repel? In a colloquial sense, magnets were known to the ancients. Recent historical studies suggest that the Olmec civilization in Central America used a magnetic rock called *lodestone* to make compasses as early 1000 BC. Chinese writings from the 4th century BC also indicate that they not only used lodestone to create compasses, but also that they knew lodestone could attract iron.

Although ancients around the world knew of magnetism, the study of this mysterious force didn't take shape until the 13th century AD when French scientists used very small iron needles to find the shape of the magnetic field around lodestone. From there the study of magnetism starts to take on a real scientific flavor. In 1750, John Mitchell determined that the attraction or repulsion of magnets followed an inverse square law (just like the electric force). Then, in 1785 Charles-Augustin de Coulomb (of Coulomb's Law fame), found that the north and south poles of magnets could not be separated – something we believe to be true to this day.

The truth is we still don't know what causes magnetism on a fundamental level. We know that magnetic fields are caused by moving charges, but just like electric fields, or even electric charge for that matter, the true fundamental mechanism is not well understood.

*Note: In introductory physics, it is common to use the term magnetic field for the field we will be talking about in this unit and the next few units. However, the quantity we call magnetic field is more precisely the **magnetic flux density**. I state this not to be confusing, but rather for any of you that might continue further in your study of physics or electric engineering.*

208-2: Moving charges and magnetic fields

Consider: *Where do magnetic fields come from?*

Now we continue our discussion of how charges change the space around them. If you remember from our introduction to charges

Charged particles produce **electric fields,**
Moving charged particles produce **magnetic fields,**
Accelerating charged particles produce **electromagnetic waves.**

We've already discussed the electric field and had an introduction to electromagnetic waves. Now it is time to delve into how moving charged particles create magnetic fields.

The main relationship between a moving charge and the magnetic field it produces was experimentally determined by *Jean-Baptiste Biot* and *Felix Savart* in 1820. This relationship is called the Bio-Savart Law after them and works well when the moving charges are not accelerating.

Biot-Savart Law for a Moving Charged Particle

$$\vec{B} = \frac{\mu_0}{4\pi} \frac{q\vec{v} \times \hat{r}}{r^2} \qquad (208\text{-}1)$$

Description – This equation describes how a charged particle with charge q, moving with velocity $\vec{v}$ creates a magnetic field, $\vec{B}$, a distance r from the source particle.

Note 1: μ_0 is the permeability of free space and is given by $\mu_0 = 4\pi \times 10^{-7} \, T \cdot m/A$

Note 2: $\hat{r}$ is directed from the source particle to the point of interest in space

Table 208-1. List of magnetic field strengths

Description	B (T)
Cell Phone at Head	2×10^{-7}
Earth (0° latitude)	31×10^{-6}
Earth (50° latitude)	58×10^{-6}
Refrigerator Magnet	5×10^{-3}
Sunspot	0.15
Neodymium Magnet	1.25
MRI Scan	$1.5 - 3$
Frog Levitation	16
Largest Created (continuous)	45

The SI unit of magnetic field is the *Tesla* (T), named after Nicola Tesla, who made major contributions to our understanding of alternating current circuits (see unit 217),

$$1T = 1\frac{N}{C\left(\frac{m}{s}\right)}.$$ (208-2)

One Tesla is a relatively large magnetic field. The magnitude of the earth's magnetic field at the surface of the planet is between 25 and 65 μT. However, typical magnetic resonance imaging (MRI) scans use either 1.5 T or 3 T magnetic fields. Table 208-1 list values for various magnetic fields.

Notice that the Bio-Savart law has a cross product in it, which makes the law inherently three dimensional! Because of this cross product, the magnetic field will always be perpendicular to the velocity of the moving charged particle. The magnetic field is also always perpendicular to a line connecting the charged particle to the point in space where we are trying to calculate the field. *This three-dimensional relationship is conceptually one of the hardest things we will do in Physics!* I would encourage you to pay close attention to the next couple of examples.

Example 208-1: First look at Magnetic Field

If an electron is moving with a speed of 2×10^5 m/s, what is the magnetic field (magnitude and direction) a distance of 0.2 meters directly above and below the electron?

Solution:

The Figure below sets up the situation, with the electron moving in the +x-direction, A is 0.2 m in the +z-direction and B is 0.2 m in the −z-direction relative to the electron.

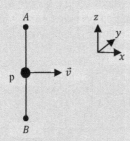

Now, consider the Biot-Savart Law

$$\vec{B} = \frac{\mu_0}{4\pi}\frac{q\vec{v} \times \hat{r}}{r^2}.$$

By definition, $\hat{r}$ points from the source charge to the location of interest, so for point A, $\hat{r}$ would be directly up in the +z-direction such that

$$\hat{r} = \begin{bmatrix} 0 \\ 0 \\ 1 \end{bmatrix}.$$

We also know that the velocity is entirely in the +x-direction, and that this distance from the charge to point A is 0.2 m ($r = 0.2m$), so that

$$\vec{B}_A = \frac{\mu_0}{4\pi}\frac{q}{r^2}\vec{v} \times \hat{r}$$

$$= \frac{4\pi \times 10^{-7} \, T \cdot \frac{m}{A}}{4\pi}\frac{1.6 \times 10^{-19}C}{(0.2 \, m)^2}\begin{bmatrix} 2x10^5 \, m/s \\ 0 \\ 0 \end{bmatrix} \times \begin{bmatrix} 0 \\ 0 \\ 1 \end{bmatrix}$$

$$= 4x10^{-2}\frac{T}{m}\begin{bmatrix} 2x10^5 \, m/s \\ 0 \\ 0 \end{bmatrix} \times \begin{bmatrix} 0 \\ 0 \\ 1 \end{bmatrix}$$

Completing this cross product will only give one component to the magnetic field in the -y-direction. Finishing the calculation, we find

$$\vec{B}_A = \begin{bmatrix} 0 \\ -8x10^{-2} \, T \\ 0 \end{bmatrix}.$$

The magnetic field at point A is pointing out of the page. Notice that this example used all three dimensions!

As for part B, the setup is very much the same, except that $\hat{r}$ is in the −z-direction, so that

$$\vec{B}_B = 4x10^{-25}\frac{T}{m}\begin{bmatrix} 2x10^5 \, m/s \\ 0 \\ 0 \end{bmatrix} \times \begin{bmatrix} 0 \\ 0 \\ -1 \end{bmatrix},$$

or

$$\vec{B}_B = \begin{bmatrix} 0 \\ 8x10^{-20} \, T \\ 0 \end{bmatrix}.$$

The magnitude of $\vec{B}$ is the same at both points; however, at point B it is into the page.

Extension:

If we tried to calculate the magnetic field at an arbitrary point relative to the proton it would always be either out of the page or into the page because of the cross product –

except exactly in front or behind he proton (where $\hat{r}$ is parallel or anti-parallel to v) where it is zero!!

Example 208-2: Field from Two

Consider two electrons, separated by 2.1 mm, traveling parallel to each other with speed $v = 3x10^3 m/s$. What is the magnetic field directly in between the two electrons?

Solution:

Conceptually, the results of the last problem tell us that the magnetic field between the two electrons should be zero, because they will offset each other. However, let's go through the process and check this.

The Figure below shows both electrons moving in the +x directions, with the point P of interest directly in between.

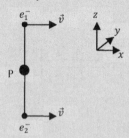

If we again consider the Biot-Savart Law

$$\vec{B} = \frac{\mu_0}{4\pi} \frac{q\vec{v} \times \hat{r}}{r^2},$$

We notice that every term in the equation will be the same for both e_1^- and e_2^-, except for their $\hat{r}$ vectors:

$$\hat{r}_1 = \begin{bmatrix} 0 \\ 0 \\ -1 \end{bmatrix} \qquad \hat{r}_2 = \begin{bmatrix} 0 \\ 0 \\ 1 \end{bmatrix}.$$

At first glance, this may seem backwards. However, remember that $\hat{r}$ always points from the charged particle to the point of interest, so for e_1^-, $\hat{r}_1$ must point *down* which is why it has the -1 in the column vector.

We can now write the equations for the magnetic field at point P due to e_1^- and e_2^-.

$$\vec{B}_{e_1^-} = \frac{\mu_0}{4\pi} \frac{q}{r^2} \vec{v} \times \hat{r} = \frac{\mu_0}{4\pi} \frac{q}{r^2} \begin{bmatrix} 3x10^3 m/s \\ 0 \\ 0 \end{bmatrix} \times \begin{bmatrix} 0 \\ 0 \\ -1 \end{bmatrix}$$

$$\vec{B}_{e_1^-} = \frac{\mu_0}{4\pi} \frac{q}{r^2} \begin{bmatrix} 0 \\ -3x10^3 m/s \\ 0 \end{bmatrix}$$

$$\vec{B}_{e_2^-} = \frac{\mu_0}{4\pi} \frac{q}{r^2} \vec{v} \times \hat{r} = \frac{\mu_0}{4\pi} \frac{q}{r^2} \begin{bmatrix} 3x10^3 m/s \\ 0 \\ 0 \end{bmatrix} \times \begin{bmatrix} 0 \\ 0 \\ 1 \end{bmatrix}$$

$$\vec{B}_{e_2^-} = \frac{\mu_0}{4\pi} \frac{q}{r^2} \begin{bmatrix} 0 \\ 3x10^3 m/s \\ 0 \end{bmatrix}$$

I did not include the values for μ_0, q or r, because they are the same for both charged particles. As for any vectors, the net magnetic field at point P is the sum of the individual magnetic fields

$$\vec{B} = \vec{B}_{e_1^-} + \vec{B}_{e_2^-} = \frac{\mu_0}{4\pi} \frac{q}{r^2} \left(\begin{bmatrix} 0 \\ -3x10^3 m/s \\ 0 \end{bmatrix} + \begin{bmatrix} 0 \\ 3x10^3 m/s \\ 0 \end{bmatrix} \right)$$

or $\vec{B} = 0$, as expected.

208-3: Electric current and magnetic field

> **Consider**: *What is electric current, and how is it related to magnetic fields?*

As soon as we start to talk about moving charges, we need to define the ***electric current***. Electric current, or just current, I, is a measure of how much charge is moving

$$I = \frac{dq}{dt}. \tag{208-3}$$

We will explore current in much greater detail in unit 213, but for now it is just important to know that if we are talking about a collection of charge moving, the current gives us a nice tidy way to talk about it. Current is measured in ***Amperes (A and Amp)***, where 1 Amp is 1 Coulomb per second (C/s) by definition.

   The great American Benjamin Franklin is responsible for deciding that the charge on an electron is negative. Unfortunately, his definition tends to cause confusion for students, since current is defined in terms of the motion of positive charge (see equation above). So, ***conventional current moves in the opposite direction to electron flow***. If we have positive charge moving, there is no problem.

   The Biot-Savart law we discussed above works well if we have individual charged particles. What if we have a stream of particles so that they are best represented by a current? Consider the combination $q\vec{v}$ in the Biot-Savart Law. We can actually rearrange this in terms of current as follows:

$$q\vec{v} = q\frac{d\vec{x}}{dt} = \frac{d(q\vec{x})}{dt} = \frac{dq}{dt}\vec{x} = I\vec{x}. \tag{208-4}$$

This little manipulation assumes, of course, that the displacement, $\vec{x}$, is constant enough to be taken out of the derivative. In practice, we think of a current, I, moving along a very short wire of length dl, that is always short enough that it can be considered *straight*. So, in practice, we take the displacement above to be a vector $d\vec{l}$, through which the current moves. With this transformation we can now write the Biot-Savart law in terms of a small piece of current as opposed to individual charges.

Biot-Savart Law for a Piece of Current

$$d\vec{B} = \frac{\mu_0}{4\pi}\frac{Id\vec{l} \, x \, \hat{r}}{r^2} \tag{208-5}$$

Description – This equation describes how a small piece of wire with length $d\vec{l}$ carrying a current, I, creates a magnetic field, $d\vec{B}$, a distance r from the wire segment.

Note 1: μ_0 is the permeability of free space and is given by $\mu_0 = 4\pi \, x \, 10^{-7} \, T \cdot m/A$

Note 2: $\hat{r}$ is directed from the source current to the point of interest in space.

Example 208-3: Magnetic field due to a long wire

What is the magnetic field 6.2 cm from a long straight thin wire carrying a current of 0.31 A?

Solution:

This is a direct application of the Biot-Savart Law:

$$d\vec{B} = \frac{\mu_0}{4\pi}\frac{Id\vec{l} \, x \, \hat{r}}{r^2}.$$

Consider the Figure below, which shows a segment of wire $d\vec{s}$ carrying a current i. First, we can use the Pythagorean theorem to relate r to other known variables

$$r^2 = s^2 + R^2,$$

where R is the distance from the wire to where we are trying to find the magnetic field.

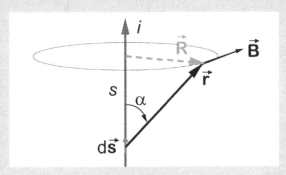

We can also use the diagram to find the cross product:

$$|d\vec{s} \, x \, \hat{r}| = \sin\theta \, ds = \frac{R}{r}ds = \frac{Rds}{\sqrt{s^2 + R^2}},$$

With these two pieces, the Biot-Savart law can be written

$$dB = \frac{\mu_0 i}{4\pi}\frac{R \, ds}{(s^2 + R^2)^{3/2}}.$$

Now we simply need to sum up all the little pieces of l by integrating over the entire length of the wire

$$B = \frac{\mu_0 i}{4\pi}\int_{-\infty}^{\infty}\frac{R \, ds}{(s^2 + R^2)^{3/2}} \cdot = \frac{\mu_0 i}{2\pi R}.$$

Finally, we can substitute in the values given in the problem to find

$$B = \frac{(4\pi \, x \, 10^{-7} \, T \cdot m/A)(0.31A)}{2\pi(0.062m)} = 1.0x10^{-6}T.$$

A couple of notes:
1) The general equation given above can be used to find the magnetic field a distance away from a long wire – you don't have to rework the problem.
2) The right hand rule – thumb along current, fingers curl along magnetic field – is used to find direction.

208-3: Current loops and magnetic moments

Consider: *How do we deal with charges moving in circles?*

Next, we're going to take a look at small current loops as shown in Figure 208-1. The outside of the loop contains a current, I, traveling counter-clockwise, enclosing an area, A. Every piece of the current loop produces a magnetic field directed upwards in the middle of the loop. As we'll see below, this arrangement comes up in many areas of magnetism, and so physicists have defined a single quantity, known as the **magnetic dipole moment**, $\vec{\mu}$, that described both the current and the size of the loop:

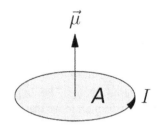

Figure 208-1. Depiction of a small current loop and the magnetic dipole moment it creates. (Need to change S to A in the file)

$$\vec{\mu} = I\vec{A}.$$ (208-6)

As you can see, the magnetic moment is a vector, and is in the same direction as the **area vector**. Area vectors are always perpendicular to the area they describe, and in this instance, that direction is defined to be in the direction of the magnetic field inside the current loop. Figure 208-2 shows the **magnetic field lines** around the current loop described in Figure 208-1. Magnetic field lines play the same role in magnetism as electric field lines do for electricity. We will see in the next unit how moving charges placed in a magnetic field respond to that field just as we saw electric charges respond to electric fields in previous units. The magnetic field of a current loop looks remarkably like the magnetic field of a permanent magnet, which will be described in the next section. The reason for this, as we'll see, is that electrons in each atom act just like little current loops.

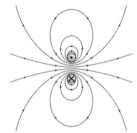

Figure 208-2. Magnetic field of a small current loop.

208-4: Magnetism in matter

Consider: *Why do some materials become magnets, but most do not?*

This entire unit has been about how moving charges create magnetic fields. We've studied both how to calculate the magnetic field of a single charged particle as well as that of a current. Both of these are generally macroscopic, meaning we were discussing how we could create a magnetic field that we could measure in some way.

Now, let's think about microscopic motion. All materials are made of atoms, with a tightly packed positively charged nucleus and an electron cloud made of at least one electron around that nucleus. On a basic level, it is easy to think of the electron as orbiting the nucleus just as a planet orbits a star. Although we now know that the electron cloud is far more complicated, the electron was historically treated in this fashion, and for many properties, this is a good model.

Well, an electron *is* a charged particle and if it is in orbit around the nucleus, it *is* moving, so it should produce a magnetic field. In fact, every electron in every atom is producing a very small magnetic field. There are even a few materials where the magnetic fields of these electrons align and create a macroscopic effect - these are the permanent magnets that we've all pinned to our refrigerators.

Let's see how this works. Consider the single electron in a hydrogen atom as shown in Figure 208-3. The Figure shows an electron orbiting the nucleus of an atom at a radius r. Note that the Figure shows how the velocity of the vector and the conventional current related to the electron are in opposite directions. Also shown in the Figure is the magnetic moment (which is opposite the angular momentum because the electron is negatively charged).

Remembering that the period of orbit is given by the circumference divided by the speed of the object, the current created by the electron as it moves is given by

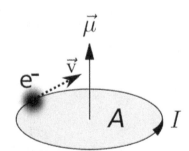

Figure 208-3. An electron orbiting a nucleus forming a magnetic moment.

$$I = \frac{e}{T} = \frac{ev}{2\pi r}.$$ (208-7)

Therefore, the magnetic moment ($\mu = IA$) is

$$\mu = IA = \frac{ev}{2\pi r}(\pi r^2) = \frac{evr}{2}. \tag{208-8}$$

We don't really know the speed (v) and orbital radius (r) of the electron, but it turns out we do know its angular momentum ($L = mvr$), which we can substitute in:

$$\mu = \frac{e}{2m}L. \tag{208-9}$$

Angular momentum is an interesting quantity. In quantum systems (such as an electron in an atomic orbit), it turns out that it is **quantized**, meaning that it can only come in specific values and not in a continuum as we are used to in classical mechanics. Angular momentum always comes in multiples of $h/2\pi$, where h is called Planck's constant and has value $h = 6.626x10^{-34} J \cdot s$. Since the value of h is so small, we never notice that L is quantized on a macroscopic scale, because every day angular momenta are huge compared to the value of h. Regardless, the smallest magnetic moment our electron in the hydrogen atom can have is

$$\mu_B = \frac{e}{2m}\frac{h}{2\pi} = 9.274x10^{-24} A \cdot m^2. \tag{208-10}$$

μ_B is called the **Bohr magneton**, and represents the smallest magnetic moment an electron can have due to its orbital motion. In addition to its motion around the nucleus, electrons also have an angular momentum similar to a spinning top, called *spin*. Although the derivation requires some quantum mechanics, the result is that the magnetic moment for this spin is almost exactly the same as the Bohr magneton ($\mu = 1.001\mu_B$),

The reason we need to know about the Bohr magneton is that a material's creation of, and reaction to magnetic fields is entirely determined by how the electrons in its atoms behave as a group. When most materials are not in an external magnetic field, the directions of the individual magnetic moments of the electrons are randomly oriented and therefore cancel out. This means that they exhibit no magnetic properties as a bulk substance outside of magnetic fields. Permanent magnets, or **ferromagnets** (discussed below), are the exception to this rule and can produce very strong magnetic fields on their own.

Diamagnetism

When the magnetic moment of an atom is placed in an external magnetic field, the moment will try to align opposite to the magnetic field (we'll see why when we talk about Faraday's law in a few units). The net effect is that the magnetic field inside the material is slightly decreased and the material itself is *repelled* by the magnetic field. This effect is called **diamagnetism**, and appears in *all* atoms. However, only a few materials are purely diamagnetic. In materials where paramagnetism or ferromagnetism (see the next couple of sections) are present, these effects overwhelm the diamagnetic effect.

The net magnetic field inside a diamagnetic material can be written

$$\vec{B}_{net} = \vec{B}_0 + \mu_0\vec{M}, \tag{208-11}$$

where $\vec{B}_0$ is the external magnetic field, $\vec{B}_{net}$ is the net magnetic field inside the material and $\vec{M}$ is the **magnetization** of the material. The magnetization gives a density measure of how many of the magnetic moments have been aligned by the external magnetic field:

$$\vec{M} \equiv \frac{\vec{\mu}_{net}}{V}. \tag{208-12}$$

In addition, we know that the magnetization of a diamagnetic material is proportional to the externally applied magnetic field

$$\vec{M} = \chi_m\frac{\vec{B}_0}{\mu_0}, \tag{208-13}$$

where χ_m is the constant of proportionality and is called the **magnetic susceptibility** of the material. If we now plug this version of the magnetization back into our equation for the net magnetic field, we get

$$\vec{B}_{net} = \vec{B}_0 + \chi_m\vec{B}_0 = \vec{B}_0(1 + \chi_m). \tag{208-14}$$

After these couple of lines of derivation, we can now see that the net magnetic field inside a diamagnetic material can be determined by just the magnetic susceptibility and the known external magnetic field. Since for diamagnetic materials we expect the net magnetic field to be smaller than the external magnetic field, the diamagnetic susceptibilities should be negative. Also, since diamagnetic effects are very small, χ_m should be small – and it is – generally on the order of 10^{-5}. χ_m for some materials can be found in Table 208-3 near the end of the unit.

It is also very important to note that all of the effects of diamagnetism disappear very quickly when the external field is removed as all of the atomic magnetic moments quickly realign due to thermal motion.

Paramagnetism

Materials that are purely diamagnetic have entirely paired electrons, which allows the net magnetic moment to be zero in the absence of an external magnetic field. Many chemical elements and compounds, however, contain unpaired electrons which leads to a small, permanent dipole moment. In **paramagnetism**, these dipole moments do not interact with each other and are therefore randomly aligned in the absence of an external magnetic field. However, once an external magnetic field is applied to a paramagnetic material, these permanent dipole moments tend to align with the magnetic field, slightly *increasing* the field in the material.

The math for paramagnetic materials is exactly the same as diamagnetic materials, so the net field in the material with an external magnetic field is

$$\vec{B}_{net} = \vec{B}_0(1 + \chi_m). \tag{210-15}$$

In paramagnetic materials, χ_m is positive unlike in diamagnetic materials where it is negative. However, like diamagnetic materials, χ_m is usually quite small (10^{-3} to 10^{-5}) for paramagnetic materials. Also like diamagnetic materials, the induced magnetic field disappears when the external field is removed.

Keep in mind that even in paramagnetic materials, the diamagnetic effect is still there, however, it is overwhelmed by the paramagnetic effect.

Ferromagnetism

Certain materials take paramagnetism to the extreme. Through two quantum mechanical process known as *Hund's Rule* and the *exchange interaction*, the first few electrons filling a shell tend to have the same *spin*. I mentioned in the section on diamagnetism that electrons have spin angular momentum as well as orbital angular momentum, and since these materials with just a few electrons in a shell have their spins in the same direction, they have a larger magnetic dipole moment. In addition, some of these materials have an odd number of electrons, leading to an orbital contribution to the magnetic dipole moment.

The net effect is that each atom in these materials have a rather large magnetic dipole moment that will strongly align when placed in an external magnetic field. Once aligned, the magnetic dipole moment of an atom in the material is strong enough to interact with the atoms around it, keeping their dipole moments aligned. This is a **permanent magnet**. Materials that exhibit this process are known as **ferromagnetic materials**.

Ferromagnetism is ubiquitous in modern technology being found in electric motors, electromagnets, hard disk drives, generators and transformers just to name a few. We'll see in the upcoming unit on Faraday's law how simple magnets can be use to transform mechanical energy to electrical energy and back – a process that drives the entire electric grid of most countries.

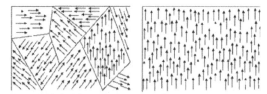

Figure 208-4. Magnetic domains of an unmagnetized ferromagnetic material (left) and one of a magnetized material (right).

The process described above is not perfect. Since the atomic interactions in ferromagnetic materials are so strong, when the material is placed in a strong magnetic field, the object magnetizes in groups called magnetic domains as can be seen in Figure 208-4. The Figure shows how when the domains in a ferromagnetic material are randomly arranged, it does not act like a permanent magnet, but when a strong magnetic field is applied to the material, it can become permanently magnetized as the domains align.

To summarize – in ferromagnetic materials, if you apply a strong magnetic field, the material will become **permanently magnetized**. What happens if we then place the material in a magnetic field that is in the opposite direction? You might guess that it would first cause the magnetization of the material to decrease and then increase in the opposite direction (in line with the new magnetic field). This is exactly what happens! The process is summarized by what is known as a **hysteresis curve**. Figure 208-5 shows a hysteresis curve and explains how the application of magnetic fields affects the magnetization of the overall object.

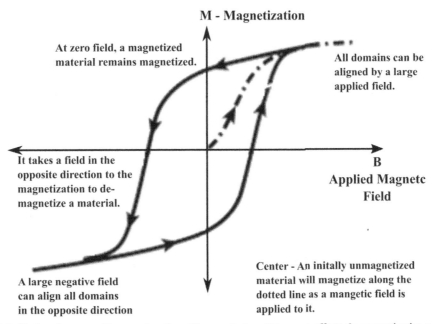

M - Magnetization

At zero field, a magnetized material remains magnetized.

All domains can be aligned by a large applied field.

B
Applied Magnetc Field

It takes a field in the opposite direction to the magnetization to de-magnetize a material.

A large negative field can align all domains in the opposite direction

Center - An initally unmagnetized material will magnetize along the dotted line as a mangetic field is applied to it.

Figure 208-5. Hysteresis curve with an explanation of how each step of the curve affects the magnetization of the material. The B axis represents the externally applied magnetic field. The inserts show overall alignment of magnetic domains at each stage of the hysteresis curve. Materials with a large area in the hysteresis curve are *called hard ferromagnetic materials* and are hard to demagnetize. Materials with a small area are called *soft ferromagnetic materials* and are easy to demagnetize.

As you can tell, magnetic materials, and especially permanent magnets, are actually quite complicated. This is made even worse by the fact that most ferromagnetic materials are not perfect. Over long periods of time, permanent magnets tend to lose their magnetization through three processes:

1) ***Thermal relaxation*** – As the atoms in each domain vibrate due to their thermal energy, there is some slight demagnetization – this process alone is quite weak and will take a long time to noticeably demagnetize a permanent magnet.

2) ***Curie temperature*** – However, there is a specific, material dependent temperature called the Curie Temperature, above which the atomic vibrations overcome the alignment of the magnetic dipoles and the magnet demagnetizes rapidly.

3) ***Physical damage*** – Permanent magnets are susceptible to physical shocks where the shock literally rearranges the magnetic domains – yes, you can demagnetize a magnet by literally hitting it!

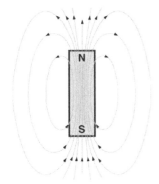

Figure 208-6. Magnetic field of a bar magnet (ferromagnet).

Although there is quite a bit to ferromagnetic materials conceptually, the math we employ is quite minimal. I mentioned earlier that we cannot use the equation for magnetic susceptibility that we used for diamagnetism and paramagnetism here. However, we can still talk about susceptibility for ferromagnetic materials that will tell us the magnetization of a permanent magnet after the external field is removed. Unlike diamagnetic and paramagnetic materials, the susceptibility for ferromagnetic materials is often quite large ($10^2 - 10^3$).

Table 208-2 summarizes the three types of magnetic materials discussed and Table 208-3 gives the magnetic susceptibility for some materials.

Table 208-3. Magnetic susceptibility for some materials.

Material	χ_m
Copper	-9.8×10^{-5}
Diamond	-2.2×10^{-5}
Gold	-3.6×10^{-5}
Lead	-1.7×10^{-5}
Silver	-2.6×10^{-5}
Aluminum	2.3×10^{-5}
Calcium	1.9×10^{-5}
Lithium	2.1×10^{-5}
Magnesium	1.2×10^{-5}
Oxygen	2.1×10^{-6}
Platinum	2.9×10^{-4}
Tungsten	6.8×10^{-5}
Iron (99.8%)	5,000
Iron (99.95%)	200,000
Nickel (99%)	600

Note: The χ_m given for ferromagnetic materials is the maximum possible value.

Table 208-2. Summary of magnetic materials.

Type	Susceptibility	Net Magnetic Field in Material	Description
Diamagnetic	$\chi_m < 0$, small	Slightly smaller than applied field	Present in all materials. Object is slightly repelled by magnetic field
Paramagnetic	$\chi_m > 0$, small	Slightly larger than applied field	Unpaired electrons required. Object is slightly attracted by magnetic fields
Ferromagnetic	$\chi_m > 0$, large	Much larger than applied field	Permanent magnets. Permanent magnets are strongly attracted to other ferromagnetic materials.

What does the magnetic field of a permanent magnet look like? Well, since it is a large number of magnetic dipoles lined up, it looks exactly the same as the magnetic field of a ***dipole***. The magnetic field of a rectangular magnet, called a bar magnet is shown in Figure 208-6. Note that the side of the magnet where the magnetic field emerges is called the ***North Pole*** of the magnet and the area where the magnetic field lines point into the magnet is called the ***South Pole*** of the magnet. Unlike poles attract, that is, the north pole of one magnet is attracted to the south pole of another magnet. Like poles repel so that two north poles repel each other as do two south poles.

One of the most interesting features of magnets is that if you split a magnet in two, you get two full magnets – each with a north and south pole. This effect happens right down to the individual electron spins. What this means is that you can never have a north pole without a south pole or vice versa. In physics speak, we say that ***there are no magnetic monopoles***.

208-5: Magnetic field of the earth

> **Consider**: *How do we describe the magnetic field of the earth? Is it like any of the fields we have discussed so far?*

Earth's magnetic field is produced by the flow of highly convective fluids in the outer core. We saw in unit 203 that both the inner and outer core of the earth are made primarily of iron alloys. The inner core is immensely hot, which heats the fluid iron alloys just outside the inner core-outer core boundary. Just as the steam from a boiling pot of water rises, the hot iron fluid then flows up towards the solid mantle above, where it cools and flows back down towards the inner core. When you throw in the overall rotation of the earth, this creates what is known as a ***dynamo***. The flow of the highly ferromagnetic iron alloys makes the earth one huge permanent magnet as can be seen in Figure 208-7.

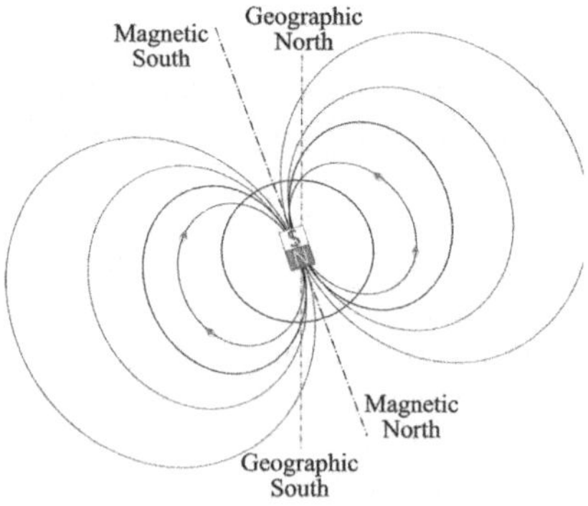

If you look very closely at Figure 208-7, you'll notice something very odd: the magnetic pole closest to the geographic North Pole is the magnetic south pole! This is mostly a historical oddity. It is the north pole of a magnet in a compass that points you *north*. This is the designation that originally gave us what we now call geographic north. Unfortunately, now that we know more about magnets, we know that the magnetic north pole in the compass would be attracted to the magnetic south pole – which is why the south magnetic pole is near the north geographic pole.

Also, notice that the geographic and magnetic poles do not exactly align! This means that a compass that is pointing 'north' is not actually pointing at the north geographic pole. Even worse – the magnetic pole keeps moving! Figure 208-8 shows the position of the south magnetic pole of the earth at different times over the last couple of centuries.

Figure 208-7. The magnetic field of earth. Note that the angle between geographic and magnetic poles is approximately 12°.

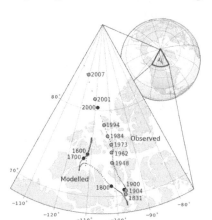

Figure 208-8. Location of the south magnetic pole (near the north geographic pole) over time.

209 – Forces in magnetic fields

In the last unit, we described how moving charges create magnetic fields. In this unit, we continue this discussion to how moving charges in a magnetic field experience a force due to that field. It is very important to keep in mind that a charged particle must be moving to experience a force in a magnetic field, unlike the electric force, where the force is present on both stationary and moving charged particles.

Integration of Ideas

Review cross products for both magnitudes and directions.
Review the idea of a vector field.
Review Newton's Second Law in 2-D and 3-D.

The Bare Essentials

- Moving charged particles feel a force in a magnetic field.

Magnetic Force on a Particle

$$\vec{F}_B = q\vec{v} \times \vec{B}$$

Description – This equation defines the magnetic force on a charged particle with charge q in terms of the speed of the particle ($\vec{v}$) and the external magnetic field ($\vec{B}$) the particle moves in.

- The magnetic force can also be written in terms of currents.

Magnetic Force on a Current

$$\vec{F}_B = I\vec{L} \times \vec{B}$$

Description – This equation defines the magnetic force on a current (I) with a given length ($\vec{L}$) in a magnetic field ($\vec{B}$).

- A loop of wire carrying a current will feel a torque when placed in a magnetic field.

Torque on a Current Carrying Loop

$$\vec{\tau} = \vec{\mu} \times \vec{B}$$

Description – This equation defines the torque ($\vec{\tau}$) on a current carrying wire in a magnetic field ($\vec{B}$)
Notes: 1) $\vec{\mu}$ is the *magnetic dipole moment* of the loop, where $\vec{\mu} = I\vec{A}$. The SI unit for the $\vec{\mu}$ is $A \cdot m^2$.

209-1: Stationary charges do not respond to magnetic fields

W E FOUND IN UNIT 208 THAT MAGNETIC FIELDS are caused by moving charged particles (including charged particles with angular momentum). We're going to find in this unit that moving charges in a magnetic field feel a force. However, stationary charged particles in a magnetic field do not feel a force. It may seem odd to start out by saying that something doesn't happen; however, one of the strongest known misconceptions related to magnetic force is that stationary particles can feel such a magnetic force.

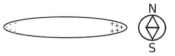

Figure 209-1. A stationary charged rod near a compass. Note that stationary charges have no effect on the compass.

As an example, what happens if a stationary, positively charged rod is near a compass, as shown in figure 209-1? When polled, many people say that the south pole of the magnet will swing and point towards the charged rod. The incorrect reasoning is that the south pole acts like a negative charge and is therefore attracted to the positively charged rod, and likewise, the north pole acts like a positive charge and is repelled by the positively charged rod. In reality, though, nothing would happen to the compass in this situation. The positive charge on the rod is creating an electric field, but not a magnetic field. In the absence of a magnetic field, the compass will not respond.

We can also think about the force on the charged rod. Newton's 3rd law tells us that the force on the compass due to the rod will have a reaction force on the rod due to the compass. We know that the permanent magnet in the compass is producing a magnetic field; however, we know from above that the stationary charges in the rod feel no force – that is, the stationary charged rod feels no force in the magnetic field. So, be careful:

Stationary charges respond to electric fields but not magnetic fields.
Moving charges respond to *both* electric fields and magnetic fields
Permanent magnets respond to magnetic fields but not electric fields.

209-2: Magnetic force on a particle

We've learned that moving charges cause magnetic fields, but it is also true that if a moving charge is in a magnetic field it feels a force (as long as it is not moving parallel to the field). The force is always perpendicular to both the velocity of the charged particle and the magnetic field itself.

Magnetic Force on a Particle

$$\vec{F}_B = q\vec{v} \times \vec{B} \qquad (209\text{-}1)$$

Description – This equation defines the magnetic force on a charged particle with charge q in terms of the speed of the particle ($\vec{v}$) and the external magnetic field ($\vec{B}$) the particle moves in.

It is instructive to compare the magnetic force to the electric force, $\vec{F}_E = q\vec{E}$. Notice that the magnetic force and electric force are close in form, except that the magnetic force requires a velocity, $\vec{v}$, as discussed. This comparison also shows why the electric field and the magnetic field must have different units: the electric field has units of N/C so that when multiplied by the charge (in Coulombs), we are left with Newtons. However, the magnetic field must not only be multiplied by the charge, but also the velocity of the particle, which is why the Tesla (as seen in Unit 208) is defined as

$$1\,T = \frac{N}{C \cdot \frac{m}{s}} \qquad (211\text{-}2)$$

Example 209-1: Magnetic Force

Near the equator, the earth's magnetic field is nearly horizontal and points north. What is the force (a) on a proton traveling due east at 1.2×10^4 m/s (b) on an electron traveling due east at 1.2×10^4 m/s (c) on a proton traveling straight down at 1.2×10^4 m/s and (d) on an electron traveling north at 1.2×10^4 m/s? You may assume that the magnitude of the earth's magnetic field is 2.5×10^{-5} T in each case.

Solution:

In each case, this problem is a direct application of the magnetic force law

(a) Using our normal coordinate system near the earth, we find

$$\vec{F}_B = q\vec{v} \times \vec{B} = 1.6 \times 10^{-1} \ C \begin{bmatrix} 1.2 \times 10^4 m/s \\ 0 \\ 0 \end{bmatrix} \times \begin{bmatrix} 0 \\ 2.5 \times 10^{-5} \ T \\ 0 \end{bmatrix}.$$

Finishing the cross produce gives us

$$\vec{F}_B = \begin{bmatrix} 0 \\ 0 \\ 4.8 \times 10^{-2} \end{bmatrix} N$$

In this case, the direction can be found by either using the right hand rule ($x \times y = z$) or using the vector-matrix form to find the final vector.

Each of the other magnetic fields can be computed in the same manner as part (a)

(b) Only a change in charge from part (a):

$$\vec{F}_B = q\vec{v} \times \vec{B} = -1.6 \times 10^{19} C \begin{bmatrix} 1.2 \times 10^4 m/s \\ 0 \\ 0 \end{bmatrix} \times \begin{bmatrix} 0 \\ 2.5 \times 10^{-5} \ T \\ 0 \end{bmatrix},$$

$$\vec{F}_B = \begin{bmatrix} 0 \\ 0 \\ -4.8 \times 10^{-2} \end{bmatrix} N.$$

(c) Change of velocity direction from (a):

$$\vec{F}_B = q\vec{v} \times \vec{B} = 1.6 \times 10^{19} C \begin{bmatrix} 0 \\ 0 \\ -1.2 \times 10^4 m/s \end{bmatrix} \times \begin{bmatrix} 0 \\ 2.5 \times 10^{-5} \ T \\ 0 \end{bmatrix},$$

$$\vec{F}_B = \begin{bmatrix} 4.8 \times 10^{-20} \\ 0 \\ 0 \end{bmatrix} N.$$

(d) $\vec{F}_B = 0$, because any time the two vectors in a cross product are parallel or anti-parallel to each other, the cross product is zero.

Note that (a)–(c) have the same magnitude, but different directions. You must be very careful with the direction of magnetic fields because of the cross-product nature!

The cross product in the magnetic force law causes an interesting effect. If you remember back to Physics I, we call a force that is always perpendicular to the velocity vector a *centripetal force* and centripetal forces cause objects to move in a circle as can be seen in figure 209.2. The figure shows an electron moving clockwise and a magnetic field directed into the page. Using the magnetic force equation, the cross product of the velocity with the magnetic field would suggest a force downwards at the point shown. However, since the electron has a negative charge, the force becomes directed upward. You should convince yourself that at every point on the circle, the cross product gives you a force directed towards the center – remember that the velocity of a particle moving in a circle is always tangent to the circle.

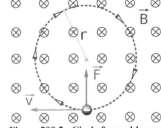

Figure 209.2. Circle formed by an electron moving in a perpendicular magnetic field.

We actually have to be careful about the circle described above. What is discussed in the last paragraph is only true if the velocity of the particle is perpendicular to the magnetic field. If the velocity is at some angle to the field, the first thing we need to do is to break the velocity into components that are parallel to the field and perpendicular to the field as shown in figure 209.3. The component of the velocity perpendicular to the magnetic field will cause a circular motion as seen in figure 209.3. However, the component of the velocity parallel to the magnetic field does not feel any force (remember, the cross product of parallel vectors is zero.). Therefore, the parallel component of the velocity is unchanged while the perpendicular component causes circular motion – which gives us a helix as can be seen in figure 209.3.

We can now relate what we know about magnetic fields to what we know about centripetal motion. We know that the magnetic force is the centripetal force in this case, and that it is the perpendicular component of the velocity that is involved in both the magnetic and centripetal forces:

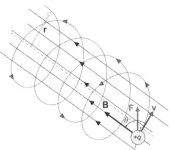

Figure 209.3. The components of $\vec{v}$ are broken into components ⊥ and ∥ to a magnetic field, $\vec{B}$. The figure then shows how these components lead to helical motion.

$$F_B = F_c \quad \rightarrow \quad qv_\perp B = \frac{mv_\perp^2}{r}. \tag{209-2}$$

This equation can be simplified some, since there is a $v_\perp$ on both sides. However, I'm not going to box this equation, because it is derived from the fact that the magnetic force is the centripetal force, which you should be able to do. Here is one version of the simplified form:

$$r = \frac{mv_\perp}{qB}, \tag{209-3}$$

which would allow you to find the radius of curvature for the circle if you know the other values. Again, there is no standard form for this equation, and it can be used to solve for any of the variables if you have the others.

Example 209-2: Mass Spectrometer

A mass spectrometer is a device used in analytical chemistry to separate charged particles/molecules of different mass. As shown in the figure to the right, the particles first enter a **velocity selector**, a region of perpendicular electric and magnetic fields that will only allow a certain velocity of particle to pass through the device undeflected. The undeflected particles then enter a region with only a magnetic field, which causes the particles to move in a circular pattern. Where the particles strike the detector then depends only on their mass and velocity.

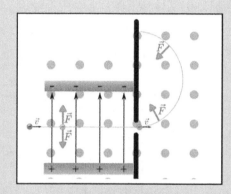

Imagine that we have an electron entering a velocity selector with a magnetic field of $4.2\text{x}10^{-5}$ T directed into the page and an electric field of $9.6\text{x}10^{-2}$ N/C directed downwards. (a) what electron velocity will make it through the velocity selector undeflected? The electron then enters a region of just magnetic field with the same field as above. (b) At what distance above the entrance to the mass spectrometer would the electron strike the detector?

Solution:

(a) When the electron enters the velocity selector, it is acted on by two forces, one electric and one magnetic:

$$\vec{F}_E = q\vec{E} \qquad \vec{F}_B = q\vec{v} \times \vec{B}.$$

Since the electron has a negative charge, the electric force will be directed in the opposite direction as the electric field – up in this case. As for the magnetic force, if the velocity is to the right and the magnetic field is into the page, the right hand rule tells us that the magnetic force is directed downwards, since the charge is negative.

Therefore, the net force can be written

$$F_{net} = 0 = F_B - F_E = qvB - qE \quad \rightarrow \quad v = \frac{E}{B},$$

since we know that particles that move through undeflected experience zero net force.

Plugging in the numbers from the problem we find that electrons will move through undeflected if they have a speed of

$$v = \frac{E}{B} = \frac{9.6\text{x}10^{-2} \text{ N/C}}{4.2\text{x}10^{-5} \text{ T}} = 2.3\text{x}10^3 m/s.$$

(b) Electrons at the speed found in (a) then enter a region of only magnetic field and therefore only magnetic force. As discussed above, this magnetic force will initially be directed upwards. However, we know that charged particles moving in a magnetic field follow a circular pattern with a radius given by

$$r = \frac{mv_\perp}{qB}.$$

Using our known values from the problem, we can find this radius of curvature to be

$$r = \frac{(9.11\text{x}10^{-31}kg)(2.3\text{x}10^3 m/s)}{(1.6\text{x}10^{-19}C)(4.2\text{x}10^{-5} \text{ T})} = 3.1 \text{ x } 10^{-4} \, m.$$

If you look again at the figure above for the mass spectrometer, you'll see that the electron will strike the detector a full diameter of the circle above where the electron enters the mass spectrometer. Therefore, the proton will strike the detector at

$$D = 6.2 \text{ x } 10^{-4} \, m.$$

209-3: Magnetic force on a wire section

In unit 208, we showed that two forms of the Biot-Savart law were the same by converting the quantity $q\vec{v}$ into the quantity $I\vec{L}$. We can do the exact same transformation on our force law to find the force on a current carrying section of wire.

Magnetic Force on a Current

$$\vec{F}_B = I\vec{L} \times \vec{B} \qquad (209\text{-}4)$$

Description – This equation defines the magnetic force on a current (I) with a given length ($\vec{L}$) in a magnetic field ($\vec{B}$).

Example 209-3: Force on a piece of wire

What is the force on a section of wire 2.3 cm long carrying a current of 6.2 μA directly northeast if it is placed in a 2.5×10^{-5} T magnetic field pointed downward.

Solution:

This is a direct application of the magnetic force on a current equation, although we must be very careful with directions. Although the problem makes it sound like the current has a direction, we must remember that current is a *scalar*; it is really the length of the wire that carries the direction information. So, we start with our basic equation:

$$\vec{F}_B = I\vec{L} \times \vec{B}.$$

Since the length of the wire is northeast, we must write it in terms of a column vector

$$\vec{L} = 0.023\ m \begin{bmatrix} \cos 45° \\ \sin 45° \\ 0 \end{bmatrix} = \begin{bmatrix} 0.016\ m \\ 0.016\ m \\ 0 \end{bmatrix}.$$

We can now substitute this into our force equation and compute the cross product:

$$\vec{F}_B = 6.2 \times 10^{-6} A \begin{bmatrix} 0.016\ m \\ 0.016\ m \\ 0 \end{bmatrix} \times \begin{bmatrix} 0 \\ 0 \\ -2.5 \times 10^{-5}\ T \end{bmatrix},$$

Which gives

$$\vec{F}_B = \begin{bmatrix} -2.48 \times 10^{-12}\ N \\ 2.48 \times 10^{-12}\ N \\ 0 \end{bmatrix}.$$

There is another way we could calculate this as well. Noting that a wire segment pointing northeast is perpendicular to a magnetic field pointed down, we could find the magnitude of the force as

$$F = ILB = 3.57 \times 10^{-12},$$

and use the right hand rule to determine that it is exactly northwest. You should convince yourself that these two answers are equivalent (up to some rounding)!

209-3: Energy stored in a magnetic field

Like electric fields, magnetic fields store energy. The energy density (energy per unit volume) of a magnetic field is given by

$$u_B = \frac{B^2}{2\mu_0}, \qquad (209\text{-}5)$$

where B is the magnetic field strength and μ_0 is the permeability of free space. If you are inside of a magnetic material and want to know the energy stored, you replace μ_0 with $\mu = \mu_0(1 + \chi_B)$, where χ_B is the magnetic susceptibility of the material.

Example 209-4: Energy stored

How much energy is stored in a cubic region of space 1 cm on a side if there is a magnetic field with magnitude 6.2×10^{-4} T passing through it?

Solution:

In order to find the energy in a certain region of space, we must first find the energy density and then multiply it by the volume of the region

$$U_B = u_B V = \frac{B^2}{2\mu_0} d^3,$$

where d is the length of one of the side of our cube.

$$U_B = \frac{(6.2x10^{-4} \ T)^2}{2\left(4\pi x 10^{-7} \frac{Tm}{A}\right)} (0.01 \ m)^3.$$

Which gives us

$$U_B = 1.53 x 10^{-7} \ J.$$

Note that the energy stored here is quite small, and that the magnetic field given in this problem is quite typical.

209-4: Torque on a magnetic dipole.

In unit 208, we looked at the magnetic field created by a current carrying loop and defined the magnetic dipole moment of the loop as

$$\vec{\mu} = I\vec{A}. \tag{209-6}$$

Remember that the current loop creates a magnetic field that closely resembles the electric field of an electric dipole, which is why $\vec{\mu}$ is called a dipole moment. Figure 209-4 shows a comparison of the electric dipole field and the magnet (dipole) field created by a current loop. In unit 207-4, we found that an electric dipole in an electric field felt a torque given by

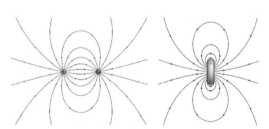

Figure 209-4. Comparison of the electric field created by an electric dipole (left) and the magnetic field created by a current loop (right).

$$\vec{\tau} = \vec{p} \ x \ \vec{E}. \tag{209-7}$$

Given the similarities in the fields, we would likewise expect to find something similar for the torque on a current loop in a magnetic field. Let's see how this works.

Consider figure 209-5, which shows a square loop of current situated in a magnetic field directed to the right. The loop is angled so that the magnetic moment of the loop makes an angle of θ relative to the magnetic field.

Using the right-hand rule, the magnetic force on the right side of the loop is up, the left side of the loop is down, the far side of the loop is into the page and the close side of the loop is out of the page. Since each side has the same current and length, the net force on the entire loop is zero. However, due to the fact that the magnetic moment makes an angle relative to the magnetic field, there will be torque on the loop as can be seen in the figure 209-5 (b). In general, torque is given by

$$\vec{\tau} = \vec{r} \ x \ \vec{F}, \tag{209-8}$$

and the magnitude can be written

$$|\vec{\tau}| = |\vec{r}||\vec{F}| \sin \theta_{rf}, \tag{209-9}$$

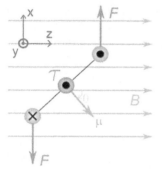

Figure 209-5. How the magnetic torque on a current carrying loop develops.

where $\vec{r}$ is the displacement from the axis of rotation to the point of action of the force, $\vec{F}$, and θ_{rf} is the angle between $\vec{r}$ and $\vec{F}$. In our case,

$$|\vec{r}| = \frac{1}{2}\ell \quad \text{and} \quad |\vec{F}| = I\ell B, \tag{209-10}$$

and there is an angle θ between $\vec{r}$ and $\vec{F}$. If you look at the force on the left-side of the loop, you will see that everything we just described for the right side is also true. Therefore, we can just double the torque from one side. Plugging all this into our torque equation, we get

$$|\vec{\tau}| = 2\left(\frac{1}{2}\ell\right) ILB \sin\theta = I\ell^2 B \sin\theta. \qquad (209\text{-}11)$$

Since the current loop is a square with length ℓ, ℓ^2 represents the area of the loop, and $I\ell^2$ is the magnetic moment of the loop. Therefore

$$|\vec{\tau}| = \mu B \sin\theta. \qquad (209\text{-}12)$$

Putting this back into vector form, we find

$$\vec{\tau} = \vec{\mu} \times \vec{B}. \qquad (209\text{-}13)$$

This has a very similar form to the torque on an electric dipole as we expected:

$$\vec{\tau}_E = \vec{p} \times \vec{E} \qquad \vec{\tau}$$
$$= \vec{\mu} \times \vec{B}. \qquad (209\text{-}14)$$

Torque on a Current Carrying Loop

$$\vec{\tau} = \vec{\mu} \times \vec{B} \qquad (209\text{-}14)$$

Description – This equation defines the torque ($\vec{\tau}$) on a current carrying wire in a magnetic field ($\vec{B}$)
Note: $\vec{\mu}$ is the *magnetic dipole moment* of the loop, where $\vec{\mu} = I\vec{A}$. The SI unit for the $\vec{\mu}$ is $A \cdot m^2$.

The torque on a current carrying loop in a magnetic field is extremely important in today's society – this is the principle that *electric motors* are built on! An electric motor uses electric energy flowing through a loop to create a mechanical torque on the loop. The torque then causes rotation in the loop and it is this mechanical motion that the motor uses to drive its applications. *Electric generators* use this principle in reverse, using current loops to turn mechanical motion into electric energy. We will study more on motors and generators in unit 216.

Example 209-5: Torque on a current loop

A wire is made into a circular coil of radius 7.8 cm with 22 turns (meaning that it is essentially 22 circles stacked up on top of each other). The coil is placed on a horizontal tabletop where there is a 0.43 T magnetic field directed to the right. What is the torque on this current loop? The current in the wire is 0.80 A, clockwise as viewed from above.

Solution:

Although this is a direct application of our torque equation, there are a number of steps. First, we must find the magnetic moment of this loop:

$$\vec{\mu} = I\vec{A}.$$

However, this equation is for one turn, so the total magnetic moment of the loop is this times the number of turns:

$$\mu = NIA = 22(0.80\ A)\pi(0.078m)^2 = 0.34\ Am^2.$$

The direction of the magnetic moment is down, using the right hand rule.

We can now calculate the torque directly. As a coordinate system, I will say to the right is positive-x and up is positive-z:

$$\vec{\tau} = \vec{\mu} \times \vec{B} = \begin{bmatrix} 0 \\ 0 \\ -0.34\ A \cdot m^2 \end{bmatrix} \times \begin{bmatrix} 0.43\ T \\ 0 \\ 0 \end{bmatrix},$$

or

$$\vec{\tau} = \begin{bmatrix} 0 \\ -0.15\ N \cdot m \\ 0 \end{bmatrix}.$$

So, the torque has a magnitude of $0.15\ N \cdot m$, and is directed along the $-y$-direction.

210 – Gauss's and Ampere's Laws

Where there is enough symmetry present in a charge or current distribution, the calculation of electric and magnetic fields can be greatly reduced by explicitly employing this symmetry. Although Gauss's Law and Ampere's Law are two of the most fundamental concepts of electricity and magnetism, they will allow us to complete these simplifications in a few, highly symmetric situations.

Integration of Ideas

Review electric fields and magnetic fields from the last few units.

The Bare Essentials

- Symmetry in electric and magnetic field lines creating a net field can often be used to simply describe the net field.

- The electric flux describes how much of an electric field pierces through a surface.

Electric Flux

$$\Phi_E = \int \vec{E} \cdot d\vec{a}$$

Description – This equation defines the electric flux, Φ_E of an electric field, $\vec{E}$, through an area, $\vec{A}$, made up of little pieces $d\vec{a}$.

- Gauss's Law describes how the electric flux through a closed surface relates to the charge enclosed in that surface.

Gauss's Law

$$\oint \vec{E} \cdot d\vec{a} = \frac{q_{encl}}{\epsilon_0}$$

Description – Gauss's Law relates the flux of an electric field through a closed surface to the net charge enclosed by that surface.
Note: The little circle on the integral reminds us that we must have a closed surface area – that is, a surface area that contains a specified volume.

- The constant of proportionality in Gauss's Law, ϵ_0, is called the **permittivity of free space** and has a value $\epsilon_0 = 8.8542 \, x \, 10^{-12} \, C^2/(N \cdot m^2)$.

- Ampere's Law describes how the current passing through a closed loop is related to the magnetic field created by the current.

Ampere's Law

$$\oint \vec{B} \cdot d\vec{l} = \mu_0 I_{encl}$$

Description – Ampere's Law relates the total current passing through a closed loop to the magnetic field created by the current.

- The constant of proportionality in Ampere's Law, μ_0, is called the **permeability of free space** and has a value $\mu_0 = 4\pi \, x \, 10^{-7} \, N/A$.

- Gauss's Law for magnetism tells us that no magnetic monopoles exist (north poles do not exist without an associated south pole and vice versa).

- Gauss's Law and Ampere's Law can also be written in differential or local form, and do not require symmetry to be solved.

210-1: Symmetry and fields

CALCULATING ELECTRIC AND MAGNETIC FIELDS due to charge distributions often requires the use of integrals as seen in units 207 (electric fields) and 208 (magnetic fields). Coulomb's Law and the Biot-Savart Law are immensely powerful and can, in principle, be used to calculate any field; however, those integrals can quickly become intractable. Physicists and engineers have two other techniques in their pockets to solve for fields around distributed objects called Gauss's Law and Ampere's Law. Both Gauss's Law and Ampere's Law are still dependent on integrals, but if enough *symmetry* exists in a system, the integrals can be removed leaving only algebra in the way of field calculations. In this unit, we will look at the laws just described and see how a couple of major symmetries can make calculating fields much easier.

210-2: Electric Flux

Consider: *How do we describe a field's interaction with surfaces?*

Let's say we want to know how much light a solar panel collects – which makes sense because the amount of energy a solar panel can produce certainly depends on how much light it collects. Well, first of all, it matters how bright the sunlight is. In unit 204, we discussed how light is made of oscillating electric and magnetic fields and that the electric field is much, much stronger than the magnetic field – so when we are collecting light, we are really collecting the oscillations of electric fields. In addition to the field, if we are trying to collect light on a panel, the size, or area, of the panel is important. The other quantity that is important is the angle between the surface of the panel and the sunlight as can be seen in figure 210-1. Remembering that the area vector for a surface is perpendicular to surface itself, the figure shows a panel at an angle θ relative to an electric field, $\vec{E}$. As θ increases, the surface area that the electric field 'sees' is reduced. Think of a piece of paper, if you hold it so that you are looking straight at the edge (so that the area vector is to your right for example), you cannot see any of the area of the paper. Now that we know the

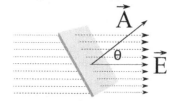

Figure 210-1. Definitions for electric flux through a flat surface

important quantities in terms of how much light the solar panel will collect, we can put them together

$$\Phi = EA \cos\theta = \vec{E} \cdot \vec{A} \qquad (210\text{-}1)$$

The quantity we just described is called the *electric flux*. The equation above works great for a flat surface; however, if the surface is curved, we look at tiny little pieces of the flux at each point and then add them up – an integral as can be seen in figure 210-2.

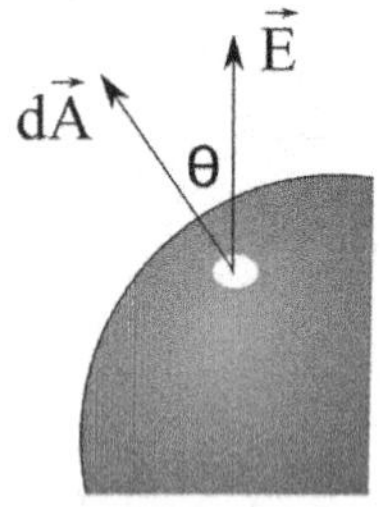

Figure 210-2. Electric flux of a small piece of area.

Electric Flux

$$\Phi_E = \int \vec{E} \cdot d\vec{a} \qquad (210\text{-}2)$$

Description – This equation defines the electric flux, Φ_E of an electric field, $\vec{E}$, through an area, $\vec{A}$, made up of little pieces $d\vec{a}$.

Example 210-1: Sunlight on earth

What is the electric flux through a flat 7.8 cm² section of earth due to sunlight, if the electric field part of the EM waves has a magnitude of 870 N/C, and is directed at an angle of 20° relative to area vector.

$$\Phi_E = \int \vec{E} \cdot d\vec{a} = EA \cos\theta,$$

since the area is flat.

Solution:

This is a direct application of the equation for electric flux:

$$\Phi_E = \left(870\frac{N}{C}\right)(7.8 \times 10^{-4} m^2)\cos 20° = 0.64 \frac{N \cdot m^2}{C}.$$

210-3: Gauss's Law

Consider: *How can symmetry be used to simplify electric field calculations?*

Take a look again at figure 210-1 and think about the flux in terms of electric field lines. If the strength of the electric field increases, the density of the field line increases and more lines would pierce through the panel. If the area of the panel increases, more lines would pierce through the surface. If the area vector and electric field vector are close to parallel, more lines pierce through the surface than if they are close to perpendicular. What I'm trying to get at is that the electric flux is related to the number of field lines that pierce through the surface.

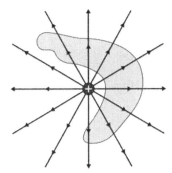

Figure 210-4. Electric field lines passing through a surface not enclosing a charge.

We are now going to relate electric field lines and flux for closed surfaces – that is a surface that completely encloses a volume. Figure 210-3 and 210-4 show electric field lines from a positive point charge passing through surfaces. Figure 210-3 depicts a charge enclosed within the closed surfaces and figure 210-4 shows a surface with the charge outside. As a convention, when discussing closed surfaces, we consider an electric field vector that pierces outward through a surface as related to a positive flux and an electric field line that pierces inwards on the surface as related to a negative flux.

Now we simply count field lines. In figure 210-3, with the charge enclosed, the surface has 12 electric field lines piercing outward and none inward (count them for yourself). This suggests that there is a net electric flux outward. The 12 electric field lines don't tell us anything quantitatively, however, because the number of lines we choose per charge is arbitrary.

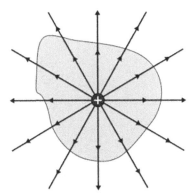

Figure 210-3. Electric field lines passing through surfaces enclosing a charge.

If we now count field lines in figure 210-4, you'll see that there are five field lines that point outward through the surface and five that point inward through the surface. Using our convention, this suggests that there is no net flux (5 outward lines – 5 inward lines = 0). Although we've only done this for a couple of surfaces, the pattern we just found is generally true – if a charge is enclosed in any closed surface, there is a net electric flux through that surface. In addition, a charge outside of a surface does not contribute any net flux through the surface. In addition, if we double the magnitude of the charge in figure 210-3, we would double the number of field lines and therefore the flux.

This is **Gauss's Law: The net electric flux through a closed surface is proportional to the charge enclosed by that surface.**

Gauss's Law

$$\oint \vec{E} \cdot d\vec{a} = \frac{q_{encl}}{\epsilon_0} \qquad (210\text{-}3)$$

Description – Gauss's Law relates the flux of an electric field through a closed surface to the net charge *enclosed* by that surface.

Note: The little circle on the integral reminds us that the we must have a closed surface area – that is, a surface area that contains a specified volume.

The constant of proportionality, ϵ_0, in Equation 210-3 is called the **permittivity of free space** and has a value $\epsilon_0 =$ 8.8542 x 10^{-1} $C^2/(N \cdot m^2)$. Technically, Gauss's law can be used in any situation; however, like Coulomb's law, the integral can be quite hard to do. Our goal is going to be to find situations in which we can get rid of the integral, and solve Gauss's law by algebra. This requires a great deal of symmetry.

So, what would be required to make the integral in Gauss's law easy? We would need the electric field to be constant everywhere on a surface. We would also need the angle between the electric field and surface area to be constant (the dot product). If this is the case, then we get

$$\oint \vec{E} \cdot d\vec{a} = \oint E \cos\theta \ da = E \cos\theta \oint da = EA \cos\theta. \tag{210-4}$$

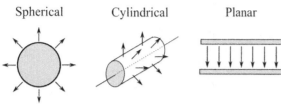

Spherical   Cylindrical   Planar

Figure 210-5. Common symmetries for Gauss's Law.

There are three main symmetries over which this works well, which are summarized in figure 210-5.

First, look at the spherical symmetry of a point charge. At all points on a sphere of radius r from the point charge, the electric field lines point radially outward. The area vector also points radially outward at all points – so the angle between the field lines and area vector is zero everywhere. These are our requirements for removing the integral in Gauss's Law, so

$$\oint \vec{E} \cdot d\vec{a} = EA \cos 0 = E4\pi r^2, \tag{210-5}$$

remembering that the surface area of a sphere is $4\pi r^2$. We can then use this in Gauss's law:

$$\oint \vec{E} \cdot d\vec{a} = \frac{q_{encl}}{\epsilon_0} \quad \rightarrow \quad E4\pi r^2 = \frac{q}{\epsilon_0}. \tag{210-6}$$

Solving for E, we find

$$E = \frac{1}{4\pi\epsilon_0}\frac{q}{r^2} = \frac{kq}{r^2}. \tag{210-7}$$

This is the exact same equation for the electric field of a point charge we found before using Coulomb's law! Consistency is good! Table 210-1 gives important information for the three types of symmetry we will use.

210-4: Electric field inside a conductor

Remember that in a perfect conductor, any excess charge is free to move within the conductor. Imagine that we have some excess electrons inside copper sphere. Those electrons want to repel each other, and will therefore get as far away from each other as possible, and will therefore will move out until they reach the surface of the conductor. Which means, there is now no excess charge inside the conductor. Gauss's Law tells us

$$\oint \vec{E} \cdot d\vec{a} = \frac{0}{\epsilon_0} \quad \rightarrow \quad E = 0. \tag{210-8}$$

> **Connection**: Lightning Storm
>
> The fact that excess charge stays on the outside of a conductor is the primary reason that you are safe in your car during a lightning storm. Even with the huge transfer of charge brought with the lightning, the charge remains on the outside surface of the car, and the electric field inside remains close to zero!

This tells us that the electric field inside a conductor is zero! In fact, this must be true for electrostatics, because *if there were a residual electric field inside the conductor, it would cause any excess charge to move! (F = qE)*. The summarize:

In electrostatics:	**Any excess charge on a conductor lies on its surface.** **The electric field *inside* a conductor is zero.**

Table 210-1. Important information for Gauss's Law

Symmetry	Description of Surface	Surface Area	Flux
Spherical	Sphere of radius r centered on charge	$A = 4\pi r^2$	$\oint \vec{E} \cdot d\vec{a} = E4\pi r^2$
Cylindrical	Cylinder of radius r and length l centered on and infinite line of charge. Note surface area contains the curved area of cylinder and end caps. End caps do not contribute to flux.	$A = 2\pi rl + 2\pi r^2$	$\oint \vec{E} \cdot d\vec{a} = E2\pi rl$
Flat Surface (plane)	Box with each face of area A placed halfway through surface. Sides do not contribute to flux.	$A = A$ (Field on one side) $A = 2A$ (Field on two sides)	$\oint \vec{E} \cdot d\vec{a} = EA$

The next section of this unit will give you a few examples of Gauss's law in action. Again, keep in mind that you must have the symmetries we've discussed in order to remove the integral in Gauss's law; if you do not have these symmetries, it is usually better to do a direct calculation of the electric field (i.e., using Equation 210-8).

210-4: Gauss's law examples

Consider: *How is Gauss's Law actually used?*

Here are a couple of tips for solving Gauss's Law problems. First, the left side of Gauss's Law (the electric flux) is all about the symmetric surface that you choose and the right side is all about the actual charge distribution enclosed by your surface:

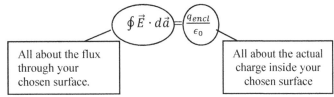

All about the flux through your chosen surface.

All about the actual charge inside your chosen surface

Also, if charge is distributed over a volume, surface, etc., make sure to use the densities we discussed at the beginning of unit 207 (see example 210-3).

Example 210-2: Electric field of a straight wire

Find the electric field a distance 8.5 meters from a very long straight wire with linear charge density $6.5 \times 10^{-12}\ C/m$.

Solution:

A very long, straight wire contains cylindrical symmetry, so we can immediate rewrite Gauss's law as

$$\oint \vec{E} \cdot d\vec{a} = \frac{q_{encl}}{\epsilon_0} \quad \rightarrow \quad E2\pi rl = \frac{q_{encl}}{\epsilon_0},$$

where r and l are the radius and length of our fictitious Gaussian cylinder.

As far as the charged enclosed in a Gaussian cylinder, this will be the charge density of the wire multiplied by the length of the cylinder:

$$q_{encl} = \lambda l.$$

Substituting this into our Gauss's law equation gives us

$$E2\pi rl = \frac{\lambda l}{\epsilon_0} \quad \rightarrow \quad E = \frac{\lambda}{2\pi \epsilon_0 r}.$$

Finally, we can substitute in our given values to find

$$E = \frac{6.5 \times 10^{-1}\ C/m}{2\pi (8.85 \times 10^{-12}\ C^2/(N \cdot m^2))(8.5m)} = 0.0138 \frac{N}{C}.$$

Example 210-3: Conducting sphere.

A metal sphere with a radius of 1.3 meters is charged with $5.4 \times 10^{-8} C$. What is the electric field in all of space due to this sphere?

Solution:

Since we are dealing with a charged sphere, we can first say that we have spherical symmetry and immediately rewrite Gauss's Law as

$$\oint \vec{E} \cdot d\vec{a} = \frac{q_{encl}}{\epsilon_0} \quad \rightarrow \quad E4\pi r^2 = \frac{q_{encl}}{\epsilon_0},$$

Where r is the radius of our fictitious Gaussian sphere.

For this problem, we must consider two regions
1) Inside the sphere
2) Outside the sphere

Inside the sphere (region 1)

Since we have a conducting sphere, all of the excess electric charge will reside on the surface of the sphere, meaning that there is *no charge inside the surface*. Another way to say this is $q_{encl} = 0$, and we can immediately say

$$E_1 = 0 \quad (r < 1.3m).$$

Outside the sphere (region 2)

Once we are outside the surface of the sphere, all of the charge on the sphere is enclosed in a Gaussian surface, so we can write

$$E_2 = \frac{q_{encl}}{4\pi \epsilon_0 r^2} = \frac{5.4 \times 10^{-8} C}{4\pi \epsilon_0 r^2} = \frac{486\ Nm^2/C}{r^2} \quad (r > 1.3m).$$

Putting it all together

So, the electric field inside the sphere is zero, and outside the sphere, the field drops off as $1/r^2$. In fact, the functional form of the electric field outside the sphere looks just like the electric field due to a point charge. The best way to visualize this effect is to plot the electric field versus distance from the center of the sphere for all of space, which is shown to the right. For this plot, the radius of the actual sphere is where the discontinuity is in the plot and is often noted as r_0 ($r_0 = 1.3m$).

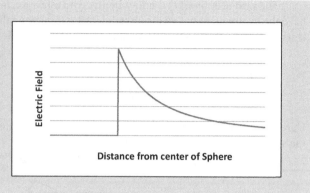

Distance from center of Sphere

Example 210-4: Sphere with uniformly distributed charge

In example 210-3, we examined a charged conducting sphere and found that the electric field inside the sphere was zero and the electric field outside the sphere dropped off as $1/r^2$, just as a point charge does. What happens if, instead of using a conducting sphere, we put the same charge on a non-conducting sphere, and distribute the charge uniformly throughout the volume? Find the electric field in all of space in this situation.

Solution:

Just as in example 210-3, we can immediately see that we have spherical symmetry and rewrite Gauss's Law as

$$\oint \vec{E} \cdot d\vec{a} = \frac{q_{encl}}{\epsilon_0} \quad \rightarrow \quad E4\pi r^2 = \frac{q_{encl}}{\epsilon_0}.$$

We will once again break this problem into two regions,
1) Inside the sphere
2) Outside the sphere

Inside the sphere (region 1)
Unlike example 210-3, there is charge distributed inside the sphere, so we must be careful when solving Gauss's Law. The charge that is enclosed within a fictitious Gaussian sphere of radius r, will have to do with the density and the volume enclosed within that sphere:

$$q_{encl} = \rho V_{encl} = \frac{Q}{4\pi R^3} 4\pi r^3 = Q\frac{r^3}{R^3},$$

where the charge density is given by the total charge of the sphere, Q, divided by the total volume of the sphere, $4\pi R^3$, and the volume enclosed is the volume out to the radius of the Gaussian sphere, r ($4\pi r^3$).

This result can now be placed back into our Gauss's law for spherical symmetry:

$$E_1 4\pi r^2 = \frac{q_{encl}}{\epsilon_0} = \frac{Q}{\epsilon_0} \frac{r^3}{R^3}.$$

Solving for E, we find

$$E_1 = \frac{Q}{4\pi\epsilon_0 R^3} r.$$

Substituting our known values in, this can be simplified to

$$E_1 = \left(221 \frac{N}{Cm}\right) r \qquad (r < 1.3m).$$

Note that this electric field grows linearly with r from the center of the sphere, whereas in example 210-3, the electric field inside the sphere was zero.

Outside the sphere (region 2)
Conceptually, when we are outside the sphere, we have the exact same situation as region 2 in example 210-3 – a total charge Q enclosed within the region of spherical symmetry. Since we have already solved this problem in the last example, we can immediately write down

$$E_2 = \frac{q_{encl}}{4\pi\epsilon_0 r^2} = \frac{5.4x10^{-8}C}{4\pi\epsilon_0 r^2} = \frac{486\ Nm^2/C}{r^2} \qquad (r > 1.3m).$$

Putting it all together

First, it is important to note that (up to some rounding) $E_1 = E_2$ ($287\ N/C$) at the boundary. This is important for continuity of the electric field at the edge of the sphere. Plotting out the above results for the electric field versus distance from the center of the sphere gives the plot below.

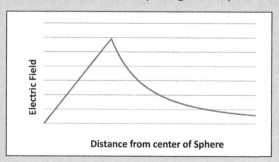

Distance from center of Sphere

Example 210-5: Electric field of a parallel-plate capacitor

Using Gauss's Law, what is the electric field between the plates of a parallel plate capacitor with charge density $52 \ pC/m^2$ on each plate?

Solution:

There are a number of ways to solve this problem; however, the question specifically asks for it to be solved using Gauss's Law. So, first, we notice that each of the plates of a parallel-plate capacitor looks like a plane, so we invoke Gauss's Law with planar symmetry:

$$\oint \vec{E} \cdot d\vec{a} = \frac{q_{encl}}{\epsilon_0} \quad \rightarrow \quad EA = \frac{q_{encl}}{\epsilon_0}.$$

For a planar surface, we choose a Gaussian shape that sticks through the plane enclosing a surface area, A of charge. Therefore, the enclosed charge is given by

$$q_{encl} = \sigma A_{encl} = \sigma A.$$

This can now be substituted back into Gauss's Law and solved:

$$EA = \frac{q_{encl}}{\epsilon_0} = \frac{\sigma A}{\epsilon_0} \quad \rightarrow \quad E = \frac{\sigma}{\epsilon_0}.$$

All that is left is to substitute in the charge density for this problem:

$$E = \frac{52x10^{-1} \ C/m^2}{8.85x10^{-1} \ F/m} = 5.88 \frac{N}{C}.$$

Note that this solution really only took three lines of math and required no calculus. If you have enough symmetry to use Gauss's Law, it greatly simplifies problems!

210-4: Ampere's Law

Consider: *How is symmetry exploited for magnetic fields?*

There is an analogous law to Gauss's law for magnetism known as Ampere's law. Ampere's law relates the current piercing through a closed loop to the magnetic field created by that current.

Ampere's Law

$$\oint \vec{B} \cdot d\vec{l} = \mu_0 I_{encl} \qquad (210\text{-}9)$$

Description – Ampere's Law relates the total current passing through a closed loop to the magnetic field created by the current.

The constant of proportionality in Ampere's Law, μ_0, is called the ***permeability of free space*** and has a value $\mu_0 = 4\pi \ x \ 10^{-7} \ N/A$.

As with Gauss's law, our goal is to find enough symmetry to remove the integral in and solve for the magnetic field using algebra. Ampere's law has a $d\vec{l}$ instead of the $d\vec{a}$ from Gauss's law. $d\vec{l}$ is a small part of a path and we need to find a path that is either parallel to the magnetic field (so that the dot product gives us a 1), and/or perpendicular to the magnetic field (so the dot product gives us a zero). In either case, we want the magnetic field to be constant along that path so that the field can be taken out of the integral.

Let's do this in the way of an example. Consider the magnetic field around a long straight current carrying wire as shown in figure 210-6. We discussed in unit 208 why the field lines are concentric circles around the wire. Along each field line, the magnetic field is the same value, so if we choose a loop (called an Amperian loop) that follows one of the field lines, we've met all of our requirements:

1) Field is constant on the loop
2) Field is parallel to the loop at all spots

Therefore

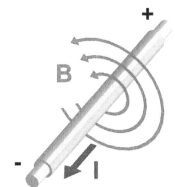

Figure 210-6. Magnetic field of a long straight wire.

$$\oint \vec{B} \cdot d\vec{l} = B \oint dl = B2\pi r, \qquad (210\text{-}10)$$

since the length around a circle is its circumference. Plugging this back into ampere's law, we get

$$\oint \vec{B} \cdot d\vec{l} = \mu_0 I_{encl} \quad \rightarrow \quad B2\pi r = \mu_0 I, \qquad (210\text{-}11)$$

which simplifies to

$$B = \frac{\mu_0 I}{2\pi r}. \qquad (210\text{-}12)$$

This is exactly what we found from many lines of integration in unit 208, and we did it with just three simple lines of algebra! Because of the symmetry requirements, there are not many situations where Ampere's law is simple to use; however, for those situations it is very powerful.

Example 210-6: Magnetic field inside of a real wire

Using Ampere's Law, find the magnetic field in all of space due to a long wire of radius 2.6 cm, carrying a current of 6.0 mA.

Solution:

Similar to example 210-4 for electric fields, we are concerned with the magnetic field inside the wire in addition to the field outside the wire. So, we first break the problem into two regions:

1) Inside the wire
2) Outside the wire.

For each case, we know we need to start with Ampere's Law, and since we still have a long wire, we can use the simplification in equation 210-12 above:

$$\oint \vec{B} \cdot d\vec{l} = \mu_0 I_{encl} \quad \rightarrow \quad B2\pi r = \mu_0 I_{encl}.$$

Inside the wire (region 1)

If we are some distance from the center of the wire, but still inside the wire, not all of the current is enclosed in the loop out of our radius, so we must write

$$I_{encl} = J A_{encl} = \frac{I}{\pi R^2} \pi r^2 = I \frac{r^2}{R^2},$$

where the current density, J, is the total current divided by the total cross-sectional area of the wire, πR^2, and the area enclosed is the cross-sectional area out to the distance we care about, r (πr^2).

Substituting this back into Ampere's Law, we get

$$B_1 2\pi r = \mu_0 I \frac{r^2}{R^2} \quad \rightarrow \quad B_1 = \frac{\mu_0 I}{2\pi R^2} r. \ (r < 2.6 \ cm)$$

Similar to the electric field inside a uniformly charged sphere, we see that the magnetic field inside the current carrying wire increases linearly with distance from the center of the wire.

Outside the wire (region 2)

Outside the wire, the entire current is enclosed and we have the exact situation described above for a long straight wire and we can immediately write down

$$B_2 = \frac{\mu_0 I}{2\pi r} \quad (r > 2.6 \ cm).$$

All that is left is to plug in the values from the description

$$B_1 = (1.78x10^{-6} \ T/m) \, r,$$
$$B_2 = (1.20x10^{-9} Tm)/r.$$

Putting it all together

Just as we did with electric fields, we can plot these magnetic fields versus distance from the center of the wire. The plot below shows this result.

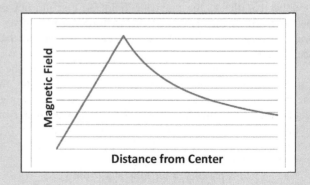

210-5: Differential forms of Gauss's and Ampere's laws

Consider: *What's wrong with Gauss's and Ampere's law? Can they be applied to very small regions to get rid of the integrals?*

There are two very important problems with Gauss's and Ampere's laws in integral form:

Integration

Both laws work well in the few situations presented in this chapter. However, most of the real world does not have the nice symmetry needed for us to reduce the calculations to algebra, and the integrals can quickly become very, very hard.

Retardation

Neither Gauss's nor Ampere's law say anything about time. Put another way, they assume that if a field is created, it is created instantly in all of space – this actually violates the fundamental principle of relativity – the cosmic speed limit of the speed of light.

Let's say the sun were to suddenly disappear. Would we know it right away? Nope. We've learned that electromagnetic waves travel at the speed of light in a vacuum, not infinitely fast. It takes about eight minutes for light from the sun to reach us here on earth (called a distance of eight light-minutes). So, what we see at any moment looking at the sun (don't actually do this, the sun is too bright!) is what happened at the sun eight minutes ago. So, if the sun were to suddenly disappear, we wouldn't know for eight minutes. The same is true of gravity – information about gravity doesn't travel infinitely fast, but rather at the speed of light. So, the earth would continue on its circular orbit for another eight minutes after the sun disappeared. This overall delay is known as *retardation*. The key here is that the same retardation effects must be true for electric and magnetic fields, but the integral forms of Gauss's and Ampere's laws do not take this into effect.

The way around both of these issues is to look at an infinitesimally small region of space instead of the large volumes we have so far in the unit. Put another way, we are going to divide up Gauss's and Ampere's law by their total volume and look at their effects on tiny spaces.

What we are going to find is two-fold

1) A point in space with a charge changes the electric field there slightly. An area of free space does not.
2) A point in space with a current changes the magnetic field there slightly. An area of free space does not.

Figure 210-6 shows the basic electric field of a point charge and magnetic field of a small straight current. Let's first assume that the region of space that contains each would have *no fields* if the charge or current we not there. That means that their presence changes the field right around them (from nothing to some field).

Consider Gauss's Law written in terms of a charge density in a tiny, tiny volume

$$\oint \vec{E} \cdot d\vec{a} = \frac{\rho \Delta V}{\epsilon_0}$$

$$\rightarrow \quad \oint \frac{\vec{E} \cdot d\vec{a}}{\Delta V} = \frac{\rho}{\epsilon_0} \tag{210-13}$$

The left side of this equation represents the net flux per unit volume. This is also called the 'divergence' of the field. If you look again at the field for a single point charge, you can see why – the fact that a point charge is at that point in space causes the electric field lines to move away, or diverge from that point. So, now we write

$$div(\vec{E}) = \frac{\rho}{\epsilon_0}. \tag{210-14}$$

This is the microscopic form of Gauss's law; however, we haven't talked about how to actually calculate the divergence yet. Consider again the left side of Gauss's law per unit volume with an electric field shown in figure 210-7:

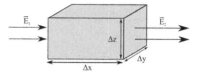

$$\oint \frac{\vec{E} \cdot d\vec{a}}{\Delta V} = \oint \frac{\Delta E \Delta z \Delta y}{\Delta x \Delta y \Delta z} = \oint \frac{\Delta E_x}{\Delta x}. \qquad (210\text{-}15)$$

Figure 210-7. A changing electric field traversing a small volume.

Now, if we let the size of the volume shrink to near zero, this gets rid of the integral because the surface area becomes quite small. Therefore

$$\oint \frac{\vec{E} \cdot d\vec{a}}{\Delta V} = \frac{\partial E}{\partial z}. \qquad (210\text{-}16)$$

Also, in this derivation, we only had an electric field in one direction. It is certainly possible that the electric field could be at an arbitrary angle, and therefore in all three dimensions, in which case the contributions from each dimension would add together. In total, this leaves us with

$$div(\vec{E}) = \frac{\partial E_x}{\partial x} + \frac{\partial E_y}{\partial y} + \frac{\partial E_z}{\partial z} = \frac{\rho}{\epsilon_0}. \qquad (210\text{-}17)$$

> **Don't get scared off by this derivation!** As can be seen in the equation to the left, the differential form of Gauss's law (and Ampere's law below) really just relate derivatives of the electric field (magnetic field) in terms of densities. All you need to do to solve these problems are take simple derivatives or complete simple integrals. See the examples in this section.

Example 210-7: Gauss's Law (differential form)

A small region in space has an electric field given by

$$\vec{E} = \begin{bmatrix} 32x + 4y \\ 2xy + z \\ 14xz + 2yz \end{bmatrix} N/C.$$

What is the charge density at this point?

Solution:

This is a direct application of the differential form of Gauss's Law. We must first find each of the partial derivatives from equation 210-13:

$$\frac{\partial E_x}{\partial x} = \frac{\partial}{\partial x}(32x + 4y) = 32,$$

$$\frac{\partial E_y}{\partial y} = \frac{\partial}{\partial y}(2xy + z) = 2x,$$

$$\frac{\partial E_z}{\partial z} = \frac{\partial}{\partial z}(14xz + 2yz) = 14x + 2y.$$

Each of these can now be combined

$$\frac{\partial E_x}{\partial x} + \frac{\partial E_y}{\partial y} + \frac{\partial E_z}{\partial z} = (32) + (2x) + (14x + 2y),$$

or

$$\frac{\partial E_x}{\partial x} + \frac{\partial E_y}{\partial y} + \frac{\partial E_z}{\partial z} = 16x + 2y + 32.$$

Now, this can be plugged into Gauss's Law to find:

$$16x + 2y + 32 = \frac{\rho}{\epsilon_0},$$

or

$$\rho = \epsilon_0(16x + 2y + 32) \, C/m^3.$$

The exact same game can be played with Ampere's law. However, if you look again at the magnetic field created by a small current element, it does not cause a divergence in the field, but rather, it causes a **curl**. The magnetic field around a small current element forms perfect circles as we saw earlier in the unit. We can rewrite Ampere's law using current density instead of total current:

$$\oint \vec{B} \cdot d\vec{l} = \mu_0 I_{encl} = \mu_0 J \Delta A \quad \rightarrow \quad \oint \frac{\vec{B} \cdot d\vec{l}}{\Delta A} = \mu_0 J. \qquad (210\text{-}18)$$

The left side of this equation is called the **curl** and is a measure of how much the current density in a region of spaces changes the way the magnetic field curls around that region. Note that the path given by $d\vec{s}$ encloses the area ΔA, so they are intimately related – if we choose to make the enclosed area smaller and smaller, we are also making the path around the area smaller as well. The only problem is that, especially as the area gets infinitesimally small, the area vector is always perpendicular to the direction of the path ($d\vec{s}$), which suggests a cross product! We can now write Ampere's law as

$$curl(\vec{B}) = \mu_0 \vec{J}. \tag{210-19}$$

Mathematically, the curl is given by

$$curl(\vec{B}) = \vec{\nabla}x\vec{B} = \begin{bmatrix} \dfrac{\partial B_z}{\partial y} - \dfrac{\partial B_y}{\partial z} \\ \dfrac{\partial B_x}{\partial z} - \dfrac{\partial B_z}{\partial x} \\ \dfrac{\partial B_y}{\partial x} - \dfrac{\partial B_x}{\partial y} \end{bmatrix}. \tag{210-20}$$

> Again, don't get scared off by the derivation. Although Ampere's law is inherently harder than Gauss's law because of the cross product, you are still just relating derivatives to densities at appoint in space.

Example 210-8: Ampere's Law (differential form)

A small region of space produces a magnetic field given by

$$\vec{B} = \begin{bmatrix} 32x10^{-6}y \\ 32x10^{-6}y \\ 32x10^{-6}y \end{bmatrix} T.$$

What is the current density that can produce this magnetic field?

Solution:

This is a direct application of the differential form of Ampere's Law. In order to solve this, we must find the individual components to

$$\vec{\nabla}x\vec{B} = \mu_0\vec{J}:$$

$$\frac{\partial B_z}{\partial y} - \frac{\partial B_y}{\partial z} = 32x10^{-6}y - 0 = 32x10^{-6}$$

$$\frac{\partial B_x}{\partial z} - \frac{\partial B_z}{\partial x} = 0$$

$$\frac{\partial B_y}{\partial x} - \frac{\partial B_x}{\partial y} = 0 - 32x10^{-6}y = -32x10^{-6}$$

Putting this all together, we see that

$$\begin{bmatrix} 32x10^{-6} \\ 0 \\ -32x10^{-6} \end{bmatrix} = \mu_0\vec{J},$$

or

$$\vec{J} = \frac{1}{\mu_0}\begin{bmatrix} 32x10^{-6} \\ 0 \\ -32x10^{-6} \end{bmatrix} A/m^2.$$

Therefore, this magnetic field can be caused by a constant current density directed 45° down from east.

211 – Electrostatic Potential

The use of electric fields gives us a powerful technique to discuss how a region of space would cause a force on a representative charged particle. There is a similar quantity related to energy called the electric potential. The potential gives us a measure of how much energy a particle would have if it were in a certain region of space. In this unit, we explore this important concept.

Integration of Ideas

Review potential energy
Review electric fields

The Bare Essentials

- Electric charges placed in electric fields contain electrostatic potential energy. As with all potential energy, this can be converted to kinetic energy as the field does work on the charged particle.

- The electric potential is the electric potential energy per unit charge at a point everywhere in space due to *other* charges. As opposed to electric and magnetic fields, the electric potential is a scalar field.

- The electric potential is related to the electric field.

Electric Potential Difference

$$\Delta V = -\int_i^f \vec{E} \cdot d\vec{l}$$

Description – This equation defines the electric potential in terms of the electric field, $\vec{E}$ and a displacement, $d\vec{l}$, from initial position, i, to final position, f.
Note 1: The electric field can also be calculated from the potential; $E_x = -\frac{\partial V}{\partial x}, E_y = -\frac{\partial V}{\partial y}, E_z = -\frac{\partial V}{\partial z}$ or $\vec{E} = -\vec{\nabla}V$.
Note 2: The SI unit for potential difference is the Volt (V). $1\ V = 1\ J/C$

- The electric potential due to a point charge depends on distance from the charge.

Electric Potential due to a Point Charge

$$V = \frac{kq}{r}$$

Description – This equation defines the electric potential due to a point charge in terms of the charge and the distance from the charge.
Note: This equation assumes that the potential at infinity is zero.

- Equipotential surfaces are perpendicular to electric field lines at all points in space and give a visual feel for electric potential

- The potential difference along a constant electric field is relatively easy to calculate.

Potential Difference along a Constant Electric Field

$$\Delta V = -Ed$$

Description: This equation relates the potential difference ΔV as we move a distance d parallel to a constant electric field E.
Note: This equation is only for constant electric fields.

- In an area of space where there is no electric field, the electric potential is constant.

- Two stationary point charges store electric potential energy that is dependent on their separation.

Potential Energy Stored by Two Point Charges

$$U_{12} = \frac{kq_1q_2}{r_{12}}.$$

Description: This equation describes the electric potential energy, U_{12}, store by two point charges, q_1 and q_1, separated by a distance r_{12}.
Note: This equation is only true for static charged particles.

- There is also a magnetic potential related to the magnetic field; however, this potential is a vector, and generally quite complicated.

211-1: Electric potential energy

Consider: *Electric fields can change the motion of a charged particle. Does that mean there can be potential energy as well?*

IN THE EXTENSION TO EXAMPLE 210-4, we found that the electric field near a very large charged plate is constant and is given by

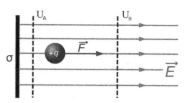

Figure 211-1. A particle with charge +q very near a large positively charged plate.

$$E_x = \frac{\sigma}{2\epsilon_0}, \qquad (211\text{-}1)$$

where σ is the surface charge density (charge per unit area) of the plate. Let's say we have a positively charged plate creating an electric field as shown in figure 211-1. What happens if we place a positively charged particle at point A near the plate? Well, we know from the electric force law,

$$\vec{F} = q\vec{E}, \qquad (211\text{-}2)$$

that the particle will experience a force to the right and will therefore accelerate to the right as it approaches point B. We could also describe this in terms of energy: the particle starts with zero kinetic energy at point A and has some kinetic energy at point B. However, conservation of energy tells us that this energy must have come from somewhere and that is usually potential energy. The Coulomb Force (like the gravitational force) is a conservative force. Remembering the relationship between force and potential energy from Physics I, we can write

$$\Delta U = -\int_A^B \vec{F} \cdot d\vec{l} = -q \int_A^B \vec{E} \cdot d\vec{l}, \qquad (211\text{-}3)$$

where ΔU is the change in potential energy, q is the charge of the particle moving through a displacement $d\vec{l}$ in the electric field $\vec{E}$. Since the change in potential energy is negative, that energy went somewhere – and in this case, it goes into the kinetic energy of the charged particle. We could also write this in terms of the work done by the electric field on the particle

$$W = \int_A^B \vec{F} \cdot d\vec{l} = q \int_A^B \vec{E} \cdot d\vec{l} = -\Delta U. \qquad (211\text{-}4)$$

For completeness, I want to remind you that the conservation of energy can be written as

$$\Delta K + \Delta U = 0 \quad \rightarrow \quad \Delta K = -\Delta U, \qquad (211\text{-}5)$$

which shows how any potential energy lost must go into kinetic energy gain.

Example 211-1: Energy in electric fields

A small particle with mass 1.2×10^{-1} *kg* and charge $+2.5 \times 10^{-15}$ *C* is placed 3.4 mm from a very large conducting plate with charge density of 2.3 x 10⁻⁵ C/m². What is the kinetic energy of the particle when it is 4.2 mm from the plate?

Solution:

This question resembles conservation of energy questions from Physics I – we are given information about initial positions and speeds and are asked about final positions and speeds. We know that the change in potential energy of a charged particle in an electric field is

$$\Delta U = -q \int_{initial}^{final} \vec{E} \cdot d\vec{l},$$

So,

$$\Delta U = -q \int_{3.4\,mm}^{4.2\,mm} \frac{\sigma}{2\epsilon_0} dl = -q \frac{\sigma}{2\epsilon_0} \int_{3.4\,mm}^{4.2\,mm} dl,$$

where I was able to get rid of the dot product since the particle will move parallel to the electric field. In addition, I was able to remove all constants from the integral.

$$\Delta U = -\frac{(2.5 \times 10^{-15}\,C)(2.3 \times 10^{-5} C/m^2)}{2(8.85 \times 10^{-12} C^2/(N \cdot m^2))}(4.2\,mm - 3.4\,mm).$$

$$\Delta U = -2.60 \times 10^{-12} J.$$

Now, we know the initial kinetic energy is zero, so

$$K_f - K_i = \Delta U \quad \rightarrow \quad K_f = 2.60 \times 10^{-12} J.$$

What is the speed of the particle when it is 4.3 mm from the charged surface?

or

Solution:

Here, we use the relationship between kinetic energy and speed

$$v = \sqrt{\frac{2(2.60 \times 10^{-12} J)}{1.2 \times 10^{-12} kg}}.$$

$$K = \frac{1}{2}mv^2.$$

Simplifying, we find

So,

$$v = 2.08 \; m/s.$$

$$v = \sqrt{\frac{2K}{m}},$$

This seems like a reasonable speed; however, if you keep in mind that the particle has only traveled 0.8 mm, you will realize that it is undergoing a very large acceleration!

As with any potential energy, you can picture what's going on here in terms of gravitational potential energy. Since a positively charged particle near a positively charged plate has a positive potential energy, this is like a ball being at the top of a hill. The particle moves away from the plate and loses potential energy, just like a ball loses potential energy as it moves down a hill. In the example above, the force was even constant, just like the gravitational force ($F=mg$ version). This analogy can be seen in figure 211-2.

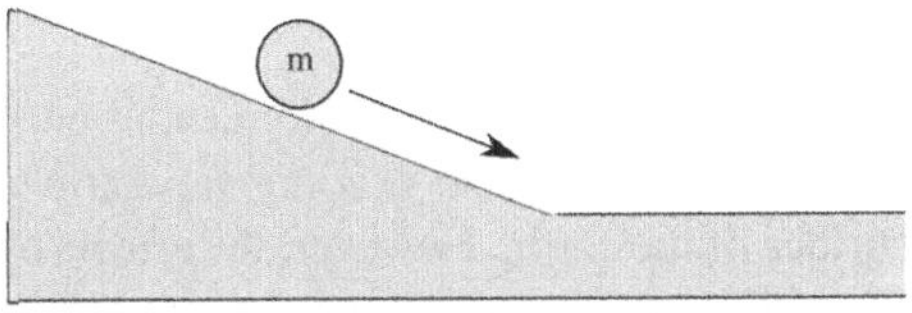

If the electric field in a given area is not constant, we have to be very careful with the gravitational analogy, because the 'hill' will not be straight! The analogy does still work; you just need to think about the shape of the potential energy curve.

Figure 211-2. Analogy of electric potential energy to gravitational potential energy for example 211-1.

211-2: Electric potential

> **Consider**: *Electric field is force per charge – is there something like that for energy – energy per charge?*

There is a very important concept related to electric potential energy, called **electric potential**. Similarly, there is a concept related to changes in electric potential energy called **potential difference**. Research in physics education has shown that the concepts of potential and potential difference are combined *the single hardest ideas for students to learn* in introductory physics.

On the flip side, electric potential is ubiquitous in our electric world. The standard unit of potential difference is the Volt. The volts in a 12-Volt battery is a measure of potential difference. If you've ever heard that you must use a converter to plug U.S. electrical equipment into electric wall sockets in the UK, this is because the U.S. uses a potential difference of 110 Volts and the UK uses 240 Volts. If you've ever taken a biology class, you may have learned that a typical nerve cell has a resting potential difference of -70 millivolts across the cell membrane.

So, potential is everywhere; but what *is* it? Electric potential is a measure of electric potential energy per unit charge:

$$V_E = \frac{U_E}{q}, \tag{211-6}$$

where V is the electric potential, U_E is the electric potential energy and q is the charge of the particle that has potential energy U_E. Potential difference, ΔV, is similarly defined:

$$\Delta V_E = \frac{\Delta U_E}{q}. \tag{211-7}$$

The SI unit for electric potential and potential difference is the **Volt (V)**. The units for volt is also defined by its relationship to potential energy and charge: 1 V = 1 J/C.

Electric potential is a **field**. In the same way that the electric field gives us a distribution of how much force an object would have if placed in the field, electric potential gives us a measure of how much potential energy a charge would have if placed in the field. One of the nice things about electric potential is that it is a **scalar field**, meaning that we do not have to worry about direction, only about a value at each point in space:

$$\vec{F}_E = q\vec{E} \qquad\qquad U_E = qV_E. \qquad\qquad (211\text{-}8)$$

Let's again make a comparison with gravity. Figure 211-3 shows a **topographic map** of two hills and a corresponding side view of the same hills. The closed lines on the topographic map show a representation of the height of the ground (above some reference point) for each point in the map. This is very useful because not only can you see how high each point is, but if you think about it, the closer the lines are, the more quickly the height is changing. So, if lines are very close together, the ground is very sloped, which we call the *gradient* of the field. Now, we know that gravitational potential energy near the surface of the earth is given by

$$U_g = mgh, \qquad\qquad (211\text{-}9)$$

where U_g is the gravitational potential energy, m is the mass of an object at height h, and g is the gravitational field strength near the surface of the earth. However, the effects of something falling down the hill are independent of its mass – meaning the speed at which something would be moving at the bottom of the hill if it slid down the hill is independent of the mass, and only depends on the height at

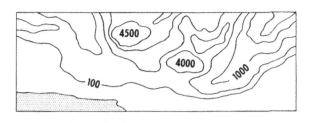

Figure 211-3. Topographic map of hills with a corresponding side view.

which it starts. So, we can define something called the **gravitational potential** that describes the important effects of the map without putting a specific mass in a specific place. Mathematically, we define the gravitational potential, V_g, in terms of gravitational potential energy

$$U_g = mV_g. \qquad\qquad (211\text{-}10)$$

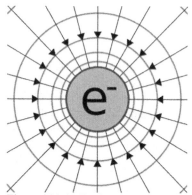

Figure 211-4. Equipotential lines and electric field lines for a negatively charged particle.

Comparing the two equations for gravitational potential energy, you can see that $V_g = gh$, so that it only has to do with a constant (g) and height (h).

Electric potential does the same thing for electric potential energy as the gravitational potential does for the gravitational potential energy. Figure 211-4 shows **equipotential lines** as well as electric field lines around a single positive charge. We will see later why these equipotential lines are perfect circles around the point charge. The equipotential lines serve the same purpose in figure 211-4 as the lines of equal height do in figure 211-3: they tell us the potential energy that a charge would have if placed at a point in the map using

$$U_E = qV_E. \qquad\qquad (211\text{-}11)$$

Hopefully, this gives you a good conceptual foundation for the electric potential. Now, we can start to describe the field mathematically by combining equation 211-11 with equation 211-3:

$$\Delta V_E = \frac{\Delta U_E}{q} = \frac{1}{q}\left(-q\int_A^B \vec{E}\cdot d\vec{l}\right) = -\int_A^B \vec{E}\cdot d\vec{l}. \qquad\qquad (211\text{-}12)$$

Notice that the dot product will make V_E a scalar as expected. This equation always works if we know the electric field, whether the electric field was found by Coulomb's Law, Gauss's Law, or any other technique.

Now, if we happen to know the potential function and want to find the electric field from it, we need to do the inverse function of equation 211-12, which is, of course, a derivative. The only problem is that we need the electric field to be a vector and now we are starting with a scalar potential. The way we do this is by using **partial derivatives**. Consider figure 211-3 again. If you were standing to the right of one of the hills and fell down, you would roll to the right (the gravitational field would cause you to fall to the right). If you were above the hills and fell down, you would roll towards the top of the map (again, the gravitational field would cause you to roll that direction). In each case, you would roll perpendicular to the lines of equal height. Now, if you are up and to the right of a hill and fell, the direction you roll would be at an angle on the map, but still perpendicular to the line of equal height (diagonal in the figure). Going back to electric potential and field, we write this relationship as

$$E_x = -\frac{\partial V}{\partial x} \quad , \quad E_y = -\frac{\partial V}{\partial y} \quad , \quad E_z = -\frac{\partial V}{\partial z}. \tag{211-13}$$

Partial derivatives are computed in a very similar fashion to regular derivatives. The only difference is that you treat the variable you are taking the derivative with respect to as a variable and all others as constants. There will be a partial derivative example soon.

Also, there is a shorthand for the partial derivatives shown in equation 211-13 using the del operator, $\vec{\nabla}$ with $\vec{\nabla} = \frac{\partial}{\partial x}\hat{\imath} + \frac{\partial}{\partial x}\hat{\jmath} + \frac{\partial}{\partial x}\hat{k}$, so that

$$\vec{E} = -\vec{\nabla}V = -\begin{bmatrix} \dfrac{\partial V}{\partial x} \\[2mm] \dfrac{\partial V}{\partial y} \\[2mm] \dfrac{\partial V}{\partial z} \end{bmatrix}. \tag{211-14}$$

$\vec{\nabla}V$ is called the **gradient of V** and is just shorthand for the derivatives shown in equation 211-13.

We now have the machinery to summarize the relationship between electric potential and electric field.

Electric Potential Difference

$$\Delta V = -\int_i^f \vec{E} \cdot d\vec{l} \qquad (211\text{-}15)$$

Description – This equation defines the electric potential in terms of the electric field, $\vec{E}$ and a displacement, $d\vec{l}$, from initial position, i, to final position, f.
Note 1: The electric field can also be calculated from the potential; $E_x = -\frac{\partial V}{\partial x}, E_y = -\frac{\partial V}{\partial y}, E_z = -\frac{\partial V}{\partial z}$ or $\vec{E} = -\vec{\nabla}V$.
Note 2: The SI unit for potential difference is the Volt (V).
1 V = 1 J/C

Example 211-2: Electric potential from field

The electric field in a region of space is given by

$$\vec{E} = \begin{bmatrix} 3.87x \\ 0 \\ 0 \end{bmatrix} \frac{N}{C}.$$

A particle moves from position $\vec{a} = [0, 4.00, 2.00]m$ to position $\vec{b} = [3.00, 0, 2.00]m$. What is the potential difference $V_b - V_a$?

Solution:

The first step in this multidimensional problem is the break the dot for the electric potential into its component form:

$$\Delta V = -\int_i^f E_x dx + E_y dy + E_z dz.$$

From here, we can split this into multiple integrals and insert the known values

$$\Delta V = -\int_0^3 3.87x\,dx - \int_4^0 0\,dy - \int_2^2 0\,dz.$$

The last two terms are zero, and the first term is

$$\Delta V = -3.87\frac{x^2}{2}\Big|_0^3 = -17.4\,V.$$

Example 211-3: Electric field from potential

What is the electric field at a point [3.0, 1.0, 6.0] m if the electric potential is given by $V = 2.32x^2yz$? (You may take x, y and z to be in meters and the units of 2.32 are V/m^4)

Solution:

To find the electric field from the potential, we must first find the partial derivative of the potential with respect to each dimension. For the x-component:

$$E_x = -\frac{\partial V}{\partial x} = -\frac{\partial}{\partial x}(2.32x^2yz) = -2.32yz\frac{\partial}{\partial x}x^2,$$

where I have explicitly noted how all variables except the one we are taking the derivative with respect to are treated as constants and can be removed from the derivative. Completing the derivative, we find

$$E_x = -4.64xyz$$

We can do the same thing for the y- and z- components

$$E_y = -\frac{\partial V}{\partial y} = -\frac{\partial}{\partial y}(2.32x^2yz) = -2.32x^2z,$$

$$E_z = -\frac{\partial V}{\partial z} = -\frac{\partial}{\partial z}(2.32x^2yz) = -2.32x^2y.$$

Finally, the values for x, y, and z at our given point can be substituted

$$E_x = -\left(4.64\frac{V}{m^4}\right)(3.0m)(1.0m)(6.0m) = -84\frac{V}{m},$$

$$E_y = -\left(2.32\frac{V}{m^4}\right)(3.0m)^2(6.0m) = -125\frac{V}{m},$$

$$E_z = -\left(2.32\frac{V}{m^4}\right)(3.0m)^2(1.0m) = 2 - 1\frac{V}{m}.$$

The entire electric field vector can then be written

$$\vec{E} = \begin{bmatrix} -84 \\ -125 \\ -21 \end{bmatrix}\frac{V}{m}.$$

211-3: Potential due to a point charge

> **Consider**: *How do you calculate the potential for a point charge?*

It is instructive to look at the potential due to a single point charge. We know the electric field of a point charge is given by

$$\vec{E} = \frac{kq}{r^2}\hat{r}. \tag{211-16}$$

Now, let's say we want to find the electric potential at a point close to the point charge. We know that the electric potential difference is given by

$$\Delta V = -\int_i^f \vec{E} \cdot d\vec{l}. \tag{211-17}$$

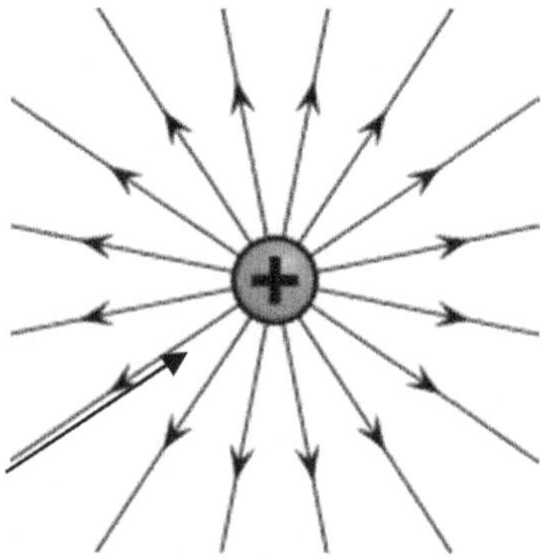

Figure 211-5. Electric field of a positive point charge. The black arrow pointing towards the charged particle represents the direction of increasing potential.

In order to determine the potential of just point charge, we will compare the potential close to the point charge to where we expect it to be zero – at infinity. Figure 211-5 shows a picture of the electric field of a positively charged point particle as well as an arrow depicting $d\vec{l}$ moving out *radially* to infinity so that dl is really a dr. Note that $d\vec{l}$ is pointing in the same direction as the electric field so that the angle between the two is zero degrees,

$$V_f - V_i = -\int_R^\infty \frac{kq}{r^2}dr\cos 0 = -kq\int_R^\infty \frac{dr}{r^2}. \tag{211-18}$$

Completing the integral, we find

$$0 - V = \frac{kq}{r}\Big|_R^\infty = 0 - \frac{kq}{R}. \tag{211-19}$$

Solving for V and replacing R with r because the point we care about could be at any radius, we get

$$V = \frac{kq}{r}.$$

(211-20)

Electric Potential due to a Point Charge

$$V = \frac{kq}{r}$$

(211-20)

Description – This equation defines the electric potential due to a point charge in terms of the charge and the distance from the charge.

Note: This equation assumes that the potential at infinity is zero.

Example 211-4: Two point charges

What is the electric potential at a point directly in between
(a) Two electrons separated by 2 μm.
(b) A proton and an electron separated by 2 μm.

Solution:

The electric due to a point charge is given by 211-19. In order to find the potential at a point in space due to many point charges, we simply find each individually and add them up. Mathematically:

$$V_{total} = V_1 + V_2 + V_3 + \cdots.$$

(a) The first thing is to note that if the electrons are separated by 2 μm, then they are each 1 μm from the point directly in between each.

Now, we can directly apply the equation for electric potential due to a point charge

$$V_{total} = \frac{kq_1}{r_1} + \frac{kq_2}{r_2}.$$

Since, $q_1 = q_2$ and $r_1 = r_2$, this can be simplified to

$$V_{total} = \frac{2kq_1}{r_1} = \frac{2(8.99x10^9 Nm^2/C^2)(-1.6x10^{-1}\ C)}{1x10^{-6}m},$$

or

$$V_{total} = -2.88\ V.$$

(b) Once again, we start with the general equation for two point charges:

$$V_{total} = \frac{kq_1}{r_1} + \frac{kq_2}{r_2}.$$

In this case, we still have $r_1 = r_2$; however, since the electron and the proton have electric charge equal in magnitude but opposite in signs, $q_1 = -q_2$, so

$$V_{total} = \frac{kq_1}{r_1} - \frac{kq_1}{r_1} = 0.$$

In fact, the electric potential at a point exactly in between charged particles with exactly opposite sign will always be zero.

211-4: Equipotential lines

Consider: *Is there a way to visualize electric potential, similar to electric field lines for electric field?*

We had a brief introduction to equipotential lines in section 211-2 when introducing the basic idea of potential and its relationship to potential energy. At that time, I mentioned that we would see why the equipotential lines of a single point charge are perfect circles, which we can now do. Consider the equation we just found for the electric potential due to a point charge:

$$V = \frac{kq}{r}.$$

(211-21)

Since this equation has a single r in the denominator, all points a distance r from the point charge will have the same electric potential. The locus of points a distance r from a point charge in a plane is a circle. In three-dimensional space, this *equipotential surface* is a sphere. Although it is important to realize that the potential field is three-dimensional, we will only consider two-dimensional potential fields, especially when drawing them.

There are a couple of important concepts for equipotential lines

1) Equipotential surfaces are everywhere perpendicular to the electric field.
2) No change in potential energy is required to move a charge along an equipotential surface.
3) Energy is converted between potential and kinetic when moving between equipotential surfaces.

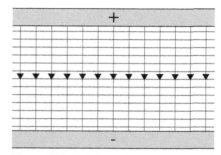

Figure 211-6. Equipotential (horizontal) and electric field lines (vertical) between two oppositely charged plates.

The equipotential lines (and electric field) between two charged plates are shown in figure 211-6. Notice that, following concept 1) above, the equipotential lines are perpendicular to the electric field at all points. This is often the easiest way to produce an equipotential map – to first draw out the electric field lines and then determine the equipotential lines from the electric field lines.

Remember, electric potential and equipotential lines are really about energy (per unit charge). The whole purpose is to let you think about how energy would be distributed on a charged particle or set of charged particles if placed in a region of space.

211-5: Two important results

Consider: *What happens if the electric field is zero?*

There are two important situations to discuss, the electric potential in a constant electric field and the electric potential in a region of no electric field. *I cannot stress enough that the results of these arrangements are very limited and should not be applied to any other situations.*

<u>Constant Electric Field</u>

If we start with our general definition of electric potential difference and break $d\vec{l}$ into components that are parallel and perpendicular to the electric field, we find

$$\Delta V = -\int_i^f \vec{E} \cdot d\vec{l} = -\int_i^f \vec{E} \cdot \left(d\vec{l}_\parallel + d\vec{l}_\perp \right) = -\int_i^f E \, dl_\parallel, \qquad (211\text{-}22)$$

because the dot product of two perpendicular vectors is zero and the dot product of two parallel vectors is given by the magnitude of the two vectors multiplied together. Now, if the electric field is constant in space, we can pull it out of the integral and find

$$\Delta V = -E \int_i^f dl_\parallel = -Ed, \qquad (211\text{-}23)$$

where d is the distance traveled parallel to the constant electric field. This is a very important result, because we will find that the electric field is constant (to good approximation) inside of a current carrying wire, and we can therefore use this result.

Potential Difference along a Constant Electric Field

$$\Delta V = -Ed \qquad (211\text{-}23)$$

Description: This equation relates the potential difference ΔV as we move a distance d parallel to a constant electric field E.

Note: This equation is only for constant electric fields.

Zero Electric Field

In a region where the electric field is zero, our equation for potential difference becomes simply

$$\Delta V = 0. \tag{211-24}$$

Be very careful with this! The fact that the change in potential is zero does not mean that the potential *is* zero. This simply means that the **potential in a region of no electric field is constant**. It is possible that the potential is zero in that region; however, all we can tell from this equation is that it is constant.

211-6: Potential energy of a set of point charges

Consider: *Since electric charges produce forces on each other, how much energy does it take it keep them close, but separated?*

Bringing the unit full circle, let's consider how much energy it takes to bring a collection of point charges together. First, consider just two point charges with charge q_1 and q_2. Let's say that q_1 produces a potential field in all of space, V_1. and then we bring in q_2 from infinitely far away. The potential energy of the system is then

$$U_{12} = q_2 V_1 = q_2 \frac{kq_1}{r_{12}} = \frac{kq_1 q_2}{r_{12}}. \tag{211-25}$$

Potential Energy Stored by Two Point Charges

$$U_{12} = \frac{kq_1 q_2}{r_{12}} \tag{211-25}$$

Description: This equation describes the electric potential energy, U_{12}, store by two point charges, q_1 and q_1, separated by a distance r_{12}.

Note: This equation is only true for static charged particles.

Next, if q_1 and q_2 are held in place and we bring in a third charged particle, q_3, the potential energy of the system would increase because of q_3's interaction with both q_1 *and* q_2:

$$U_{13} = \frac{kq_1 q_3}{r_{13}} \qquad \text{and} \qquad U_{23} = \frac{kq_2 q_3}{r_{23}}. \tag{211-26}$$

where r_{13} is the final distance between q_1 and q_3 and r_{23} is the final distance between q_2 and q_3.

Now, the total potential energy of the system is

$$U = U_{12} + U_{13} + U_{23} = \frac{kq_1 q_2}{r_{12}} + \frac{kq_1 q_3}{r_{13}} + \frac{kq_2 q_3}{r_{23}}. \tag{211-27}$$

The next example will further show how to find the energy of a set of point charges.

Example 211-5:   Potential energy of a triangle

How much potential energy is stored in an equilateral triangle with an electron at each vertex, if the sides of the triangle have a length of 12 nm?

Solution:

We start this problem with a clean slate, meaning that it costs us no energy to bring in the first electron from very far away.

When we bring in the second electron, we have to do work against the first electron, which will, in turn, store potential energy:

$$U_{12} = \frac{kq_1q_2}{r_{12}} = \frac{ke^2}{r},$$

So

$$U_{12} = \frac{(8.99x10^9\, Nm^2/C)(1.6x10^{-19}C)^2}{12x10^{-9}m} = 1.9\, x10^{-20}J.$$

When we bring in the third electron, we have to work against both electron 1 and electron 2, storing energy in each case. Note, however, that the magnitude of all of the charges and all of the distances are the same in this problem, so

$$U_{13} = \frac{ke^2}{r} = 1.9\, x10^{-20}J,$$

$$U_{23} = \frac{ke^2}{r} = 1.9\, x10^{-20}J,$$

Therefore, the total work we must do (and therefore total energy stored) is

$$U = U_{12} + U_{13} + U_{23} = 5.7x10^{-20}J.$$

Although this problem used similar charges and distances, the same process could be used for any number of charged point particles.

211-7: The magnetic vector potential

> **Consider**: *We talked quite a bit about the electric potential. Is there something similar for magnetic fields?*

It turns out that all vectors fields have a potential associated with them. In this unit, we have seen how both the gravitational field and the electric field have potentials that can be used to directly relate to energy. The magnetic field is no different; however, the three-dimensional nature of the field makes the potential considerably harder to deal with. In fact, unlike the gravitational and electric potentials, the potential associated with the magnetic field is a ***vector potential***. The magnetic vector potential is usually denoted by $\vec{A}$ and is given by

$$\vec{B} = \vec{\nabla}\, x\, \vec{A}. \tag{211-28}$$

Compare this to the related electric potential

$$E = -\vec{\nabla}V, \tag{211-29}$$

and you can see that the vector nature of the magnetic vector potential and the fact that it is related by a cross product make it considerably harder to deal with mathematically We will not do anything related to the magnetic vector potential in this course – I just mention it for completeness with the other fields we are studying.

212 – Capacitance and Capacitors

Electric potential and electric potential energy give us a way to describe how electric charges are distributed and how they store electric potential energy. In a specific situation where we have two surfaces that contain the same magnitude but opposite sign of charge, we call the element a capacitor, and define its capacitance to simplify calculations and later use the element in circuits.

The Bare Essentials

- The capacitor is an electric element designed to store energy in an electric field between two equal, but oppositely charged plates.

> **Definition of Capacitance**
>
> $$C \equiv \frac{q}{\Delta V}$$
>
> **Description** – This equation defines the capacitance of a capacitor as the amount of charge, q, one plate of the capacitor can hold to produce a potential difference between the plates of the capacitor ΔV.
> **Note 1:** The unit of capacitance is the Farad (F), with 1 F = 1 C/V
> **Note 2**: A Farad is a huge capacitance. More common units are mF, μF and nF.
> **Note 3**: Capacitance should only depend on geometry and not on other factors such as charge or fields.

- One important geometry for capacitance is the parallel plate capacitor where two flat plates are held parallel and separated.

> **Parallel Plate Capacitor**
>
> $$C = \frac{\epsilon_0 A}{d}$$
>
> **Description** – This equation describes the capacitance of a parallel plate capacitor in terms of the area of one plate of the capacitor, A, and the distance between the plates, d.
> **Note 1:** ϵ_0 is the permittivity of free space.
> **Note 2**: This equation only holds for parallel plate capacitors.

- A capacitor's job is to store electric potential energy.

> **Energy Stored in a Capacitor**
>
> $$U_e = \frac{1}{2} q\Delta V = \frac{1}{2} C\Delta V^2 = \frac{q^2}{2C}$$
>
> **Description** – These equations define the electric potential energy stored in a capacitor in terms of the charge on one plate of the capacitor, q, the capacitance, C, and the potential difference, V, developed across the plates.
> **Note 1:** The three equations are equivalent (related by $q = C\Delta V$. You choose the best equation based on the quantities you have.

- Dielectrics increase the usefulness of capacitors by increasing the amount of charge they can hold for a given potential difference.

212-1: Why Should You Care?

SOMETIMES YOU NEED to temporarily store electric energy. One way to do that is with the use of a *capacitor*. As we'll see, capacitors are designed to hold charges in specific geometries, allowing energy to be stored in electric fields. Capacitors are ubiquitous in electric circuits. In addition to energy storage, they can be used to control the flow of electrons in circuits and even control whether circuits allow high frequency or low frequency oscillations in the circuit (called filtering). In order to understand these properties later, we must first discuss how capacitors are set up. Oh, and we'll also discuss why you get shocked in the winter when you try to touch a doorknob.

212-2: Capacitors

Consider: *what is a capacitor?*

A **capacitor** is an electric element made up of two conductors separated by a non-conductive region, which for now we'll take to be either air or a vacuum. Capacitors can be made in many geometries; however, the easiest to picture and understand is known as the parallel-plate capacitor. Figure 212-1 gives a visual representation of a parallel plate capacitor – note the similarities used in describing both the electric field between two parallel conducting plates and the potential difference between two such plates in the last couple of units.

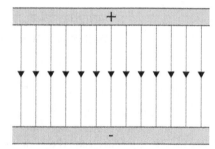

Just to recap, since the top plate contains positive charge and the bottom plate contains negative charge, there is an electric field between the two plates, pointing from the positive charges towards the negative. We discussed how symmetry makes the electric field uniform and constant between the two plates. Now, in addition, we found in the last unit that since we have an electric field extended over a distance, there is a potential difference, or voltage, developed between the two plates.

It turns out that there is a relatively simple equation that will help us determine the ability of this setup to hold charge, create electric fields and voltage and therefore store energy that only depends on geometry, and this is called the *capacitance*.

Figure 212-1. Diagram of a parallel plate capacitor

Definition of Capacitance

$$C \equiv \frac{q}{\Delta V} \qquad (212\text{-}1)$$

Description – This equation defines the capacitance of a capacitor as the amount of charge, q, one plate of the capacitor can hold to produce a potential difference between the plates of the capacitor ΔV.

Note 1: The unit of capacitance is the *Farad* (*F*), with $1\ F = 1\ C/V$

Note 2: A Farad is a huge capacitance. More common units are mF, μF and nF.

Note 3: Capacitance should only depend on geometry and not on other factors such as charge or fields.

Please note that there are one-plate capacitors, where the second plate is considered to be at infinity and the potential difference in calculated assuming the position is zero at infinity.

Why, exactly, the capacitance will only depend on geometry is relatively complicated; however, if you think about the potential due to a point charge

$$\Delta V = \frac{kq}{r}, \qquad (212\text{-}2)$$

you can see that if you divide q by ΔV, you will be left with a quantity that only has to do with distance, r, and a constant, k. It turns out this trend is true for all capacitor geometries.

There is a standard set of steps you must go through to determine capacitance:

> 1) Use either Gauss's Law or the equation for electric field to find the field between the two plates of the capacitor,
> 2) Use the relationship between electric field and electric potential to find the electric potential difference,
> 3) Use the electric potential in the equation for capacitance to find the capacitance.

We will see in the next two sections how this plays out for a couple of geometries.

212-3: Parallel-Plate Capacitors

Consider: *What is the simplest type of capacitor and how can it be used?*

Let's now return to the parallel plate capacitor of figure 212-1. In order to find the capacitance, we first need to find the potential difference or voltage between the plates. In order to find the voltage, we need to know something about the electric field. Luckily, we've done that before! Using Guass's Law, we found that the magnitude of the electric field between the two oppositely charged plates is given by

$$E = \frac{\sigma}{\epsilon_0},$$

(212-3)

where σ is the charge density (total charge divided by total area) of *one* the plates. Also, since we know that the positive charges are on the top plate, the direction of the electric field must be down.

Now, to find the electric potential (again, a review), we start with the integral form:

$$\Delta V = -\int_{r_i}^{r_f} \vec{E} \cdot d\vec{l} = -\int_{r_i}^{r_f} \frac{\sigma}{\epsilon_0} dl \cos\theta = \int_{r_i}^{r_f} \frac{\sigma}{\epsilon_0} dl$$

(212-4)

where I have inserted what we know about the electric field and the fact that if we move along an electric field line from the negative plate towards the positive plate, the angle between the electric field and $d\vec{l}$ is 180 degrees ($\cos 180 = -1$).

Since both quantities inside the integral are constant, we can take them out of the integral and then we have the simplest of integrals left. Therefore

$$\Delta V = \frac{\sigma d}{\epsilon_0},$$

(212-5)

where d is the distance between the two plates of the capacitor.

Now, we are left with applying our definition of capacitance. The only problem is that the equation for capacitance has charge in it and our equation for potential difference has area charge density. So, first we rewrite the potential difference in terms of charge and using the definition of capacitance,

$$\Delta V = \sigma \frac{d}{\epsilon_0} = \frac{q}{A}\frac{d}{\epsilon_0} \quad \rightarrow \quad C \equiv \frac{q}{\Delta V} = \frac{q}{\frac{q}{A}\frac{d}{\epsilon_0}} = \frac{\epsilon_0 A}{d}$$

(212-6)

Parallel Plate Capacitor

$$C = \frac{\epsilon_0 A}{d}$$

(212-6)

Description – This equation describes the capacitance of a parallel plate capacitor in terms of the area of one plate of the capacitor, A, and the distance between the plates, d.

Note 1: ϵ_0 is the permittivity of free space.

Note 2: This equation only holds for parallel plate capacitors with a vacuum between the plates.

Connection – iPhone touch screens

How does an iPhone, iPad or other device know where you are touching the screen? Under the glass exterior is transparent conducting material known as ITO (indium tin oxide). When you place your finger on the screen, you form a capacitor with the ITO layer at that spot. Small capacitance meters built into the screen can sense where a capacitance has formed and tells the electronics below where your finger is on the screen! This is why only certain materials work on the screen – they must be conductors!

How much charge

You have a parallel plate capacitor in which the plates have an area of 12 cm² and a separation of 1.7 cm.
  (a) What is the capacitance of this capacitor
  (b) What happens to the capacitor if you shrink the separation by half?
  (c) What happens to the capacitance if you shrink the area of each plate by half?

Solution:

Each of these parts is a direct application of the parallel plate capacitor equation.

(a) We can find the capacitance directly:

$$C = \frac{\epsilon_0 A}{d} = \frac{(8.85 x 10^{-12} \, F/m)\,(12 x 10^{-4} m^2)}{0.017 \, m}.$$

This gives us

$$C = 6.24 \, x 10^{-13} \, F$$

(b) Since the separation of the plates is in the denominator of the capacitance equation, we should expect the capacitance to double if we shrink the separation by half:

$$C = \frac{\epsilon_0 A}{(1/2)d} = 2\frac{\epsilon_0 A}{d}.$$

Therefore the capacitance would be $1.25 \, x 10^{-12} F$.

(c) Since the area of the parallel plate capacitor is in the numerator of our capacitance equation, we should expect the capacitance to decrease to one-half its original value:

$$C = \frac{\epsilon_0 (1/2)A}{d} = \frac{1}{2}\frac{\epsilon_0 A}{d}.$$

Therefore the capacitance would be $3.13 \, x 10^{-13} F$.

You can see each of these transformations is relatively straight forward, but you must be careful as to whether the quantity is in the numerator or the denominator.

212-4: Example of another Geometry

This section only contains an example of calculating capacitance for a geometry that is different than the parallel plates.

Cylindrical Capacitance Example.

Determine the capacitance of the cylindrical capacitor shown on the right. The capacitor has a length, l. The inner conductor is positively charged and has radius R_1, and the outer conductor is negatively charged and has radius R_2.

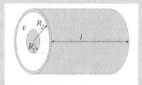

Solution:

We must follow the steps prescribed above for finding the capacitance of a capacitor.

We found the electric field in a cylindrical geometry using Gauss's Law in unit 210 (example 210-2):

$$E = \frac{\lambda}{2\pi\epsilon_0 r} = \frac{Q}{2\pi\epsilon_0 r l},$$

where Q is the charge in a length l of the cylinder and r is the distance from the center of the cylinder.

The next step is to find the potential difference between the inner and outer conductor of the capacitor due to this electric field

$$\Delta V = -\int_{R_2}^{R_1} \vec{E} \cdot d\vec{l}.$$

Since the electric field in between the conductors is radially outward, we take our direction of integration radially inward (from negative to positive plate), and get

$$\Delta V = -\int_{R_2}^{R_1} \frac{Q}{2\pi\epsilon_0 l}\frac{dr}{r} = -\frac{Q}{2\pi\epsilon_0 l}\ln r \Big|_{R_2}^{R_1}.$$

Simplifying,

$$\Delta V = \frac{Q}{2\pi\epsilon_0 l}\ln\frac{R_2}{R_1}.$$

The last step is to substitute this into the capacitance equation

$$C \equiv \frac{Q}{\Delta V} = \frac{Q}{\frac{Q}{2\pi\epsilon_0 l}\ln\frac{R_2}{R_1}},$$

Which can be simplified to

$$C = \frac{2\pi\epsilon_0 l}{\ln R_2/R_1}.$$

212-5: Energy Stored in a Capacitor

Consider: *How much energy is stored in a capacitor?*

So, let's think about how a capacitor is formed – and I suggest you use the picture of the parallel plate capacitor as we work through this. Initially, the plates of the capacitor are uncharged. We first need to move a small charge from one plate to the other (we get to move this for free since we are not working against a potential), and once this charge is in place, it creates a small potential difference across the plates given by $\Delta V = q/C$.

The next little bit of charge (dq) we want to bring to the other side of the capacitor now takes a little bit of work (dW), since we are now trying to bring the new charge closer to a charge that is already there:

$$dW = \Delta V \, dq = \frac{q}{C} \, dq. \tag{212-7}$$

Now imagine trying to continuously transfer small charges until a total charge Q has been transferred to the other side. This is an integral since we are constantly adding one little bit of charge to what has already been transferred, and the potential we are working against has to do with how much is already there:

$$W = \int_0^Q dW = \int_0^Q \frac{q}{C} \, dq = \frac{1}{C} \int_0^Q q \, dq, \tag{212-8}$$

Where I was able to take out the capacitance since I know it is a constant for a given geometry. We now have one of the easiest integrals to compute:

$$W = \frac{1}{C} \frac{q^2}{2} \Big|_0^Q = \frac{Q^2}{2C}. \tag{212-9}$$

Then, we can simply say that if this was the amount of work that it took to complete the system, then it is also the potential energy that the system holds (assuming all the work was done with conservative forces).

$$U_e = \frac{Q^2}{2C}. \tag{212-10}$$

This equation is great if we know the charge stored and capacitance of a capacitor, but what if we know the voltage and charge, or the capacitance and charge? Fortunately, we can use the fact that, for all capacitors, $Q = C\Delta V$ to manipulate our equation for potential energy:

$$U_e = \frac{Q^2}{2C} = \frac{Q^2}{2\left(\frac{Q}{\Delta V}\right)} = \frac{1}{2} Q\Delta V = \frac{1}{2}(C\Delta V)\Delta V = \frac{1}{2} C\Delta V^2. \tag{212-11}$$

Any of these versions can be used interchangeably – it literally only depends on what quantities you have on hand.

Energy Stored in a Capacitor

$$U_e = \frac{1}{2} q\Delta V = \frac{1}{2} C\Delta V^2 = \frac{q^2}{2C} \quad (212\text{-}11)$$

Description – These equations define the electric potential energy stored in a capacitor in terms of the charge on one plate of the capacitor, q, the capacitance, C, and the potential difference, V, developed across the plates.

Note 1: The three equations are equivalent (related by $q = C\Delta V$. You choose the best equation based on the quantities you have.

Energy Storage Problem

A parallel plate capacitor has plates of area 1.6 m^2 and separation of 17 cm. How much energy does this capacitor store when each plate holds a charge with magnitude 65 mC?

Solution:

Although this problem will be a direct application of the energy stored in a capacitor equation, we must first find the capacitance from the geometry:

$$C = \frac{\epsilon_0 A}{d} = \frac{(8.85x10^{-12}\ C/V \cdot m)\ (1.6\ m^2)}{0.17\ m},$$

which simplifies to $C = 8.33\ x10^{-11}F$.

The energy storage equation can now be employed

$$U = \frac{q^2}{2C} = \frac{65x10^{-3}C}{2(8.33\ x10^{-1}\ F)} = 3.90\ x10^8 J.$$

This is a very large amount of energy; however, realize that this capacitor is very large (1.6 m^2), and carries a large amount of charge (65 mC).

212-6: Dielectrics

Consider: *What happens to electric fields inside matter?*

There is a problem with air-gap capacitors – the amount of charge they can hold is limited to something called **dielectric breakdown** – and this is something you have all experienced. We've discussed the situation before where you walk across a rug in wool socks on a cold winter's night and get shocked when your hand approaches a doorknob. This is actually dielectric breakdown. As the excess charges in your hand approached the doorknob, charges of the opposite sign moved to the edge of the doorknob (a conductor) by simple electric attraction. In essence, though, your hand and the doorknob have now become a capacitor, as can be seen in figure 212-2.

Figure 212-2. Charge distribution just before you get shocked by a door handle.

The problem is that **air is not a perfect insulator**. Once the electric field between two charged surfaces separated by air gets to about three million volts per meter (3,000,000 V/m), air starts to act like a conductor and electric current flows between your hand and the doorknob. *Zap.* This is exactly the same process that causes lightning. Friction within storm clouds causes a separation between positively and negatively charged particles within the cloud that then separate. The situation is shown in figure 212-3. Now that the bottom of the cloud is negatively charged, it causes surface positive charges on objects on the ground. Once the charge separation gets large enough...zap! So, the shock you feel when reaching for the doorknob is actually a tiny little lightning bolt – on a very different scale though!

Figure 212-3. Distribution of charges just before a lightning strike.

The same is true with any air-gap capacitor, so the amount of charge that it can hold is limited. As an example, let's ask how much charge a parallel plate capacitor with area of 1 cm^2 (square plates with 1 cm on a side) can hold. We know that the magnitude of the electric field between two parallel plates is given by

$$E = \frac{\sigma}{\epsilon_0} = \frac{Q}{\epsilon_0 A}. \qquad (212\text{-}12)$$

Solving for Q and substituting, we find

$$Q = \epsilon_0 EA = \left(\frac{8.85x10^{-12}F}{m}\right)\left(3,000,000\ \frac{V}{m}\right)(1\ x\ 10^{-4}m^2) \qquad (212\text{-}13)$$
$$= 2.66x10^{-9}C.$$

So, this capacitor could only hold 2 nC of charge before it would undergo breakdown and have a spark cross the plates. This is not good!

To combat this, we can place different **dielectrics** between the plates of the capacitor. A dielectric is simply an insulating material that will increase the dielectric breakdown point of

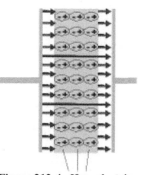

Figure 212-4. How electric dipoles change the net electric field. The internal dipoles create a field that adds to the external field.

the capacitor. Most dielectrics do have a breakdown point, but most are drastically above that of air. On the other hand, the dielectric also changes the electric field between the plates of the capacitor; here's why.

Consider figure 212-4, which shows what happens to atoms inside the dielectric when an electric field is applied. The atoms form dipoles (which is where the term dielectric comes from), which means that there will now be a second electric field inside the dielectric formed simply by the dipoles, and this field will be in the opposite direction to the applied field!

The net electric field is always in the same direction as the applied field (the internal dipoles do not overcome the applied field), but the overall field inside the dielectric is *much smaller* than the external field. But this is good!! If the net field inside the dielectric is smaller, the chances of dielectric breakdown occurring has been drastically reduced – i.e., we can apply a very large external field (have large amounts of charge on the plates of a capacitor) and not worry about breakdown. Of course, most dielectric materials do still have a breakdown point – but this is not usually a concern because it is so high.

The crux of the whole situation is that if you fill a capacitor completely with a dielectric, the capacitance changes:

$$C_{diel} = \kappa C_{vacuum} \qquad (212\text{-}14)$$

where C_{diel} is the capacitance with the dielectric, C_{vacuum} is the capacitance in vacuum and κ is known as the **dielectric constant**, a constant that is material dependent and relates how strongly it changes the capacitance. Table 212-1 gives the dielectric constants for a number of materials.

There is another shortcut that can be useful when working with dielectrics. If we've determined a value for a region of space with no dielectric material such as voltage or electric field, we can immediately write down that same quantitity if the space is filled with dielectric by replacing

$$\epsilon_0 \rightarrow \kappa\epsilon_0. \qquad (212\text{-}15)$$

Table 212-1. κ for various materials

Material	Dielectric Constant (κ)
Air (1 atm)	1.00054
Polystyrene	2.6
Paper	3.5
Pyrex	4.7
Mica	5.4
Silicon	12
Germanium	16
Ethanol	25
Water (20°)	80.4
Strontium	310

This works because the capacitance of a material can always be written in the form $C = \epsilon_0 \mathcal{L}$, where $\mathcal{L}$ is some geometry with units of length (A/d for the parallel plate, or ab/(a-b) for the sphere, etc.). So, replacing ϵ_0 with $\kappa\epsilon_0$ always has the same effect as replacing C with κC.

Example 212-4: Water filled capacitor

A parallel plate capacitor filled with air can hold 23.4 nC of charge on each plate before dielectric breakdown occurs. How much charge could the capacitor hold if it were filled with water?

Solution:

This is the exact reason dielectrics are used in capacitors: it allows us to increase the total charge held on each plate. We know that for any capacitor

$$C_0 \equiv \frac{q_0}{\Delta V} \quad \rightarrow \quad q_0 = C_0 \Delta V.$$

Let's say that we connect the capacitor to a voltage supply and charge up the plates just until breakdown would occur (23.4 nC). If we maintain connection to the battery, the potential difference across the capacitor will continue to be the voltage of the power supply. Therefore, the charge on each plate will increase if we fill the capacitor with water

$$q = C\Delta V = \kappa C_0 \Delta V = \kappa q_0.$$

Therefore, the charge on the plates will increase by a factor of kappa (80.4 for water) to

$$q = \kappa q_0 = (80.4)(23.4\ nC) = 1.88\ \mu C.$$

Extension:

What would happen to the capacitor above if we charged it up, but then filled it with water after disconnecting the capacitor from the voltage supply?

Well, if the capacitor is disconnected, the charge on each plate cannot change. Hoswever, when we fill it with water, we know that our capacitor equation must still hold:

$$C \equiv \frac{q}{\Delta V}.$$

In this case, it is the potential difference between the plates that changes due to a change in C (again, since the charge must remain the same):

$$\Delta V_0 = \frac{q}{C_0} \quad \rightarrow \quad \Delta V = \frac{q}{\kappa C_0} = \frac{1}{\kappa}\Delta V_0.$$

Therefore, the voltage decreases by a factor of kappa.

213 – Electric Currents

Earlier in the course, we discussed how moving electric charges create electric currents. In this unit, we explore the idea of electric current in more detail and set ourselves up for the discussion of electric circuits to come in the next few units.

Integration of Ideas

Review the definition of current from Unit 208.

The Bare Essentials

- In many common household applications, batteries create a current in a circuit.

- The direction of an electric current in a wire is in the opposite direction to the motion of the electrons in the wire.

- Current is related to the drift speed of the electrons in a wire.

Current and Drift Speed

$$I = nqAv_d$$

Description – This equation defines the current in a wire in terms of the number density of charge carriers (n), the charge of the carriers (usually electrons), the cross-sectional area of the wire (A) and the drift speed of the charge carriers, v_d.
Note: Electric current is a scalar and therefore does not include direction.

- The current density describes how the current is distributed over the cross-sectional area of a wire.

Current Density

$$\vec{J} \equiv \frac{I}{A}\hat{v} = nq\vec{v}_d$$

Description – This equation defines the current in a wire in terms of the number density of charge carriers (n), the charge of the carriers (usually electrons), and the drift speed of the charge carriers, v_d.
Note 1: The SI unit of current density is A/m^2.
Note 2: Current density is a vector, so be careful about the direction!

- Ohm's Law relates the current density in a conductor to the electric field created by a potential difference across the wire.

Ohm's Law

$$\vec{J} = \sigma\vec{E}$$

Description – Ohm's law relates the current density in a conductor to the electric field set up in the conductor. The constant of proportionality in this equation (σ) is called the conductivity of the material.
Note 1: The SI unit of conductivity is Siemens/meter (S/m)
Note 2: Materials that follow Ohm's Law (i.e. have a constant conductivity over a given temperature range) are known as Ohmic materials.

- The resistivity of a material is defined as the inverse of the conductivity ($\rho = 1/\sigma$).

- The resistance of an object depends on its resistivity (or conductivity) and its geometry

---
Resistance-Resistivity Relationship

$$R = \frac{L}{\sigma A} = \frac{\rho L}{A}$$

Description – This equation defines the electrical resistance of a specific object in terms of its conductivity (σ) or resistivity (ρ), its length (L) and its cross-sectional area.
Note 1: The SI unit of resistance is the Ohm (Ω).

- Ohm's Law can be also be written as $\Delta V = IR$.

- The power dissipated in an electrical element is related to the voltage drop, current and/or resistance.

Electrical Power

$$P = I\Delta V = I^2 R = \frac{\Delta V^2}{R}.$$

Description – This equation describes the power dissipated in a circuit element in terms of the current (I) through the element, the voltage drop (ΔV) through the element and the resistance of the element (R).
Note 1: The SI unit of power is the watt (W)
Note 2: The power equation can also be used to determine the power added to a system by a battery.

213-1: Electric current and batteries

Consider: *What are currents in everyday life?*

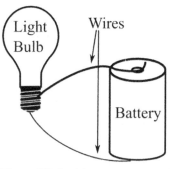

Figure 213-1. A battery connected to a light bulb with two wires.

What happens if you connect a battery to a small light bulb with two wires as shown in figure 213-1? Well, if you connect it correctly, the light bulb lights up. Let's first take a look at what we mean by connecting it *correctly*. Notice that one wire is connected to the + side of the battery (with the little bump) and this connects to the side of the bulb (where the bulb usually screws into the socket). The second wire then goes from the bottom of the bulb back to the – (negative) side of the battery (the flat side). The idea here is to create a *closed circuit* where charges leaving one side of the battery can travel to the light bulb, through the filament, out the other side of the bulb and the back to the battery. As we discussed before, the *electric current* is the rate at which charges move. In this case, electrons emerge from the – (negative) side of the battery and flow through the bulb until they return to the positive side of the battery. This is called the *electron current*. However, we all know that electrons have a negative electric charge. The *conventional current* is defined as if positive charge was moving through the wires. Unfortunately, this means that the conventional current (the current everyone talks about) is in the opposite direction as the electron movement!

> Conventional current moves in the opposite direction of the electron current.

The battery is a source of energy, converting chemical energy to electric energy. Such *electrochemical batteries* have three main components – an anode, a cathode and an electrolyte in which the other components are immersed. When the terminals of the battery are connected together the anode and cathode undergo chemical reactions with the electrolytic solution. The anode undergoes an oxidative reaction that produces electrons, and the cathode undergoes a reduction reaction that absorbs electrons. Although the net effect is that the same number of electrons are produced as absorbed, the free electrons can traverse through a circuit (such as our light bulb above) on their way from the anode to the cathode.

Today, we use four main types of batteries in everyday life. You can see the main properties of these batteries in table 213-1.

Table 213-1. Types of common electrochemical batteries

Battery Type	Use	Anode	Cathode	Electrolyte
Zinc-carbon	Cheap household batteries (AAA, AA, C, D)	Zinc	Manganese Dioxide	Ammonium chloride
Alkaline	Better household batteries (AAA, AA, C, D)	Zinc Powder	Manganese Dioxide mixture	Potassium hydroxide
Lithium-ion	Rechargeable household batteries (cell phones, laptops, etc.)	Carbon	Lithium cobalt oxide	Lithium salts in organic compounds
Lead-acid	Rechargeable car batteries	Lead Dioxide	Lead	Sulfuric acid

One other important characteristic of batteries is their *capacity*. Most of the common household batteries (AAA, AA, C, D) have a nominal potential difference across their terminals of *1.5 Volts* (soon, we will call this the *emf* of the battery). So, why the different sizes? It's all about the total charge they can transfer, or their *capacity*. A D-cell battery can transfer considerably more charge and therefore holds more energy than a AAA battery, meaning that it will last longer for the same application. Capacity is usually measured in a weird unit, the milliamp-hour ($mA \cdot h$), which is a unit of charge since it is Amps multiplied by time. The capacity of standard batteries can be seen in table 213-2.

As you can see in the table, an alkaline D-cell battery has more than ten times the capacity relative to an alkaline AAA battery, which is why they are used in heavier applications. Also, notice that alkaline batteries tend to have twice the capacity of the cheaper zinc-carbon batteries in all sizes. Be careful when you are purchasing batteries. Don't get fooled into buying something that is only half as good for 75% the price!

Table 213-2. Capacity of household batteries in $mA \cdot h$ ($1 \, mA \cdot h = 3.6 \, C$)

Size	Capacity (zinc-carbon)	Capacity (alkaline)
AAA	540	860-1,200
AA	400-1,700	1,800-2,600
C	3800	8,000
D	8000	12,000-18,000

213-2: Electric fields create electric current

Consider: *How are electric currents created?*

Figure 213-2. Free electrons in a copper lattice. Each copper atom is noted as a positive ion that gives away a loosely bound electron (small dots) when an electric field is present.

We're now going to switch to thinking about what happens inside the wires of figure 213-1. As conductors, the wires are made of some sort of metal, often copper. The copper atoms are held together rather tightly by bonds; however, each copper atom has at least one very loosely held electron. These loosely held electrons are called *free electrons* because almost any force placed on them will free them from their atoms. This picture can be seen in figure 213-2. When we take all of the free electrons together, we call this an *electron sea*. The crux of this argument is that a perfect metal can be treated as if it has an almost unlimited number of mobile electrons that can be used to create an electric current.

Now, let's say we put an electric field across the wire such that all of the electrons are immersed in the field. They will each feel a force, $F = qE$, in the opposite direction of the electric field (since the electrons are negative). We would expect them to accelerate in this field. There is a problem though – they will keep running into the positively charged metal ions as they try to move through the wire. If you have ever seen the game *Plinko* on *The Price is Right*, this is exactly the situation – as the disk tries to slide down the Plinko board, it continuously hits pegs that randomly change its motion. Same thing for electrons. So, although they are trying to accelerate down the wire, they are interrupted. As can be seen in figure 213-3, this leads to a small average speed known as a *drift speed*.

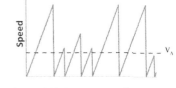

Figure 213-3. How the drift speed of an electron arises from collisions with ions. $V_{Average}$ is the overall average speed.

As the electrons accelerate in an electric field, they undergo a change in momentum given by

$$\vec{F}_{net} = \frac{\Delta \vec{p}}{\Delta t}. \tag{213-1}$$

In our current case, I will take the force, field, momentum and velocity to all be along the wire, so I will neglect the vector symbols from here on out. We know that the net force in the wire is given by an electric field, so we can write

$$\Delta p = p - 0 = qE\Delta t, \tag{213-2}$$

where I have assumed that the initial speed of the electron is zero. From here, I can relate this to the speed of the electron by using the relationship between momentum and speed, $p = mv$, so

$$v = \frac{p}{m} = \frac{qE\Delta t}{m}. \tag{213-3}$$

If you again look at figure 213-3 relating the drift speed to collisions, you will see that the average speed has to do with the time between collisions – if there is a greater time between collisions, the average speed will get to be larger. So, what we do is say that Δt is that time between when an electron first starts moving and when it hits an ion, and if we average the time between collisions, we find

$$v_d = v_{ave} = \frac{qE\Delta t_{ave}}{m}, \tag{213-4}$$

Where v_d is the drift speed of electrons in the wire. What's interesting is that an electron in an electron sea with no electric field across it will have speeds due to thermal motion of 10^6 m/s; however, drift velocities in wires tend to be only around 10^{-4} or 10^{-5} m/s!

Remember that current is defined as the charge per second that flows through a given area:

$$I = \frac{dq}{dt}. \tag{213-5}$$

We can now use the drift velocity to write this in terms of electrons flowing through a wire. In order to do this, we have to define a density of charges, n, for a given metal. This is simply the number of mobile charged particles per volume. Using this, the charge per unit volume is given by nq, where q will

Connection: Instant light.

Why is it that lights come on immediately when you flip a light switch? *The electrons do not push each other through the circuit.* When you flip the switch, the electric field propagates through the entire wire at the speed of light. Therefore, the entire sea of electrons starts to feel a force and move almost instantly!

be the charge on an electron. If we now multiply the drift velocity by this quantity, nq, we wind up with something of units Amperes per meter squared, which is called the ***current density***.

$$J = nqv_d. \qquad (213\text{-}6)$$

The current density is actually a vector and its direction is related to the drift velocity:

$$\vec{J} = nq\vec{v}_d. \qquad (213\text{-}7)$$

It is again very important to keep in mind that the ***charge on an electron is negative***, so that the conventional current density in a wire is opposite to the direction of the drift velocity!

Current Density

$$\vec{J} = nq\vec{v}_d \qquad (213\text{-}7)$$

Description – This equation defines the current density in a wire in terms of the number density of charge carriers (n), the charge of the carriers (usually electrons) and the drift speed of the charge carriers.

Note 1: The SI unit of current density is A/m^2.

Note 2: Current density is a vector, so be careful about the direction!

Now, if we know the cross-sectional area of a wire, we can also define the total current in the wire in terms of the current density and the area.

Current and Drift Speed

$$I = nqAv_d \qquad (213\text{-}8)$$

Description – This equation defines the current density in a wire in terms of the number density of charge carriers (n), the charge of the carriers (usually electrons), the cross-sectional area of the wire (A) and the drift speed of the charge carriers, v_d.

Note: Electric current is a ***scalar*** and therefore does not include direction.

The model described above for electrons in metals is called the ***Drude model*** after Paul Drude who suggested it in 1900.

Example 213-1: Something about current and charge

The density of free electrons in copper is approximately $8.5 \times 10^{28}\, electrons/m^3$ (there is approximately one free electron per atom in copper). What is the drift velocity of electrons in a copper wire of radius 0.96 mm, carrying a current of 0.87 A.

Solution:

This is a direct application of the drift speed equation. Rearranging the equation to solve for drift velocity:

$$v_d = \frac{I}{nqA}$$

This gives a drift velocity of

$$v_d = \frac{0.87A}{(8.5 \times 10^{28}\, e/m^3)(1.6 \times 10^{-19}C)\pi(0.96 \times 10^{-3}m)^2},$$

or

$$v_d = 2.2 \times 10^{-5}\, m/s.$$

Note that this is an incredibly small speed. The electrons in a standard wire are moving quite slowly. However, information (electricity) transfers very quickly because of the electric fields in those wires.

Example 213-2: Current Density

What is the magnitude of the current density in the wire described in example 213-1?

$$J = \frac{I}{A} = \frac{I}{\pi r^2} = \frac{0.87\,A}{\pi(0.96 x 10^{-3}m)^2} = 3.0 x 10^5 A/m^2.$$

Solution:

There are two ways to solve this problem:
1) Current density is current divided by cross-sectional area
2) Current density can be found using the boxed equation above.

2) Use the current density equation

$$J = nqv_d = (8.5 x 10^{28}\,e/m^3)(1.6 x 10^{-1}\,C)2.2 x 10^{-5}\,m/s)$$

or

$$J = 3.0 x 10^5 A/m^2.$$

Note that both methods gave the same answer.

1) First, we can find the current density directly

213-3: Electric fields allow current to take corners

Consider: *Do electrons just bump into walls inside of wires?*

What produces the electric field that causes electrons to flow through the entire wire? It can't be just the battery, because if it were, the electric field would be stronger near the battery than far away – said another way, our light bulb in figure 213-1 would be brighter with shorter wires, which we know is just not the case!

The answer is surface charges on the wires. Gauss's law told us that we can't have a net excess charge inside a conductor (and remember, our sea of electrons is balanced by the positive ions inside a wire!). If there were an excess charge, the particles would repel each other and wind up on the surface of a conductor, and this is exactly what happens. Figure 213-4 shows the approximate surface charge distribution around a simple circuit. Notice that the surface charges are most prevalent near the battery (positive near the positive end of the battery and negative near the negative end of the battery). The net result of the surface charges is that the electric field throughout the entire loop is the same! This is quite important for conservation of energy to hold for the loop.

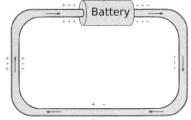

Figure 213-4. Distribution of surface charges around a simple circuit. Note that arrows show the direction of conventional current.

The other thing that surface charges allow for is the current to take corners. Consider the upper-left corner of figure 213-4. As the electrons move up through the left branch (remember, they are moving in the opposite direction of the electric field), they would pile into the corner since Newton's 1st law tells us that they want to move in a straight line. Well, as they accumulate on the surface at the corner, they will repel other electrons as they approach and push them around the corner. This setup can be seen in figure 213-5. The figure is actually quite an exaggeration. Since the mass and drift velocity of electrons is quite small, it is often only one electron that needs to build up in the corner to cause the current to turn.

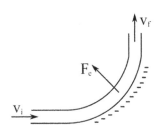

Figure 213-5. Depiction of surface charges accumulated at a corner of a wire.

213-4: Ohm's Law relates the current density to the electric field

Consider: *How do we know how much current will be formed by an electric field?*

We now know that the electric field inside the wire of our simple circuit is constant. Again, this is important because it means that the drift velocity of the electrons and therefore the current density will be the same throughout the wire (as long as the material doesn't change). Now, consider again, our definition of current density above. We can use the definition of drift velocity to directly relate the current density to the drift velocity

$$\vec{J} = nq\vec{v}_d = nq\frac{q\vec{E}\Delta t_{ave}}{m}.$$ (213-9)

This equation can be simplified slightly to

$$\vec{J} = \frac{nq^2 \Delta t_{ave}}{m} \vec{E}.$$ (213-10)

The constants of proportionality between the current density and the electric field are based on properties of the electron and material properties (density of charge carriers and time between collisions). Taken together, this quantity is called the conductivity, σ, and is a measure of how easily electrons will flow in a material

$$\sigma = \frac{nq^2 \Delta t_{ave}}{m}.$$ (213-11)

This relationship between current density and electric field is known as Ohm's Law.

Ohm's Law

$$\vec{J} = \sigma \vec{E}$$ (213-12)

Description – Ohm's law relates the current density in a conductor to the electric field set up in the conductor. The constant of proportionality in this equation (σ) is called the conductivity of the material.
Note 1: The SI unit of conductivity is Siemens/meter (S/m)
Note 2: Materials that follow Ohm's Law (i.e. have a constant conductivity over a given temperature range) are known as Ohmic materials.

Many of you may know Ohm's law in another form, and we'll get to that shortly. One of the key things here is that Ohm's law is not perfect. The conductivity of many materials depends on temperature and as current flows through materials, friction causes their temperature to increase. Ohm's law does tend to work well over small ranges of current densities, in which case the material is called *ohmic*.

It is also important to note that conductivity is a material property, meaning that it has to do with the *type* of material you are using and *not the shape of size* of the object.

213-5: Resistivity, resistance and the other Ohm's law

Consider: *Do all materials conduct the same amount of current?*

We can find the potential difference along a piece of our wire using the definition from unit 211,

$$\Delta V = -\int \vec{E} \cdot d\vec{l}.$$ (213-13)

We just found that the electric field inside the wire is constant and directed along the length of the wire. Since $\vec{E}$ and $d\vec{l}$ are parallel to each other, we do not have to worry about the dot product, and since $\vec{E}$ is constant, it can be removed from the integral. Therefore

$$\Delta V = -EL,$$ (213-14)

where L is the total length of wire we traverse through. Inserting the relationship between electric field and current density gives us

$$\Delta V = -\frac{J}{\sigma}L.$$ (213-15)

Finally, we can write this in terms of the current as opposed to the current density by employing I = JA:

$$\Delta V = -\frac{I}{\sigma A}L = -\frac{L}{\sigma A}I.$$ (213-16)

The constant of proportionality between ΔV and I is called the **resistance**. Resistance is literally a measure of how much a material resists the flow of current for a given potential difference across its ends. One of the first things to notice is that the ΔV is negative, meaning that any element with resistance is going to cause a **potential drop**. This is often just assumed for resistive materials and so the negative sign is not usually written.

In addition, we can introduce a quantity called the **resistivity**, which is the inverse of the conductivity:

$$\rho = \frac{1}{\sigma}. \qquad (213\text{-}17)$$

A highly conductive material will have a high conductivity and a low resistivity and a material that is not a good conductor will have a low conductivity and a high resistivity. Which value is used is usually just a matter of convention since they essentially give the same information. The conductivity and resistivity of various materials can be found in Table 213-3.

With all of this together, we can now very precisely define the resistance of an object as shown in the box below.

Table 213-3. Conductivities and resistivities of various materials.

Material	Conductivity $(\Omega \cdot m)^{-1}$	Resistivity $(\Omega \cdot m)$
Nichrome	6.7×10^5	1.5×10^{-6}
Copper	5.7×10^7	1.8×10^{-8}
Gold	4.1×10^7	2.4×10^{-8}
Aluminum	3.6×10^7	2.8×10^{-8}
Iron	1.0×10^7	1.0×10^{-7}
Lead	4.6×10^6	2.2×10^{-7}
Carbon	4.9×10^4	2.0×10^{-5}
Sea Water	4.0	0.25
Pure Water	4.0×10^{-8}	2.5×10^7
Glass	$\sim 10^{-12}$	$\sim 10^{12}$
Rubber	$\sim 10^{-13}$	$\sim 10^{13}$

Resistance-Resistivity Relationship

$$R = \frac{L}{\sigma A} = \frac{\rho L}{A} \qquad (213\text{-}18)$$

Description – This equation defines the electrical resistance of a specific object in terms of its conductivity (σ) or resistivity (ρ), its length (L) and its cross-sectional area, A.

Note 1: The SI unit of resistance is the Ohm (Ω).

Again, note that the conductivity and resistivity are material properties; that is, they depend only on the type of material. The resistance, on the other hand, depends on the type of material **and** its geometry.

Example 213-3: Resistance of something

The resistivity of pure gold is $2.44 \times 10^{-8}\Omega \cdot m$. What is the resistance of a 5.2 meter gold wire with a radius of 2.3 mm?

Solution:

This is a direct application of the resistance-resistivity relationship.

or

$$R = \frac{\rho L}{A} = \frac{(2.44 \times 10^{-8}\Omega \cdot m)(5.2m)}{\pi(2.3 \times 10^{-3}m)^2},$$

$$R = 7.63 \times 10^{-3}\Omega.$$

This is a very low resistance, which is expected since gold is a good conductor and therefore makes very efficient wires.

Using our definition of resistance, we are now ready to write down the more well-known version of **Ohm's law for a resistive element**:

$$\Delta V = \frac{L}{\sigma A}I \quad \rightarrow \quad \Delta V = IR. \qquad (215\text{-}19)$$

So, what is a resistor? A resistor is any element that has a much higher resistance than the near-ideal wires that connect it. One example is the tungsten filament in the light bulb of figure 213-1. The tungsten filament is very thin so it has a relatively high resistance. Ohm's law for resistive elements tells us that the filament will also have a large potential difference when compared to the wires. One

> Wires in circuits are usually considered ideal, meaning that we treat them as if they have zero resistance.

important idea is that wires connecting resistive elements are often considered as ideal, that is, they have no resistance themselves. This is not completely true, but in many situations, it is a very good approximation.

213-6: Power in batteries and resistors

Consider: *What sort of energy transfer is occurring inside batteries, wires and the like?*

Power is the rate at which energy is transferred into or out of a system. When related to what we've been discussing in this unit, the power of a battery is the rate at which it adds electric energy to a system and the power dissipated by a resistor is the rate at which a resistor transfers energy out of a system.   Starting with our basic definition of power,

$$P = \frac{dW}{dt} = \frac{dq\Delta V}{dt} = I\Delta V. \tag{213-20}$$

Thus, the power a battery adds is the potential difference across its terminals times the current through the battery. Similarly, the power dissipated by a resistor is the current through the resistor multiplied by the potential difference developed across it. Using Ohm's law for resistive elements, the power can also be written in terms of resistance and current or potential difference and current as well.

Electrical Power

$$P = I\Delta V = I^2 R = \frac{\Delta V^2}{R}. \tag{213-20}$$

Description – This equation describes the power dissipated in a circuit element in terms of the current (I) through the element, the voltage drop (ΔV) through the element and the resistance of the element (R).

Note 1: The SI unit of power is the watt (W)

Note 2: The power equation can also be used to determine the power added to a system by a battery.

Example 213-4: Electric heaters

Electric heating elements in portable heaters, toaster ovens, hair dryers, etc., make use of high-resistivity materials to turn electric energy into thermal energy. One of the most common materials used for heating elements is nichrome (an alloy of nickel, chromium and often iron), with a resistivity of $100x10^{-8}\Omega \cdot m$. How much thermal energy per second (power) is produced if a 3.2 meter nichrome wire with a radius of 0.55 mm carries a current of 1.9 Amperes?

Solution:

Since we are given current and resistivity in the problem and are looking for power, we will need to use

$$P = I^2 R.$$

However, in order to do this, we must first find the resistance of the wire from its resistivity and geometric measurements:

$$R = \frac{\rho L}{A} = \frac{(100x10^{-8}\Omega \cdot m)(3.2m)}{\pi(0.55x10^{-3}m)^2},$$

which gives us

$$R = 3.37\Omega.$$

The power dissipated is then given by

$$P = (1.9A)^2(3.37\Omega) = 12.2 \; W.$$

This is definitely enough energy per second to heat up a thermal element!

214 – Basic Circuits

We now have the understanding of electric current and resistance to explore what happens inside of basic circuits – circuits that are constructed of batteries and resistors in simple geometries. Many important electric systems are made of circuits similar to those described in this unit, and therefore this forms an important first step to understanding modern electronics.

Integration of Ideas

The ideas of current and resistance from the unit 213.
Conservation of charge and energy.

The Bare Essentials

- Batteries use chemical energy to create an EMF and terminal voltage across the battery.

- Basic direct current (dc) circuits contain three elements: batteries, wires and resistors

- A *series connection* is a connection where elements have no branches between them. Such elements must all have the same *current* through them by the conservation of charge.

- Resistors in series add directly

Resistors in Series

$$R_{series} = R_1 + R_2 + R_3 + \cdots$$

Description – This equation describes the equivalent resistance of resistors in series as the direct sum of the resistances.

- A *parallel connection* is a connection where elements start at the same node and end at the same node. Such elements must have the same *voltage drop* by the conservation of energy.

Resistors in Parallel

$$R_{Parallel} = \left(\frac{1}{R_1} + \frac{1}{R_2} + \frac{1}{R_3} + \cdots \right)^{-1}$$

Description – This equation describes the equivalent resistance of resistors in parallel as the reciprocal of the sum of reciprocals of the resistances.
Note: All elements in parallel must have the same voltage drop.

- Two meters are commonly used to test circuits and circuit elements -
  - *Voltmeters* measure the voltage drop in a circuit between two points. Voltmeters must be connected in parallel to the circuit section of interest.
  - *Ammeters* measure the current in part of a circuit. An ammeter must be connected in series with the circuit section of interest.
  - *Multimeters* combine the abilities of both voltmeters and ammeters and often contain further functionality.

- The power delivered by a circuit element or dissipated by a circuit element is related to the voltage across and current through that element.

Power in Electric Circuits

$$P = I\Delta V$$

Description – This equation describes the power delivered to or dissipated by a circuit in terms of the current and voltage of the circuit element.
Note: For resistors, Ohm's law can be used to write the power equation in a few different ways:
$$P = I\Delta V = I^2 R = \Delta V^2 / R.$$

214-1: What is a basic circuit?

Our discussions of current and resistance in the last unit were very general and were meant to describe the motion of any electric current through any material. In this case, we could be describing the current that jumps from your finger to a doorknob as you get a shock on a cold winter's morning or even the electric impulse that travels through your body to tell your brain that your finger just touched something hot. In this unit, we are going to be more specific and focus on ***electric circuits***; that is a closed loop where some source of electricity (battery, power supply) cause a current that flows through other electric elements (resistors, capacitors, transistors, etc.). Electric circuits are all around us in our modern world from the computer I'm writing this unit on to your cell phones, to the inner workings of all living things – each of these are composed of circuits of varying complexity. What makes a circuit a circuit is our ability to identify the electric sources and other elements and therefore find a way to analyze them as we'll see in the next couple of units.

214-2: Batteries, Wires and Resistors.

Consider: *What are the basic components that make up a circuit?*

In order to discuss even the most basic electric circuits, we must have a few definitions down. There are three basic elements that we will consider - batteries, wires and resistors. Batteries supply a voltage difference across their terminals, essentially adding energy to the system. Batteries are described in detail in the next section. Wires are lengths of wire that connect different elements. In general we will assume that wires are perfect conductors and have no resistance. Finally, resistors are circuit elements that tend to resist the flow of electrons (thus their name) and therefore remove energy from the circuit. Common resistors found in basic circuits are light bulbs and heating elements.

Each of our three basic elements has a small diagram that is used so that we can visually recognize them in a circuit:

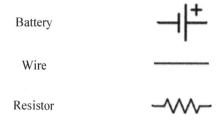

The simplest circuit that we can then write down contains a battery connected by wires to a single resistor, as can be seen in figure 214-1. There are a couple of very important concepts to discuss that are true from this very basic circuit as well as the more complicated circuits we will discuss later. First, note that this is a ***complete or closed circuit***, meaning there are no breaks as you move from the battery through the resistor and back to the battery. This is very important because if there is a break in an area of a circuit, current cannot flow. If there is a break such that we do not have a complete circuit, we then have an ***open circuit***. Open circuits actually happen all the time – usually when you turn something off using a power button, you are creating an open circuit, thereby preventing the flow of electrons and powering off the instrument.

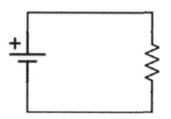

Figure 214-1. A simple circuit containing a battery and a resistor connected by wires.

It turns out that both conservation of charge and conservation of energy play very important roles in the analysis of circuits. Consider an electron leaving the negative terminal of the battery – remember that, unfortunately, the conventional current and the actual motion of electrons are in the opposite direction from each other. Anyway, the wire in which the electron is trying to move is filled with electrons, such that as it tried to move through the wire, it pushes on electrons in front of it, which push on electrons in front, etc., etc., until we return all the way back to the battery. You can visualize this similar to peas moving in a straw – if you were to take a drinking straw, fill it with peas and then try to put another pea in the end, one would have to come out the other side because only so many peas can fit in the straw. This 'conservation of peas' has to work for each section of wire, meaning that each small section of our simple circuit has to have the same motion of electrons – or rather, the same current. Note that later when we allow wires to split and recombine, we're going to have to be much more careful about this.

What about energy? As one of our electrons leaves the negative terminal of the battery, it has kinetic energy and potential energy. Now, when it returns later to the battery, it still has to have the same kinetic energy (remember all of our electrons are moving at the same rate), however, it has lost all of its potential energy. As an analogy, you can think of this as water in a pipe flowing down a hill. If the water is contained in a pipe where the radius does not change, the continuity equation (equation 202-1) tells us that the water at the bottom of the hill has to have the same speed as the water at the top, however, it has lost its potential energy as it flowed down the hill. In this analogy, the battery is a pump that moves the water up the hill and the resistors represent where the water flows down the hill.

Battery

Let's make this more precise. Consider a small group of electrons with charge dq moving through the battery in a time dt. The battery does work, dW, on the charged particles increasing their potential energy. We define the *emf* of the battery as the amount of work it does per unit charge,

$$\mathcal{E} = \frac{dW}{dq},\tag{214-1}$$

where the symbol $\mathcal{E}$ represents the *emf* of the battery. *emf* is measured in units of volts and is often called the **voltage** of the battery. Be very, very careful with the use of *emf* – the term was originally an abbreviation for *electromotive force*; however, this is very much a misnomer, because it is *not* a force, but rather a potential difference. Instead, consider the following definition:

> The **emf** ($\mathcal{E}$) of a battery is the potential difference (voltage) developed between the terminals of the battery when there is no current flowing through the battery.

Wires

After leaving the battery, our charged particles travel through a section of ***ideal wires***. Ideal wires are considered to have zero resistance therefore there is no voltage drop according to Ohm's Law ($V = IR$). Thus, ideal wires act as if they have no effect on the circuit at all, except to connect other circuit elements. In reality, wires do have a small, but measureable resistance. However, that resistance is usually much smaller than the other elements of the circuit and it is therefore a good approximation to consider them ideal.

Resistors

As the charged particles travel through the resistor they do experience a loss in potential energy. Of course, the energy cannot simply disappear; it is converted to other forms of energy such as thermal energy, sound or light (see the section on power in the previous unit). As we learned in our unit on electric potential, the work done on a set of charged particles with charge dq, as they move through a potential difference V is given by

$$dW = Vdq.\tag{214-2}$$

However, we already know from Ohm's Law that the potential drop across a resistor is given by $V = IR$, therefore

$$V = \frac{dW}{dq} = IR.\tag{214-3}$$

Entire Circuit

Let's now apply conservation of energy to the entire circuit (as seen in Figure 214-1) – meaning that the net change in energy around the entire circuit must be zero. Consider the work done by each element as we move around the circuit. I like to start at the battery, but you can start at any point in the circuit,

$$W_{battery} + W_{wire} + W_{resistor} + W_{wire} = 0.\tag{214-4}$$

We can also write this in terms of the work per unit charge of each element, which gives us

$$\frac{dW}{dq}_{battery} + \frac{dW}{dq}_{wire} + \frac{dW}{dq}_{resistor} + \frac{dW}{dq}_{wire} = 0.\tag{214-5}$$

We can now replace each of these terms based on our discussion above, noting that the battery gives energy to the system and the resistor removes energy from the system,

$$\mathcal{E} + 0 - IR + 0 = 0,\tag{214-6}$$

which simplifies to

$$\mathcal{E} - IR = 0. \tag{214-7}$$

To put this simply, the voltage added to the circuit must equal the voltage removed from the circuit. This is known as **Kirchhoff's Voltage Law (KVL)** (also known as **Kirchhoff's Loop Rule**), which is a very powerful way to solve complex circuits.

It is also very important to note that the last equation above can be solved for I, giving us the current which flows through the resistor in our example

$$I = \frac{\mathcal{E}}{R}. \tag{214-8}$$

Example 214-1: Electric Current

Consider figure 214-1. If the battery has $\mathcal{E} = 1.5V$ and the resistor has a resistance of $R = 10\,\Omega$, what is the current through the resistor?

Solution:

Using the derivation above, we know that (for a single battery and resistor setup) we have

$$I = \frac{\mathcal{E}}{R}.$$

Therefore

$$I = \frac{1.5\,V}{10\,\Omega} = 0.15\,A.$$

This is a very reasonable answer. Please note that most standard batteries (AA, AAA, C, D) have an *emf* of 1.5 V.

214-3: Internal Resistance of Batteries

> **Consider**: *Is there friction inside of batteries? How can this be accounted for?*

The ideal EMF of a battery is the voltage across the terminals of the battery when there is no current flowing through it. Because of the nature of the acids creating the voltage difference, when current is flowing though the battery, there is an **internal resistance** that must be accounted for. The internal resistance acts just like a small resistor external to the battery in the way it effects circuits. Therefore, Ohm's law can be used to described the overall effect:

$$\Delta V_{Real\ Battery} = \mathcal{E} - IR_{internal}, \tag{214-9}$$

where $\Delta V_{Real\ Battery}$ is the voltage across the terminals of a battery with EMF $\mathcal{E}$ and internal resistance $R_{internal}$, when a current I flows through it. In most situations, we will treat batteries as ideal and not include this internal resistance; however, it is important to realize that in the real word, there is a small resistance associated with the battery itself.

214-3: Elements in Series

> **Consider**: *What are elements 'in series'? What does this mean for current?*

Circuit elements are said to be **in series** if they are directly connected together by wires and there are no branches in the wires that connect them. Figure 214-2 shows three resistors in series (a) and three batteries in series (b). Note that each part of figure 214-2 represent just part of a circuit and to complete a circuit they would need to be connected to other elements.

Let's first consider the three resistors in series connected to a battery as shown in figure 214-3. The current through each of the resistors must be the same because they are connected similarly to our elements in the last section – that is, since there are no branches in the wire, conservation of charge says that the motion of charges through each element must be the same, i.e., they have the same current. Using *Kirchhoff's Voltage Law*, we find

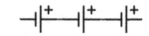

Figure 214-2. (a) three resistors connected in series. (b) Three batteries connected in series.

$$V_1 - IR_1 - IR_2 - IR_3 = 0. \tag{214-10}$$

Since each of these currents are the same, we can simplify the above equation to

$$V_1 - I(R_1 + R_2 + R_3) = 0, \tag{214-11}$$

which we can then write

$$V_1 - I(R_{eq}) = 0, \tag{214-12}$$

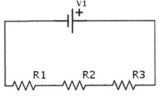

where $R_{eq} = R_1 + R_2 + R_3$ is known as the **equivalent resistance** and represents how the group of resistors acts when considered together, or rather, how the group of series resistors acts as one single *equivalent* resistor.

Figure 214-3. Three resistors in series connected to a battery.

Resistors in Series

$$R_{series} = R_1 + R_2 + R_3 + \cdots \tag{214-13}$$

Description – This equation describes the equivalent resistance of resistors in series as the direct sum of the resistances.
Note: The current in each element is the same.

If batteries are connected in series as shown in figure 214-2(b), the *emf* of the equivalent battery is equal to the sum of the *emf* of all batteries.

Batteries in Series

$$\mathcal{E}_{series} = \mathcal{E}_1 + \mathcal{E}_2 + \mathcal{E}_3 + \cdots \tag{214-14}$$

Description – This equation describes the equivalent *emf* of batteries in series as the direct sum of the individual *emf*s.

214-3: Elements in Parallel.

> **Consider**: *What are elements 'in parallel'? What does this mean for current?*

Circuit elements are said to be **in parallel** if there is a single junction which then goes through each element and then recombines as can be seen in figure 214-4(a). Figure 214-4(b) shows the same set of parallel resistors connected to a battery. An important thing to note is that the parallel connection inherently creates a more complex circuit because it is a **multi-loop** connection. If you consider the two circuits we've discussed so far in this unit (Figures 214-1 and 214-3), each of these circuits consisted of a single-loop. One of the major differences is that when an electron enters a junction (as seen in figure 214-4) it has a choice of which resistor to go through, and this choice is what defines a junction.

In order to analyze the parallel connection we must once again turn to the conservation of charge. As a current enters the junction on the left side of figure 214-4(a), the current will split into three parts, one that goes though R1, one that goes though R2 and one that goes through R3. Call these currents I_1, I_2 and I_3 respectively. The conservation of charge says that the total charge that enters the junction per unit time must equal the charge that leaves the junction per unit time, or

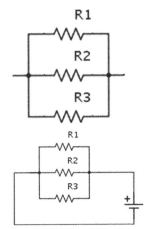

Figure 214-4. (a) A set of three parallel resistors. (b) Three parallel resistors connected to a battery.

$$I_{in} = I_{out} \tag{214-15}$$

This is a statement of **Kirchhoff's' Current Law** (KCL), which states that the sum of the currents into a junction must equal the sum of the currents leaving the junction.

In our specific situation, KCL gives us

$$I_{in} = I_{battery} = I_1 + I_2 + I_3, \tag{214-16}$$

where $I_{battery}$ is the current through the battery.

We can now apply *Kirchhoff's Voltage Law* to each of the possible loops in figure 214-4; that is, we consider the loops with just a battery and R_1, the battery and R_2 and the battery and R_3. In each case, we have a single loop to work with, giving us

$$\mathcal{E} - I_1 R_1 = 0 \qquad\qquad I_1 = \frac{\mathcal{E}}{R_1},$$
$$\mathcal{E} - I_2 R_2 = 0 \quad \Rightarrow \quad I_2 = \frac{\mathcal{E}}{R_2}, \tag{214-17}$$
$$\mathcal{E} - I_3 R_3 = 0 \qquad\qquad I_3 = \frac{\mathcal{E}}{R_3}.$$

If we again employ the *Kirchhoff's Current Law*, we find

$$I_{in} = I_1 + I_2 + I_3, \tag{214-18}$$

$$I_{battery} = \frac{\mathcal{E}}{R_1} + \frac{\mathcal{E}}{R_2} + \frac{\mathcal{E}}{R_3} = \mathcal{E}\left(\frac{1}{R_1} + \frac{1}{R_2} + \frac{1}{R_3}\right) = \frac{\mathcal{E}}{R_{parallel}}, \tag{214-19}$$

where $R_{parallel}$ is the equivalent resistance of the parallel system. Simplifying this expression leads to

$$\frac{1}{R_{parallel}} = \left(\frac{1}{R_1} + \frac{1}{R_2} + \frac{1}{R_3}\right). \tag{214-20}$$

It is extremely important to note that this equation is for the reciprocal of the equivalent resistance and not the equivalent resistance itself. Therefore, I will write this expression slightly different in the box below.

Resistors in Parallel

$$R_{Parallel} = \left(\frac{1}{R_1} + \frac{1}{R_2} + \frac{1}{R_3} + \cdots\right)^{-1} \tag{214-21}$$

Description – This equation describes the equivalent resistance of resistors in series as the reciprocal of the sum of reciprocals of the resistances.

Note: All elements in parallel must have the same voltage drop.

Example 214-1: Power Example

What is the equivalent resistance of a 2.3 Ω resistor connected in parallel with a 7.4 Ω resistor?

Solution:

This is a direct application of the equation for equivalent resistance for resistors in parallel.

$$R_{Parallel} = \left(\frac{1}{R_1} + \frac{1}{R_2}\right)^{-1} = \left(\frac{1}{2.3\ \Omega} + \frac{1}{7.4\ \Omega}\right)^{-1},$$

giving us

$$R_{Parallel} = 1.75\ \Omega.$$

Example 214-2: Series/Parallel Example 1

Consider the simple circuit below consisting of three resistors connected to a single battery.

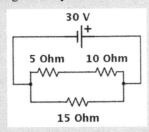

Determine the current through and voltage drop for each resistor.

Solution:

This problem will make repeated use of the equations for series and parallel resistors as well as Ohm's law.

Inspecting the resistors, the 5 Ohm and 10 Ohm resistors are in series and then that combination is in parallel with the 15 Ohm resistor. In doing this type of problem, we always start reducing the circuit from the 'inside-out,' meaning that I will start with the two resistors in series and then move on.

Since we have a 5 Ohm and 10 Ohm resistor in series with each other, their equivalent resistance is

$$R_{12} = 5\ \Omega + 10\ \Omega = 15\ \Omega.$$

If we now replace the two series resistors with this one equivalent resistor, the circuit can be drawn as shown to the right. The two resistors that remain are connected in parallel, so they can be combined to a single equivalent resistor:

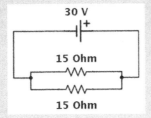

$$\frac{1}{R_{eq}} = \frac{1}{15\ \Omega} + \frac{1}{15\ \Omega} \rightarrow R_{eq} = 7.5\ \Omega.$$

In reality, we didn't even need this last step, because we know that resistors in parallel must have the same voltage, so if the 7.5 Ω equivalent resistor has a voltage drop of 30 V, so do the two 15 Ω resistors in the figure above. The current through each of the 15 Ω resistors is given by Ohm's Law

$$I = \frac{\Delta V}{R} = \frac{30V}{15\Omega} = 2A.$$

Since the 15 Ω resistor in the bottom branch is a real resistor from our first diagram, we already know all the important information for this resistor:

$$A_{15} = 2A, \quad V_{15} = 30V.$$

The 15 Ω resistor on the top branch is the equivalent resistnace for two resistors in series (the 5 Ω and 10 Ω resistors). Since these resistors are in series, they must have the same current and we know now that it is 2 A. From this, we can use Ohm's Law to find the voltage drop across each of these resistors:

$$A_5 = 2A \quad \rightarrow \quad V_5 = I_5 R = (2A)(5\Omega) = 10\ V,$$

$$A_{10} = 2A \quad \rightarrow \quad V_{10} = I_{10} R = (2A)(10\Omega) = 20\ V.$$

We now have the current through and voltage drop across each resistor, as asked. Notice that we first went through and found equivalent resistances for all sets of series and parallel resistors and then went back and used Ohm's Law to find current (when we know voltage) or voltage (when we know current). This is the best strategy for purely series/parallel arrangements.

133

Example 214-3: Series/Parallel Example 2

Consider the simple circuit shown to the right. All resistors have a resistance of 4.0 Ω. What is the current through R1?

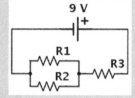
9 V

R1
R3
R2

$$\frac{1}{R_{12}} = \frac{1}{R_1} + \frac{1}{R_2} = \frac{1}{4.0\Omega} + \frac{1}{4.0\Omega} \quad \rightarrow \quad R_{12} = 2.0\Omega$$

R_{12} is now in series with R_3, so the equivalent resistance of the circuit is $R_{eq} = R_{12} + R_3 = 6.0\Omega$. Since this equivalent resistance is connect to a 9V battery, the current through it is $I = V/R = 9V/6\Omega = 1.5\,A$. Since R_{12} and R_3 are in series, they both have the same current (1.5A) and we can find the voltage drop across R_{12}: $V_{12} = I_{12}R_{12} = (1.5A)(2.0\Omega) = 3V$. Finally, since R_1 and R_2 are in parallel, they must have the same voltage (3 V) and we can find the current through R_1: $I_1 = V_1/R_1 = 3V/4\Omega = 0.75\,A$.

Solution:

Although we are only asked for the current through R1, we really need to complete all of the same steps from example 214-2. First, R1 and R2 are in parallel, so their equivalent resistance is

214-4: Measurement in Electrical Circuits.

Consider: *How are current and voltage measured in a real circuit?*

There are two common meters used when making measurements in electrical circuits – voltmeters to measure voltage drops and ammeters to measure currents.

Voltmeters

A voltmeter is used to measure potential difference **between two points**, or **across an element**. We learned earlier that any elements connected in parallel must have the same potential difference (voltage). Therefore, if we connect a meter capable of measuring voltage in parallel with any element, it is guaranteed to give us the voltage drop of that element. Voltmeters are designed to have very high resistance so that little current flows through them, and they therefore have a minimal effect on the circuit. The symbol for a voltmeter is shown in figure 214-5.

Figure 214-5. Symbol for a voltmeter

If you can remember that **voltmeters measure voltage between two points** (beginning and end of an element, for example), it can be easier to remember that they **must be placed in parallel** connected at the two points of interest!

Ammeters

Ammeters are meters that are designed to **measure the current through** an element or branch. Again, we learned earlier that all elements connected in series must have the same current flowing through them. So, if we connect an ammeter in series with an element, it will tell us the current that flows through that element. The symbol for an ammeter is shown in figure 214-6. Ammeters are designed to have very low resistance so that they do not have an appreciable voltage drop across them and therefore did not disrupt the circuit.

Figure 214-6. Symbol for an ammeter.

If you can remember that **ammeters measure current through an element**, it can help remind you they must be connected in **series with that element**. The circuit in figure 214-7 shows the correct positioning of an ammeter and voltmeter to measure the current and voltage, respectively, across a capacitor. Note again that the ammeter is connected in series with the capacitor and the voltmeter is connected in parallel.

Multimeters

It is not uncommon to find meters that are designed to act both as voltmeters and ammeters. These devices are called (not too surprisingly), multimeters. A picture of one type of multimeter can be seen in figure 214-8.

As you can see, the multimeter has a number of **ports**, or holes that can accept standard wire connectors (known as banana connectors). One of these ports, COM, will always be used no matter in which mode the multimeter will be used. The other

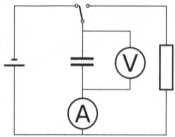

Figure 214-7. A simple circuit with a battery and a capacitor, showing how the correctly connect an ammeter and voltmeter to the capacitor.

Figure 214-8. A multimeter.

port that is used depends on what type of measurement you plan to make. Commonly, there are holes labeled mA or µA for measurement of milliamps or microamps of current, one labeled A for amps of current and one labeled V for voltage measurements. Once the wires are connected to the correct holes, you must also turn the dial on the multimeter to the correct setting to tell the internal circuit which ports to use and how to show the measurements on the screen.

Multimeters come in many types and they all vary some. For example, some meters also have the ability to measure resistance in a circuit, and some even have the ability to measure capacitance. For all of these measurements, the basic operating procedure is the same – find the correct ports and turn the dial to the correct setting.

214-5: Some Final Notes

> **Consider**: *Are there any rules of thumb that can be used to understand basic circuits conceptually?*

There are a couple of other takeaways for series and parallel resistors:

1) The equivalent resistance of resistors in series is *always* larger than the resistance of any of the individual resistors.
2) The equivalent resistance of resistors in parallel is *always* smaller than the resistance of any of the individual resistors.

It is, of course, more complicated when you combine both series and parallel systems together as seen in examples earlier in the unit.

Using these two rules-of-thumb above, it is often possible to determine what will qualitatively happen to a circuit if different elements are removed. For example, let's say three identical light bulbs (circuit element shown in figure 214-9) are connected in parallel and then the set is connected to a battery, the light bulbs will light up. What happens if you remove one of the light bulbs? Qualitatively, we know that the equivalent resistance of the circuit will increase; meaning that less current will leave the battery. However, the current now must split between only two branches as opposed to three and the overall current through each bulb is unchanged. These types of qualitative questions can help you test your understanding of basic circuits.

Figure 214-9. The circuit diagram of a light bulb.

Example 214-4: Conceptual Circuit Questions

Three identical light bulbs each with resistance, R, are connected to a battery as shown in the figure to the right.

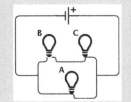

  (a) How does the brightness of each branch compare?
  (b) What happens to bulbs B and C if bulb A is removed?
  (c) What happens to bulbs A and B if bulb C is removed (and the slot is left open)?

Solution:

For this problem, we need to think conceptually about what happens to the light bulbs since we are given neither the voltage of the battery, or the resistance of the bulbs.

(a) Bulb A is in parallel with the branch containing bulbs B and C (which are in series with each other). We know that each branch of a parallel system must have the same voltage drop across it (which would be the voltage of the

battery in this case). Since the resistance of the upper branch (R+R = 2R) is twice as large as the resistance of the lower branch (R), bulb A will have twice the current flowing through it as bulbs B and C. Therefore bulb A will be brighter than bulbs B and C, which will have the same brightness (same current since they are in series).

(b) Each branch of the parallel system must have the same voltage drop, which in this case is the voltage of the battery. Before bulb A is removed, the voltage drop across bulbs B and C together must therefore be equal to the voltage of the battery. After bulb A is removed, the same must be true - the voltage drop across bulbs B and C must still be equal to the voltage of the battery. Therefore, there is no change in the brightness of bulbs B and C when bulb A is removed.

(c) If bulb C is removed, we can immediately say that bulb B will go dark, since it is in series with bulb C and no current can flow through the open slot where bulb C was previously. Similarly to the argument for part (b), the voltage drop across bulb A must still be the voltage of the battery, and the brightness will therefore not change.

215 – Complex Circuits

We saw in unit 214 that resistive circuits that contain series and parallel connections can be dealt with directly by finding equivalent resistances, currents and voltages. However, not all circuits are made of parallel and series connections. In fact, most circuits are constructed of more elements than just batteries and resistors. In this unit, we begin our explanation of how Kirchhoff's laws can be used to analyze more complex circuits.

Integration of Ideas

Review the definition of current from Unit 213.
Review the ideas of parallel and series circuits from Unit 214.

The Bare Essentials

- Complex circuits can be solved using Kirchhoff's Laws.

- Following the conservation of charge, Kirchhoff's Current Law (KCL, also known as the node rule) states the net current entering a node must equal the net current leaving a node.

Kirchhoff's Current Law (KCL)

$$\sum I_{in} = \sum I_{out}$$

Description – KCL states that the sum of the currents entering the node of a circuit must equal the sum of the currents leaving the same node.

- Derived from the conservation of energy, Kirchhoff's Voltage Law (KVL, also known as the voltage rule) states that the net voltage drop around any closed loop in a circuit must be zero.

Kirchhoff's Voltage Law (KVL)

$$\sum_{loop} V = 0$$

Description – KVL states that the net voltage around a closed loop in a circuit must be zero.
Note: KVL works for any and all loops in a circuit as long as the loops do not cross themselves.

- In order to solve complex circuits:
  - Use the same number of KCL and KVL equations as you have unknowns;
  - In most situations, you can only use n-1 KCL equations, where n is the number of nodes (junctions). This is to avoid redundancy.

- Capacitor charging and discharging in a circuit can be described by simple differential equations.

Charging Capacitor

$$q(t) = C\mathcal{E}\left(1 - e^{-\frac{t}{RC}}\right)$$

Description – This equation describes the charge on a charging capacitor with capacitance, C, being charged by a battery with *emf*, $\mathcal{E}$, in a series circuit with resistance, R.

Discharging Capacitor

$$q(t) = q_0\left(e^{-\frac{t}{RC}}\right)$$

Description – This equation describes the charge on a discharging capacitor with capacitance, C, and initial charge, q_0, in a series circuit with resistance, R.

- Note that the quantity RC has units of time and is sometimes called the *time constant* (τ) for a discharging capacitor; it represents the time for the charge on the capacitor to drop to 1/e of its initial value

215-1: Kirchhoff's Laws

Consider: *How are conservation of charge and conservation of energy applied to electric circuits?*

Unfortunately, not all circuits are made of parallel and series configurations of resistors and simple battery setups. In fact, most circuits contain more elements than just our simple batteries and resistors. There is a technique that can be used to solve just about any circuit, and we've already introduced it to you – Kirchhoff's Laws:

Kirchhoff's Current Law (KCL)

$$\sum I_{in} = \sum I_{out} \qquad (215\text{-}1)$$

Description – KCL states that the sum of the currents entering the node of a circuit must equal the sum of the currents leaving the same node.

Kirchhoff's Voltage Law (KVL)

$$\sum_{loop} V = 0 \qquad (215\text{-}2)$$

Description – KVL states that the net voltage around a closed loop in a circuit must be zero.
Note: KVL works for any and all loops in a circuit as long as the loops do not cross themselves.

We used Kirchhoff's Laws in the last unit when discussing how to find the relationship for series and parallel resistors, but now it is time to generally put them to use and see how they can be used to solve seemingly very complex circuits.

Consider the circuit shown in figure 215-1. This circuit cannot be a simple series and parallel circuit because of the branch containing R2. As an example, you might be tempted to say that the combination of R3 and R1 are in parallel with R2; however, since there is another branch between R1 and R3 that then goes though R5, this branch destroys the series/parallel nature of the circuit. In this case, we **must** use Kirchhoff's Laws. Learning how to use Kirchhoff's Laws is also very valuable since they work on all electric circuits no matter how complicated. Remember, the two laws are based entirely on the conservation of charge and the conservation of energy, so any deviation from these laws would violate one of these fundamental principles.

In order to analyze the circuit below using Kirchhoff's Laws, we have a couple of steps of setup to do. First, for each element, we must define a current. For batteries, the direction of the current is well defined, since we know the direction that the battery is *trying* to move the current. For each resistor, we also need to define the current; however, for resistors you have full discretion in deciding which direction the current should travel – you only need to be consistent with it moving forward. As you can see in figure 215-2, I have labeled the current for each circuit element with a name and an arrow. By convention, I like to name the currents with the

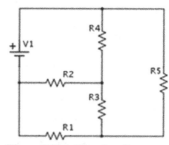

Figure 215-1. The circuit

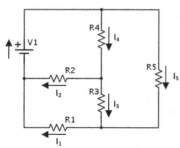

Figure 215-2. Circuit with currents defined.

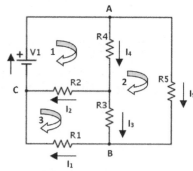

Figure 215-3. Circuit with currents, nodes and loops defined.

138

same subscript as the resistors for consistency, but your choice of naming convention is up to you.

The next step is to choose nodes and loops over which to apply Kirchhoff's Laws. A **node** is any place where multiple branches come together. Our **loops** are any closed loop around the circuit which does not cross itself. In general, we need the same number of equations as we have unknowns in our circuit. Our unknowns in figure 2I5-2 are the current through each of the five resistors and the current through the battery. So, we have six unknowns and will therefore need six equations.

Figure 215-3 adds the nodes (A, B and C) and loops (1,2 and 3 noted by the curvy arrows) to the circuit we've been discussing. Although the diagram becomes very busy with all of these added elements, they are all quite important to forming our required equations.

Now, let's start with *Kirchhoff's Current Law*. This rule states that the total current entering a node must be equal to the current leaving the node (conservation of charge). Consider node A – the arrow for the battery is pointing into the node and the arrows for currents I4 and I5 are pointing away from the node, so I would write the node equation for junction A as

$$Node\ A:\ \ I_{batt} = I_4 + I_5. \tag{215-3}$$

Similarly, for nodes B and C, we find

$$node\ B:\ \ I_3 + I_5 = I_1, \tag{215-4}$$

$$node\ C:\ \ I_1 + I_2 = I_{batt}. \tag{215-5}$$

Now we have three equations that include all six of our currents. We still need three equations so that we can solve for each of the currents in our circuit and these come from using Kirchhoff's Voltage Law. For each of our chosen loops, we will make our way around the loop using the following conventions

For sources (batteries, power supplies, etc.)
> If the current and loop are in the same direction, the voltage for that element is positive.
> If the current and loop are in the opposite diection, the voltage for that element is negative.

For other circuit elements (resistors, light bulbs, capacitors, etc.)
> If the current and loop are in the same direction, the voltage for that element is negative.
> If the current and loop are in the opposite direction, the voltage for that element is positive.

Consider loop 1 in figure 215-3. Starting with the general equation for Kirchhoff's Loop Rule, if we were to circle the loop and insert the voltage for each element, we would find

$$\sum_{loop} V = 0. \tag{215-6}$$

You can start at any element in the loop you wish. For this example, I will start with the battery in loop one. Looking at the diagram, you can see that the direction of the current arrow of the battery and the direction of the loop at the position of the battery are in the same direction. Using the conventions above, the voltage of the battery in our KVL equation will be positive, and is given in the diagram as V1. Following the loop, the next element we find is R4. Again, the direction of the current we chose and the direction of the loop at that position are the same. Since this element is a resistor, our convention tells us that the voltage for this element will be negative. Using Ohm's Law, we will write this as $V_{R4} = -I_4 R_4$. Finally, the last element in loop1 is R2. Again, the current we chose for R2 is in the same direction as loop 1 at that point, so the voltage will be negative.

Putting this all together, we find

$$V_{bat} + V_{R4} + V_{R2} = 0, \tag{215-7}$$

which becomes

$$V_1 - I_4 R_4 - I_2 R_2 = 0. \tag{215-8}$$

Now consider loop 3, starting at R_1. For R1, the current we chose and the direction of the loop coincide, so, as before, the voltage will be negative. However, for R2, the direction of the current is opposite the direction of the loop at that point, so we have to use a positive voltage for R2. Putting this all together, our KVL equation for loop to is

$$-I_1 R_1 + I_2 R_2 - I_3 R_3 = 0. \tag{215-9}$$

Using the same process, please convince yourself that the KVL equation for loop 2 is

$$I_4 R_4 + I_3 R_3 - I_5 R_5 = 0. \tag{215-10}$$

We now have our six equations for our six unknown currents. Summarizing:

$$I_{batt} = I_4 + I_5 \qquad\qquad V_1 - I_4 R_4 - I_2 R_2 = 0$$

$$I_3 + I_5 = I_1 \qquad\qquad -I_1 R_1 + I_2 R_2 - I_3 R_3 = 0 \tag{215-11}$$

$$I_1 + I_2 = I_{batt} \qquad\qquad I_4 R_4 + I_3 R_3 - I_5 R_5 = 0$$

Unfortunately, this now represents a system of six equations and six unknowns. There are multiple ways to complete this task, including substitution, elimination of variables and Cramer's Rule. You are probably most familiar with substitution, which always works, but can be very time consuming. An introduction to Cramer's Rule is given at the end of this unit for those that may be interested.

Example 215-1: Substitution.

Consider the circuit shown in figures 215-1 through 215-3. If each of the resistors has a known resistance of 3.0 Ohms, the battery has a voltage of 18 Volts and a current of 6 Amps through it, what is the current through each branch of the circuit?

Solution:

Since the above text already gave the general equations for the circuit, we can start by substituting the known values in, simplifying where possible:

(1) $6 = I_4 + I_5$ (4) $18 - 3I_4 - 3I_2 = 0$

(2) $I_3 + I_5 = I_1$ (5) $I_1 = I_2$

(3) $I_1 + I_2 = 6$ (6) $I_4 + I_3 - I_5 = 0$

We can find a few things relatively quickly. First, equation (5) tells us that currents 1 and 2 must be the same. Substituting this into equation 3:

$$I_2 + I_2 = 6 \quad \rightarrow \quad I_2 = 3 = I_1.$$

This result can now be substituted into equation (4):

$$18 - 3I_4 - 3(3) = 0 \quad \rightarrow \quad I_4 = 3.$$

We now have all the currents except I_3 and I_5. If we use our result for I4 in equation (1), we get

$$6 = 3 + I_5 \quad \rightarrow \quad I_5 = 3.$$

Finally, equation (6) gives us

$$6 + I_3 - 6 = 0 \quad \rightarrow \quad I_3 = 0.$$

This result for I_3 is curious. However, it does have a direct explanation: For the given circuit, if all the resistances are equal, no current flows through R_3. Our circuit is symmetric in this case, that is, if you remove R_3, each of the branches left are exactly the same, so the current has no reason to flow through R_3.

To summarize our results:

$$I_1 = 3\ A$$

$$I_2 = 3\ A$$

$$I_3 = 0$$

$$I_4 = 3\ A$$

$$I_5 = 3\ A$$

Problem Solving Strategy (Kirchhoff's Laws):

Using Kirchhoff's Laws is very powerful and relatively formulaic. Here is a general list of steps to take when solving circuit problems in this manner

1) Assign a current direction and label each circuit element. Your direction does not have to be correct; if you are consistent and you chose wrong, the current will come out negative.
2) Assign labels to the node that you will use to apply the KCL. Be careful that two nodes do not give you the same information. To avoid redundancy, we can only use n-1 KCL equations where n is the number of nodes.
3) Assign and label loops along which you will apply KVL.

4) The total number of nodes and loops used must equal the number of branches   of your circuit.
5) Apply KCL and KVL to your circuit and solve the equations.

The following couple of examples will hopefully expound on this process.

Example 215-2: Circuit 2.

For the circuit shown below, determine the current through each of the three branches.

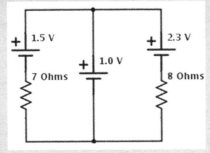

Solution:

Following the steps above, we must first assign and label currents, loops and nodes, as shown below.

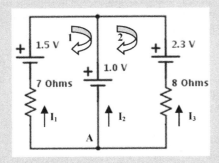

First, note that only the lower node was assigned (as node A). If we apply KCL to the upper node, we would get the same result as for node A because the same branches are involved. Applying KCL, we find

$$(1) \quad 0 = I_1 + I_2 + I_3.$$

Following all of our conventions, KVL for loop 1 and loop 2 give us, respectively,

$$(2) \quad 1.5 - 1.0 - 7I_1 = 0,$$

$$(3) \quad 1.0 - 2.3 + 8I_3 = 0.$$

Solving equation (2) gives us $I_1 = 0.07\,A$, and equation (3) gives us $I_3 = 0.16\,A$.

Finally, we can use our results in equation (1) to find $I_2 = -0.23\,A$. Again, note that since I2 wound up being negative, the real current goes in the opposite direction. It is important to note that even though there is no resistor in that branch, there is a finite current because of the overall resistance in the circuit.

In summary:

$$I_1 = 0.07\,A, I_2 = -0.23\,A, I_3 = 0.16\,A.$$

Tip: I would also leave I_3 as negative to be consistent with our drawn diagram.

I also want to introduce the idea of a **Solutions Table** for a circuit problem. Once we've found the current in each resistor, we can then also find the voltage drop across the resistor and the power dissipated by the resistor. Similarly, if we know the current through a battery, we can also find the power delivered by the battery to the circuit. So, we would expect our solutions table that have these quantities. You can see an example of how a solutions table would be set up using the results of example 281-2 in table 215-1.

Table 215-1. *A Solutions Table for example 215-2 above. Notice that once you fill in known values, the values in the next column to the right are just the previous two columns multiplied together (bold values).*

Name and R(Ω)	I (A)	V (V) = IR	P (W) = IV
$R_1 = 7\,\Omega$	0.07 A	**-0.49 V**	**-0.034 W**
$R_3 = 8\,\Omega$	0.16 A	**-1.28V**	**-0.205 W**
Battery 1	0.07 A	1.5 V	**0.105 W**
Battery 2	-0.23 A	1.0 V	**-0.230 W**
Battery 3	0.16 A	2.3 V	**0.368 W**

Filling out a solutions table for a circuit tells you just about everything you might want to know about each element in the circuit. For example, R_1 in the above table is dissipating 0.034 Watts, meaning that 0.034 Joules of electrical energy per second are being transformed to other types of energy which we may measure as heat, light, sound, etc.

Example 215-3: Light bulb power

Consider the following circuit diagram. For this problem we will assume that all resistors are light bulbs.

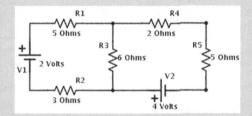

Find the voltage drop across each resistor as well as the power delivered or dissipated by each element.

Solution:

The first step in this problem is to realize that R1 and R2 are in series as are R4 and R5. We can use what we know about resistors in series to reduce the circuit to the following diagram.

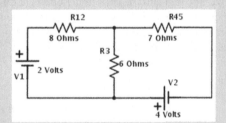

Always use the simple techniques to reduce complex problems!

Following our general steps, we can define currents in each branch, a node, and loops with directions.

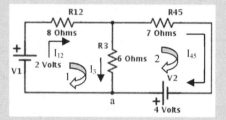

Using our general conventions, KCL and KVL give us:

(1) *Node a* $I_3 + I_{45} = I_{12}$
(2) *Loop* 1: $2 - 8I_{12} - 6I_3 = 0$
(3) *Loop* 2: $4 + 6I_3 - 7I_{45} = 0$

Although only three equations, we must do some manipulation to get a solution. First, substitute equation (1) into equation (2) to eliminate I_{45}.

$$2 - 8(I_3 + I_{45}) - 6I_3 = 0.$$

This equation and equation (3) are both in terms of I_3 and I_{45}. They can be solved together to find $I_3 = -0.123$ A, and $I_{45} = 0.466$ A. Then, using equation (1), we find $I_{12} = 0.342$ A. Again, note that we chose I_3 in the wrong direction.

The last step is to put together a solutions table. It is very important to remember that (from our original diagram), $I_1 = I_2 = I_{12}$ and $I_4 = I_5 = I_{45}$

Solutions Table:

Name and R(Ω)	I (A)	V (V) = IR	P (W) = IV
$R_1 = 4\ \Omega$	0.342 A	1.37 V	0.469 W
$R_2 = 3\ \Omega$	0.342 A	1.03 V	0.352 W
$R_3 = 6\ \Omega$	-0.123 A	-0.738 V	0.091 W
$R_4 = 2\ \Omega$	0.466 A	0.932 V	0.434 W
$R_5 = 5\ \Omega$	0.466 A	2.33 V	1.09 W
Battery 1	0.342 A	2.0 V	0.684 W
Battery 2	0.466 A	4.0 V	1.86 W

Please note that (up to some rounding) the power put into the system by the batteries is the power removed from the system by the light bulbs – conservation of energy. So, which light bulb is the brightest? Well, assuming that all of the bulbs are the same type, R5 would be the brightest. In fact, it would be approximately twice as bright as the next brightest bulb. Remember that the power of the light we see from a bulb is not exactly the electrical power that goes into the resistor/bulb. Some energy is lost to other forms; however, if the bulbs are the same type, this analogy works.

2I8-2: Capacitors and Circuits

Consider: *How do capacitors react inside of a circuit?*

So far, we have only talked about circuits that contain voltage sources and resistors. Another important circuit element is the capacitor (introduced in unit 212). Remember that capacitors give us a way to store energy in electric fields between two oppositely charged surfaces. There are two common circuit element diagrams for capacitors, as shown in figure 215-4. In the figure, C1 represents a non-polarized capacitor, meaning that the capacitor can be connected to the circuit in either direction. C2, on the other hand, is polarized capacitor and must be connected in one direction, or you will find unexpected results. If a one or more capacitors are placed in a circuit, the circuit can still be analyzed using Kirchhoff's Laws. In order to do this, we must remember that the voltage drop across a capacitor is given by

Figure 215-4. Schematic diagram for a capacitor.

$$V_C = \frac{Q}{C}.$$ (215-12)

Consider the **series RC circuit** shown in figure 215-5. The elements with arrows at the top of the diagram are known as switches, and can be set to either be closed, completing the circuit (known as a **closed switch**, which is shown on the left), or open, leaving a break in the circuit so that no current can flow through that branch (known as an **open switch**, which is shown on the right). Imagine that we start with the switches as shown in figure 215-5. The open switch on the right means that no current will flow through that branch, meaning that it can essentially be ignored. In the left branch, we can use KVL to find the voltage developed across the plates of the capacitor:

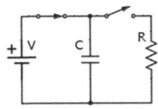

$$V_B - V_C = 0 \quad \rightarrow \quad V_C = V_B,$$ (215-13)

Figure 215-5. Charging the capacitor in an RC circuit.

or rather that the capacitor will be charged by the battery until the voltage across the plates of the capacitor is equal to the voltage of the battery.

If we now simultaneously open the left switch and close the right switch, we have the situation shown in figure 215-6, with the capacitor having an initial voltage of V_C. Once the right switch is closed, the charged capacitor will now act as the voltage source, and if we apply KVL, we find

$$V_C - V_R = 0 \quad \rightarrow \quad V_C = V_R,$$ (215-14)

which, substituting what we know about the voltage of each element, we can write

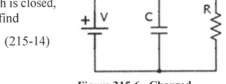

Figure 215-6. Charged capacitor connected to a resistor in an RC circuit.

$$\frac{Q}{C} = -IR,$$ (215-15)

where Q is the charge needed on the capacitor to give a voltage between its plates of V_C. We know that current is the rate at which charge flows, meaning that the current is the derivative of the charge with respect to time, we find

$$\frac{Q}{C} = -\frac{dQ}{dt}R,$$ (215-16)

which can be rearranged to

$$\frac{dQ}{dt} = -\frac{Q}{RC}.$$ (215-17)

The above equation represents a differential equation that is relatively easy to solve if we isolate our variables, Q and t. In order to do this, I will divide both sides of the equation by Q and multiply both sides of the equation by *dt*, leaving us

$$\frac{dQ}{Q} = -\frac{dt}{RC}.$$ (215-18)

We can now solve this by simple integration:

143

$$\int \frac{dQ}{Q} = -\frac{1}{RC} \int dt. \tag{215-19}$$

The integral on the left gives us the natural logarithm and the integral on the right is just t by definition:

$$\ln(Q) = -\frac{1}{RC} t + K, \tag{215-20}$$

where K is the constant of integration (I used K as the constant so that it is not confused with the capacitance, C). We can now solve for the charge, Q, as a function of time

$$Q = e^{-\frac{1}{RC}t+K} = e^K e^{-\frac{1}{RC}t} = Q_0 e^{-\frac{t}{RC}}, \tag{215-21}$$

where the constant of integration e^K is taken as the initial charge Q_0 (charge at time t = 0). What this tells us is that the charge on an initially charged capacitor placed in a series RC circuit will decrease exponentially with time.

Discharging Capacitor

$$q(t) = q_0 \left(e^{-\frac{t}{RC}} \right) \tag{215-21}$$

Description – This equation describes the charge on a discharging capacitor with capacitance, C, and initial charge, q_0, in a series circuit with resistance, R.

Note that RC has units of time and is sometimes called the ***time constant*** (τ) for a discharging capacitor; it represents the time for the charge on the capacitor to drop to 1/e of its initial value.

Another potentially important quantity would be the current created in the RC circuit. Since the current is the derivative of the charge with respect to time, we find

$$I(t) = -\frac{q_0}{RC} \left(e^{-\frac{t}{RC}} \right) = -\frac{V_C}{R} \left(e^{-\frac{t}{RC}} \right). \tag{215-22}$$

In describing how the capacitor above became initially charged, we didn't discuss any dependence on time. Mostly, I did this because of a bit of mathematical complexity which will be easier to describe now that we solved for the discharging capacitor. Consider figure 215-7 with an initially uncharged capacitor. Let's apply KVL shortly after the switch is closed

$$V_B + V_C + V_R = 0, \tag{215-23}$$

where V_B, V_C, and V_R represent the voltage across the battery, capacitor and resistor, respectively. We can now substitute in what we know about these voltages

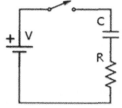

Figure 215-7. A series RC circuit.

$$V - \frac{Q}{C} - IR = 0 \quad \rightarrow \quad V - \frac{Q}{C} - \frac{dQ}{dt} R = 0, \tag{215-24}$$

where V is used more generally for the battery voltage. As you can see, this equation is very similar to the KVL equation for a discharging capacitor, except that we have an extra constant, V. With what we did for the discharging capacitor, we can now solve this by analogy. We know the solution must be exponential, but that the charge on the capacitor starts at zero and ends at its maximum, which suggests

$$q = q_0 \left(1 - e^{-\frac{t}{RC}} \right). \tag{215-25}$$

What about q_0 (the max current)? By analogy, we know that the max voltage across the capacitor should equal the *emf* of the battery. Since, for a capacitor, $q = CV$, we can write

$$q = C\mathcal{E} \left(1 - e^{-\frac{t}{RC}} \right). \tag{215-26}$$

Charging Capacitor

$$q(t) = C\mathcal{E}\left(1 - e^{-\frac{t}{RC}}\right) \qquad (215\text{-}26)$$

Description – This equation describes the charge on a charging capacitor with capacitance, C, being charged by a battery with *emf*, $\mathcal{E}$, in a series circuit with resistance, R.

Note thatWhen you consider that the argument of an exponential function must be unitless, we know that the quantity RC must have units of time (so that it cancels the t in the numerator of the exponential). Therefore, RC is sometimes called the ***time constant*** for the RC circuit. In one time constant of time, the charge (or current) of the RC circuit changes by 1/e of its original value. Although this seems like an arbitrary measure, it is used because it falls out of the equation so readily.

Example 215-4: Capacitor

A 2.00-MΩ resistor is connected in series with a 6.00-μF capacitor. When a switch is thrown, these elements are connected to a 12-V battery with negligible internal resistance. The capacitor is initially uncharged. (a) What is the maximum charge on the capacitor? (b) What is the maximum current in the circuit? (c) What is the maximum power provided by the battery? (d) How much energy is stored in the capacitor at t = 50 ms?

Solution:

(a) The charge on the capacitor is given by

$$q(t) = C\mathcal{E}\left(1 - e^{-\frac{t}{RC}}\right).$$

Since the maximum value of this function would only occur when the term in parenthases is 1, this would be

$$q_{max} = C\mathcal{E} = (6\,\mu F)(12\,V) = 72\,\mu C.$$

(b) Current is the rate at which charge flows (dq/dt), so first, we must take the derivative to the charging capacitor equation with respect to time

$$I(t) = \frac{dq}{dt} = \frac{d}{dt}\left(C\mathcal{E}\left(1 - e^{-\frac{t}{RC}}\right)\right),$$

or

$$I(t) = \frac{-C\mathcal{E}}{-RC}e^{-\frac{t}{RC}} = \frac{\mathcal{E}}{R}e^{-\frac{t}{RC}}$$

The maximum of this equation will happen at t = 0, so

$$I_{max} = \frac{\mathcal{E}}{R} = \frac{12\,V}{2x10^6\Omega} = 6\,\mu A.$$

(c) In general, the power provided by a battery is given by P=IV, so

$$P(t) = I(t)V(t) = \left(\frac{\mathcal{E}}{R}e^{-\frac{t}{RC}}\right)\left[\mathcal{E}\left(1 - e^{-\frac{t}{RC}}\right)\right],$$

which simplifies to

$$P(t) = \frac{\mathcal{E}^2}{R}\left(e^{-\frac{t}{RC}} - e^{-\frac{2t}{RC}}\right).$$

Each of the terms in parentheses are a max at t = 0. so the max power is then becomes:

$$P_{max} = \frac{\mathcal{E}^2}{R} = \frac{(12\,V)^2}{2x10^6\Omega} = 72\,\mu W.$$

(d) The energy stored in a capacitor is

$$U = \frac{q^2}{2C} = \frac{1}{2C}\left[C\mathcal{E}\left(1 - e^{-\frac{t}{RC}}\right)\right]^2,$$

which gives us

$$U = \frac{(12\,V)^2(6x10^{-6}F)}{2(6\,x\,10^{-6}\,F)}\left(1 - e^{-\frac{50x10^{-3}s}{2x10^6\Omega(6x10^{-6}F)}}\right),$$

or

$$U = 7.47x10^{-9}J.$$

2l8-3: Row Reduction (optional)

A common technique for solving systems of linear equations is known as **row reduction**, or **Gaussian elimination**. Row reduction is very methodical, and can be readily applied to small and enormous systems alike. This process does take space and use quite a bit of paper, but once you get the hang of it, you can solve any system of linear equations in this fashion.

In example 215-3, we used simple substitution to solve for the currents in each branch of the circuit. Let's now return to that example and apply Gaussian elimination. The three equations we found from KCL and KVL are

$$
\begin{aligned}
&(1) \quad Node\ a\text{:} && I_3 + I_{45} = I_{12} \\
&(2) \quad Loop\ 1\text{:} && 2 - 8I_{12} - 6I_3 = 0 \\
&(3) \quad Loop\ 2\text{:} && 4 + 6I_3 - 7I_{45} = 0
\end{aligned}
\tag{215-27}
$$

The first thing we need to do is to rewrite these equations so that the variables (the currents) are all in the same order in the equations. I will do this by putting them in order of their numbering on the left side and moving any constants to the right side of the equation.

$$
\begin{aligned}
I_{12} - I_3 - I_{45} &= 0 \\
-8I_{12} - 6I_3 + 0I_{45} &= -2 \\
0I_{12} + 6I_3 - 7I_{45} &= -4
\end{aligned}
\tag{215-28}
$$

Note also that I made sure each variable was listed in each equation, even in the coefficient is zero. Now, I'm going to form an **augmented matrix**, by placing the coefficients of each variable in place and the constants on the right side:

$$
\left[\begin{array}{ccc|c}
1 & -1 & -1 & 0 \\
-8 & -6 & 0 & -2 \\
0 & 6 & -7 & -4
\end{array}\right].
$$

The overall idea is to get the 3x3 section of the matrix (the part to the left of the vertical line) to look like this

$$
\begin{bmatrix}
1 & 0 & 0 \\
0 & 1 & 0 \\
0 & 0 & 1
\end{bmatrix},
$$

so that we can get the solutions easily.

In order to pull this off, we will do a number of **row operations**. For example, the first thing we would like to do is to get a zero in the first column of the second row. In order to do this, we should multiply the first row by 8 and then add the two rows. In equation form, this would look like:

$$
8\ x\ (I_{12} - I_3 - I_{45} = 0) \qquad \rightarrow \qquad 8I_{12} - 8I_3 - 8I_{45} = 0,
\tag{215-29}
$$

which we then add to row number 2:

$$
\begin{aligned}
8I_{12} - 8I_3 - 8I_{45} &= 0 \\
+ -8I_{12} - 6I_3 + 0I_{45} &= -2 \\
\hline
0I_{12} - 14I_3 - 8I_{45} &= -2
\end{aligned}
\tag{215-30}
$$

We now replace row two of our enhanced matrix with our solution from above:

$$
\left[\begin{array}{ccc|c}
1 & -1 & -1 & 0 \\
0 & -14 & -8 & -2 \\
0 & 6 & -7 & -4
\end{array}\right].
$$

There is a simplified notation for writing out the operation we just performed. $8R_1 + R_2 \rightarrow R_2$, meaning what we did was multiply row 1 by 8, added it to row 2 and replaced row 2 with the result.

Since every term in the second row of this matrix is even, I'm going to divide that row by two make the numbers a bit smaller

$$\frac{R_2}{2} \to R_2 \qquad \begin{bmatrix} 1 & -1 & -1 & | & 0 \\ 0 & -7 & -4 & | & -1 \\ 0 & 6 & -7 & | & -4 \end{bmatrix}. \qquad (215\text{-}31)$$

Next, I want to get a zero in the second column of row three. If we multiply the third row by 7/6, the second column will match the coefficient in the second column of the second row, which we can then add to together to get the desired result:

$$R_2 + \frac{7}{6} R_3 \to R_3 \qquad \begin{bmatrix} 1 & -1 & -1 & | & 0 \\ 0 & -7 & -4 & | & -1 \\ 0 & 0 & -\frac{73}{6} & | & -\frac{34}{6} \end{bmatrix}. \qquad (215\text{-}32)$$

The next step is to multiply row three by -6/73 so that the only coefficient on the left side of the vertical line is a 1:

$$R_3 x \frac{-6}{73} \to R_3 \qquad \begin{bmatrix} 1 & -1 & -1 & | & 0 \\ 0 & -7 & -4 & | & -1 \\ 0 & 0 & 1 & | & 0.466 \end{bmatrix}. \qquad (215\text{-}33)$$

The last line of the matrix would be read, $0I_{12} + 0I_3 + I_{45} = 34/73$, which is the same as saying $I_{45} = 0.466$, which is the same solution we got in example 215-3!

We now start moving backwards up the matrix. Since the last row gives us the answer for I_{45}, we can substitute this into the second row, which would then be read

$$-7I_3 - 4I_{45} = -1, \qquad (215\text{-}34)$$

however, since we now have a value for I_{45}, we can substitute this in to find

$$-7I_3 - 4(0.466) = -1 \quad \to \quad I_3 = -0.123. \qquad (215\text{-}35)$$

Remembering that this solution is the same as saying $0I_1 + I_2 + 0I_3 = -0.123$, we can substitute this into our augmented matrix in row 2:

$$\begin{bmatrix} 1 & -1 & -1 & | & 0 \\ 0 & 1 & 0 & | & -0.123 \\ 0 & 0 & 1 & | & 0.466 \end{bmatrix}. \qquad (215\text{-}36)$$

Finally, we can use the same steps as above to substitute I_3 and I_{45} into the first row of the matrix, giving us our final solution (check for yourself):

$$\begin{bmatrix} 1 & 0 & 0 & | & 0.342 \\ 0 & 1 & 0 & | & -0.123 \\ 0 & 0 & 1 & | & 0.466 \end{bmatrix}. \qquad (215\text{-}37)$$

This final form, with just 1's along the diagonal and everything else zero to the left of the vertical line is known as ***echelon form***, directly gives us the solution to our problem.

I did out this problem in all its gory detail. The power of the row reduction method is that it works for any sized set of equations you can come up with, and it is very formulaic – at each step you simply do one row manipulation to find a leading zero along the diagonal and then use the results of the last line of the matrix to substitute into every row above.

In reality, the row reduction method *is* the substitution method you have used many times, except that it takes away the guessing and gives you a set of steps to follow in order to solve the problem. If you do a couple of problems with it and get used to the method, it will *always* work for you!

147

216 – Faraday's Law and Inductors

Gauss's Law and Ampere's Law gave us a way to calculate the electric and magnetic fields, respectively, of static arrangements – that is of static electric charges or non-changing currents. Once we allow these distributions to vary in time, we find that changing magnetic fluxes cause electric fields and that changing electric fields cause magnetic fields. In this unit, we explore this interrelationship in greater detail.

Integration of Ideas

Review the basics of electric and magnetic fields from the last few units.

The Bare Essentials

- Our discussion of electric and magnetic fields so far have assumed *static* fields. These equations must be modified if the fields are allowed to be time dependent.

- Time varying magnetic fields create electric fields that have a different character (curly electric fields) than electric fields created by discrete electric charges.

- *Faraday's Law* of Induction relates these curly electric fields to induced EMFs in a circuit immersed in the fields

---

Faraday's Law

$$\mathcal{E}_{ind} = \oint \vec{E}_{ind} \cdot d\vec{l} = -\frac{d\Phi_B}{dt}$$

Description – Faraday's Law relates a changing magnetic flux, Φ_B, to the EMF, and therefore electric field it induces in a closed conducting loop.
Note 1: $\Phi_B = \int \vec{B} \cdot d\vec{a}$ is the magnetic flux, analogous to electric flux but for magnetic fields.
Note 2: Faraday's Law may be written in local form as $\vec{\nabla} \times \vec{E} = -\frac{\partial \vec{B}}{\partial t}$

- *Lenz's Law*, the minus sign in Faraday's Law and the most famous minus sign in all of physics, states that the direction of the EMF induced by a changing magnetic flux is such that it opposes the change in flux.

- Inductors are circuit devices that regulate the rate of change of current due to Faraday's Law.

Self Inductance

$$\mathcal{E}_{ind} = -L\frac{dI}{dt}$$

Description – This equation relates the back voltage created in an inductor with self-inductance, L, with a current changing at a rate given by dI/dt.
Note 1: The SI units of self inductance is the Henry ($1\,H = \Omega \cdot s$)

- The energy stored in an inductor is given by

$$U_L = \frac{1}{2}Li^2$$

- Magnetic fields are created by changing electric fields (in addition to currents). The correction to Ampere's Law to accommodate this is known as the *Ampere-Maxwell relationship*.

Ampere-Maxwell Relationship

$$\oint \vec{B} \cdot d\vec{l} = \mu_0 I_{enc} + \mu_0 \epsilon_0 \frac{d\Phi_E}{dt}$$

Description – The Ampere-Maxwell relationship relates the creation of a magnetic field to both the current flowing through a closed loop and the rate of change of the electric flux through that same closed loop.
Note 1: $\Phi_E = \int \vec{E} \cdot d\vec{a}$ is the electric flux through the closed loop
Note 2: The local form of the Ampere-Maxwell relationship may be written as $\vec{\nabla} \times \vec{B} = \mu_0 \vec{J} + \mu_0 \epsilon_0 \frac{\partial \vec{E}}{\partial t}$

216-1: Static Fields

S O FAR, WE HAVE DISCUSSED ONLY *STATIC* ELECTRIC and magnetic fields, meaning that we had fields that did not change with time. Some interesting things start to happen if we allow the fields to change with time. First, however, I want to note that this is a very important topic for modern electronics and communications. Most of the electricity grids in the western hemisphere are built on alternating current, or AC circuits. We'll go more into the why's and how's of this in the next chapter. For now, though, if you have a current that is changing, that means that you have a magnetic field that is changing...so, no longer a static field. Also, let's say we are in the process of charging or discharging a capacitor. While the charge is changing on the plates of the capacitor, the electric field between the plates is changing...so, no longer a static field.

What we're going to find is that ***changing electric fields cause magnetic fields*** and ***changing magnetic fields cause electric fields***. We introduced all of electricity and magnetism by studying electromagnetic waves – waves where the electric field oscillations and magnetic field oscillations were intimately tied together. Therefore, it shouldn't be too much of a stretch to suggest that as one changes it is related to the other. In this unit, we will start to see how the two basic EM fields relate to each other.

216-2: Faraday's law and curvy electric fields

Consider: *Why does a changing magnetic field cause an electric field?*

Physicists love symmetry. For example, we made heavy use of symmetry when using Gauss's and Ampere's laws to determine the electric and magnetic fields around various distributions. Symmetry doesn't only come about when talking about geometry either. For example, we have found that electric currents cause magnetic fields. A symmetry question would then be, can magnetic fields cause current?

In unit 209 on magnetic forces, we started off by saying that a stationary charge in a magnetic field does not feel a force. A quick review of this would suggest that no, a magnetic field does not cause a current because there already needs to *be* a current present for there to be a magnetic force. However, it turns out that the story is much more interesting.

As expected, if you take a magnet and hold it near a loop of wire, nothing happens. However, if you now take that magnet and move it towards or away from the loop, a current does form in the loop, as can be seen in figure 216-1.

Interestingly, if you then try to find the magnetic field in the center of the loop created by the current, its direction always *opposes the change* of the magnetic field in the loop, meaning that if the magnetic field through the loop is increasing (bringing the magnet closer to the loop), the *induced* magnetic field is in the opposite direction to the field (see figure 216-1). If the magnetic field in the loop is decreasing (moving the magnet away), the *induced* field tries to help out the decreasing field. Initially, this type of experiment baffled physicists, but it is also quite exciting because it showed that there is a symmetry between creating fields and currents.

Here are the overall observations found in the type of experiment described

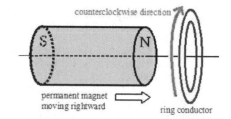

Figure 216-1. A magnet moving towards a ring of conductor causes a current to be induced in the ring.

- An induced field is only found when there is relative motion between the loop and magnetic, *or* if the loop and magnet are stationary and the area of the loop changes
- Faster motion or changes in area create larger currents
- The current direction depends on which magnetic pole is facing towards the loop and the motion of the loop.

The first bullet-point above suggests that it is the ***magnetic flux*** that is the most important quantity. If you remember from our discussion of Gauss's law, the flux of a field is a measure of how much of a field goes perpendicularly through a given area:

$$\Phi_B = \int \vec{B} \cdot d\vec{a}. \tag{216-1}$$

The SI unit of magnetic flux is the Weber (Wb), although $T \cdot m^2$ and $V \cdot s$ are also acceptable, equivalent units.

The second bullet point suggests that it is the rate of change of this flux that is important, and the third bullet-point really just tells us we have to remember that the magnetic field is a vector. Putting this all together, we can write

$$\varepsilon = -\frac{d\Phi_B}{dt}. \tag{216-2}$$

This equation is the equivalent of saying that there is an induced EMF in an enclosed loop through which there is a changing magnetic flux. This is **Faraday's Law**. The minus sign tells us that the magnetic field created by the induced current always opposes the change in the magnetic flux. This is probably the most famous minus sign in all of physics, and has its own name: **Lenz's Law**.

We learned in unit 213 that an EMF in a wire creates an electric field that then creates an electric current. We can make this connection directly in Faraday's law by inserting the relationship between EMF and electric field

$$\varepsilon = \oint \vec{E} \cdot d\vec{l} = -\frac{d\Phi_B}{dt}. \tag{216-3}$$

Faraday's Law

$$\mathcal{E}_{ind} = \oint \vec{E}_{ind} \cdot d\vec{l} = -\frac{d\Phi_B}{dt} \tag{216-3}$$

Description – Faraday's Law relates a changing magnetic flux, Φ_B, to the EMF and therefore electric field, it induces in a closed conducting loop.

Note 1: $\Phi_B = \int \vec{B} \cdot d\vec{a}$ is the magnetic flux, analogous to electric flux but for magnetic fields.

Note 2: Faraday's Law may be written in local form as

$$\vec{\nabla} \times \vec{E} = -\frac{\partial \vec{B}}{\partial t} \quad \rightarrow \quad curl(\vec{E}) = -\frac{\partial \vec{B}}{\partial t}$$

There are a couple of important practical points:

1) Since the magnetic flux is made up of magnetic field, area and angle, changes in any or all of them will cause a change in flux. In most situations we will deal with, you can remove the integral in the flux equation.

2) Remember that the electric field in a wire is constant. If you are trying to find the induced electric field, you can often remove the integral on the field.

3) Faraday's law is defined above for one turn (wire going once around). If the loop has multiple turns (N turns) of wire, each turn will contribute to the flux, i.e.

$$\varepsilon = -N\frac{d\Phi_B}{dt} \tag{216-4}$$

Figure 216-2. Right-hand rule for a current loop.

Lenz's Law

Lenz's law requires a bit more explanation from a practical perspective. What do we mean by the induced magnetic *field opposes the change in magnetic flux*? Let's start off by remembering the right-hand rule for the magnetic field direction in the center of a loop. If you curl your fingers around in the direction of the current around the loop, your thumb points in the direction of the magnetic field through the loop, which is the same direction as the magnetic dipole moment. This can be seen in figure 216-2.

Put simply, if that magnetic flux through the loop is decreasing, the induced current is such that it is in the same direction as the external magnetic field. If the magnetic flux is increasing, the induced magnetic field is in the opposite direction to the external field.

Curly fields

The nature of electric fields created by changing magnetic fields is very different from that of a point charge. The electric field created by a point charge is called a **coulomb field** and shows the divergence that we saw when discussing Gauss's law in unit 210. In contrast, the electric

Figure 216-3. The induced curly electric field around a solenoid with increasing current.

field created by changing magnetic fields does not resemble the coulomb field at all, and in fact looks much more like the magnetic fields we found for static currents – that is a *curly electric field*.

As a quick example, consider a solenoid. Using Ampere's law, we found that the magnetic field outside a very long solenoid is zero. However, if the current is changing in the wire of the solenoid, there is an induced electric field outside the solenoid that circles around it. Consider figure 216-3 which shows the curly electric field outside a long solenoid with increasing current. Notice that the magnetic field inside the solenoid is to the right, and the circulation of the induced electric field is in the opposite direction (opposing the change).

The curly nature of the induced electric field is also why it is so efficient at inducing currents around closed loops. The curvy nature of the electric field and the loop nature of the wires complement each other perfectly!

216-3: Faraday's law examples

Consider: *So, how do I actually do Faraday's law problems?*

The following are three *classic* examples of Faraday's law in introductory physics. You should study these extensively.

Example 216-1: Changing B-field

A circular wire loop with radius 2.5 cm sits in the x-y plane and is situated in a magnetic field whose strength changes with time given by $\vec{B} = 0.052t^2\ T\hat{z}$. What is the induced EMF in the loop due to this changing magnetic field?

Solution:

This is a relatively straightforward application of Faraday's law:

$$\mathcal{E}_{ind} = -\frac{d\Phi_B}{dt}$$

Key words that should make you think of Faraday's law are changing magnetic field and induced electric field.

The first step is to find the magnetic flux through the loop

$$\Phi_B = \int \vec{B} \cdot d\vec{a} = BA\cos\theta,$$

where I could eliminate the integral because although the magnetic field is changing in time, it is not changing through the area and A and θ are also constants.

Since the loop is situated in the x-y plane, the area vector (which is perpendicular to the surface) must be along the z-axis. This means that the area vector and the magnetic field vector are parallel to each other, so $\theta = 0$. Therefore, we can find the magnetic flux:

$$\Phi_B = (0.052t^2\ T)\pi(0.025m)^2\cos 0,$$

which simplifies to $\Phi_B = 1.02x10^{-4}t^2\ Tm^2$.

We can now substitute this flux into Faraday's law to find

$$\mathcal{E}_{ind} = -\frac{d\Phi_B}{dt} = -\frac{d}{dt}[(1.02x10^{-4}\ Wb)t^2],$$

which reduces to

$$\mathcal{E}_{ind} = -2.04x10^{-4}t\ V.$$

The induced EMF just found acts as a time-dependent battery in the loop. If we knew the resistance of the wire, we could even calculate the current!

Extension: What is the direction of the current in the loop in the example above?

Lenz's law tells us that the direction of the induced EMF will be to oppose the change of the magnetic field. In example 216-1, the magnetic field is pointed upward (+z-direction) and increases. Therefore, the induced magnetic field should point downwards. Using the right hand rule for current loops, a magnetic field downwards corresponds to a *clockwise* current if looking from above. Note that it is important that we say we are looking down on the loop as well.

Now, if the magnetic field is upwards and *decreasing*, the induced field would need to be in the same direction as the original field (so that the induced field opposes the decrease in the original field). In this case, the magnitude of the induced EMF and current would be the same, but they would be in the opposite direction, i.e. the current would flow counterclockwise in the loop.

Example 216-2: Motional EMF

A metal bar of length $l = 0.67\,m$ slides along a stationary u-shaped metal as shown to the right. Assuming that the magnetic field in the region of the setup has a magnitude of 0.27 T, and the bar moves with a speed of 1.8 m/s, what is the EMF induced in the bar.

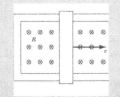

Solution:

This problem is direct application of Faraday's law. Key terms that should lead you to this determination are magnetic field and induced EMF.

Starting with Faraday's law,

$$\mathcal{E} = -\frac{d\Phi_B}{dt},$$

we must first find the magnetic flux through the area enclosed by the bar and the u-shaped metal. The magnetic flux is given by

$$\Phi_B = \int \vec{B} \cdot d\vec{a} = BA\cos\theta,$$

Where I could remove the integral because the magnetic field and angle (dot-product) are constant, which leaves

$$\int \vec{B} \cdot d\vec{a} = B\cos\theta \int da = BA\cos\theta,$$

The area enclosed within the metal bar and rod is a rectangle and is therefore given by

$$A = lw = lvt,$$

Where l is the length of the setup and w is the width of the rectangle which changes with time as the bar moves ($w = vt$). Putting this all together in Faraday's law yields

$$\mathcal{E} = -\frac{d\Phi_B}{dt} = -\frac{d}{dt}Blvt = -Blv\frac{d}{dt}t = -Blv.$$

All that is left is to substitute in our known values:

$$\mathcal{E} = -(0.27\,T)(0.67\,m)(1.8\,m/s) = 0.33\,V.$$

Example 216-3: Electric generator (rotating loop)

A circular loop of wire with radius 1.5 m is immersed in a magnetic field with magnetic field strength 0.090 T. Initially, the wire loop is oriented so that the area vector of the loop is parallel to the magnetic field (i.e., the loop is perpendicular to the field). If the loop is now rotated end over end with a frequency of 60 Hz, what is the EMF induced in the loop?

Solution:

This problem is a direct application of Faraday's Law. Key words that help lead to this decision are magnetic field and induced current.

Starting with Faraday's law,

$$\mathcal{E} = -\frac{d\Phi_B}{dt},$$

we must first find the magnetic flux through the area enclosed by the bar and the u-shaped metal. The magnetic flux is given by

$$\Phi_B = \int \vec{B} \cdot d\vec{a} = BA\cos\theta,$$

where the integral can be removed because the variables do not change over the area (although they may change in time).

The angle, θ, between B and A is time dependent while the loop is spinning. Since we are given a frequency of 60 Hz and we know that units inside of the sine function must be radians, we can first convert the frequency to angular frequency and then multiply by time to get the angle:

$$\theta = \omega t = 2\pi f t = 2\pi(60\,Hz)t = 120\pi t.$$

We can now apply this to Faraday's law:

$$\mathcal{E} = -\frac{d\Phi_B}{dt} = -\frac{d}{dt}BA\cos(\omega t) = BA\omega\sin(\omega t).$$

Finally, substituting in our known values:

$$\mathcal{E} = (0.090\,T)\pi(1.5\,m^2)(120\pi)\sin(120\pi t),$$

which gives
$$\mathcal{E} = 240\,V\sin(120\pi t).$$

This is the setup for a basic *electric generator*, and the values in the problem give a standard 240 Volt, 60 Hz supply to a U.S. Household.

216-3: Inductors

Consider: *How can Faraday's law be used in circuits?*

Connection – Traffic Lights

Loops of wire placed in the pavement of roads are often used as sensors for traffic lights. When your car drives over the loop, the metal causes a change in the inductance of the loop. A small circuit senses this difference in inductance and sends a signal to the traffic light causing to change the traffic pattern.

An interesting effect happens if you place a loop of wire in a circuit. If the current in the circuit changes, Faraday's law kicks in for the loop of wire and creates a **back-EMF** that opposes the current change. The mechanism seems like circular reasoning – the current in the loop causes a magnetic field, as the current itself tries to change, faraday's law produces an EMF that opposes the change in current. However, this is a real

Figure 216-4. Circuit symbol for an inductor.

effect and is called **self-inductance**. Today, circuit designers routinely exploit this effect and the circuit elements that work off of faraday's law are called **inductors**. Figure 216-4 shows the circuit symbol for an inductor.

The inductance of a coil is defined as

$$L \equiv \frac{N\Phi_B}{I}, \tag{216-5}$$

Connection – Metal Detectors

Standard metal detectors in airports work the same way as the traffic light detectors described to the right - any metal passing through the metal detectors is noted as the inductance of the loop changes. This process will only detect metal, however, which is why more sophisticated technology is often used these days.

where L is the inductance, N is the number of turns of the coil, Φ_B is the magnetic flux through the inductor, and I is the current through the inductor. The inductance of a coil is similar to the capacitance of a capacitor meaning that although the definition has current in it, the overall inductance is only dependent on the geometry of the inductor. This is possible because the magnetic flux will be dependent on the current and therefore the current will cancel in the end.

In practice, the inductance of an inductor in a circuit is given on the inductor and is therefore a known numerical value

Example 216-5: Inductance of a solenoid

Find the self-inductance of a solenoid with radius, r, and length, l, with N turns when the solenoid carries a current, I.

$$\Phi_B = BA\cos\theta = \left(\frac{\mu_0 NI}{l}\right)\pi r^2 \cos 0,$$

Solution:

This is a direct application of the self-induction equation (because it tells us). Starting with the equation for self-inductance:

$$L \equiv \frac{N\Phi_B}{I},$$

You can see that we must first find the magnetic flux through one loop of the solenoid.

Where I have used the magnetic field inside a solenoid from Unit 208 and the fact that the magnetic field inside the solenoid is parallel to the axis ($\theta = 0$).

This can now be used in the equation for self-inductance:

$$L \equiv \frac{N\Phi_B}{I} = \frac{N(\mu_0 NI)\pi r^2}{Il} = \frac{\mu_0 N^2 \pi r^2}{l}.$$

Although this result is in algebraic form, it could now be used for any solenoid.

The EMF produced in an inductor can be rewritten in terms of the inductance from Faraday's law

$$\mathcal{E}_L = -N\frac{d\Phi_B}{dt} = -\frac{d}{dt}(N\Phi_B). \tag{216-6}$$

Using the definition of inductance above, this can be rewritten

$$\mathcal{E}_L = -\frac{d}{dt}(LI) = -L\frac{dI}{dt}. \tag{216-7}$$

Self Inductance

$$\varepsilon_{ind} = -L\frac{dI}{dt} \qquad (216\text{-}7)$$

Description – This equation relates the back voltage created in an inductor with self-inductance, L, with a current changing at a rate given by dI/dt.

Note 1: The SI units of self-inductance is the Henry $(1\ H\ =\ \Omega \cdot s)$

Again, the potential difference across an inductor will be most useful when analyzing alternating current circuits in the next unit; however, this will give you the induced EMF in the loop due to a changing current.

Example 216-6: EMF induced in a solenoid.

The current through a solenoid varies with time given by

$$I = 12\ A \cos (120\pi t).$$

What is the EMF induced in the solenoid at time, t = 6.8 x 10^{-6} s, if the solenoid has a 175 turns, a radius of 2.45 cm and a length of 56 cm?

Solution:

In reality, this is a Faraday's law problem, because the magnetic flux through the solenoid is changing with time. However, we just found a way to more easily find the induced EMF if we know the geometry (self-inductance) and how the current changes.

Start with our equation for EMF with self-inductance:

$$\varepsilon_{ind} = -L\frac{dI}{dt}.$$

The self-inductance of this solenoid and the rate of change of current are given by

$$L = \frac{\mu_0 N^2 \pi r^2}{l} \approx \frac{\mu_0 (175)^2 \pi (0.0245\ m)}{(0.56\ m)} = 4.12 x 10^{-5} H,$$

$$\frac{dI}{dt} = \frac{d}{dt}(12\ A \cos 120\pi t) = 12(120\pi)A \sin(120\pi t).$$

The derivative of the current at 2.3 s is therefore

$$\frac{dI}{dt} = 12(120\pi)A \sin(120\pi\ 5 = 6.8 x 10^{-6}\ s) = -116\frac{A}{s}.$$

These can then be combined to find the induced EMF:

$$\varepsilon_{ind} = -L\frac{dI}{dt} = -(4.12 x 10^{-5} H)\left(-116\frac{A}{s}\right) = 0.0048\ V.$$

Again, this could also be done directly with Faraday's law; however, the current method is easier with the quantities given in the problem.

Just capacitors store energy in the electric field contained between capacitor plates, inductors store energy in the magnetic field created by the current of the inductor. For an inductor with inductance L, carrying a current, i, the energy stored in the magnetic field of the inductor is given by

$$U_L = \frac{1}{2}Li^2. \qquad (216\text{-}8)$$

216-4: The Ampere-Maxwell law and induced magnetic fields

Consider: *How does a changing electric field cause a magnetic field?*

Until now, the entire unit has been about induced electric fields due to a changing magnetic flux. As noted in the introduction, physicists love symmetry, and it turns out there is symmetry when inducing fields, meaning that in addition to electric fields being induced by changing magnetic flux, magnetic fields are also produced by changing electric fields.

Consider the capacitor connected to two wires shown in figure 216-5. If there is no current in the wires, then the capacitor is not charging or discharging and there is an electric field between the plates of the capacitor, but there is no magnetic field around either the wires or in the capacitor.

Now, if the capacitor is charging or discharging, there is a magnetic field around the wires that can be found using Ampere's law as in unit 210. However, experimentally, it was found that there is also a magnetic field outside the region of the capacitor where there is no current, but in this case, there is a changing electric field; the magnetic field at that point is the exact same as that caused by the current carrying wire that is charging or discharging the capacitor! This is called the **displacement current (I_D)**.

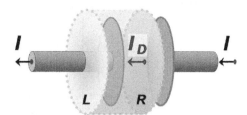

Figure 216-5. Induced magnetic field due to a changing electric field in a capacitor.

What's more, if you try to measure the magnetic field inside the capacitor while charging or discharging, the magnetic field is the same as that given inside a current carrying wire as seen in example 213-5. It is as if the current simply expanded and acted like a thick current carrying wire, except that there was no current there, just a changing electric field! This also means that what is important is not the total electric field but rather the electric flux out to the region in which we care about.

This suggests that we can simply add to Ampere's law a term that accounts for changing electric flux creating magnetic fields. That is just what James Clerk Maxwell did, and we wind up with the Ampere-Maxwell law.

Ampere-Maxwell Relationship

$$\oint \vec{B} \cdot d\vec{l} = \mu_0 I_{enc} + \mu_0 \epsilon_0 \frac{d\Phi_E}{dt} \qquad (216\text{-}9)$$

Description – The Ampere-Maxwell relationship relates the creation of a magnetic field to both the current flowing through a closed loop and the rate of change of the electric flux through that same closed loop.

Note 1: $\Phi_E = \int \vec{E} \cdot d\vec{a}$ is the electric flux through the closed loop

Note 2: The local form of the Ampere-Maxwell relationship may be written as $\vec{\nabla} \times \vec{B} = \mu_0 \vec{J} + \mu_0 \epsilon_0 \frac{\partial \vec{E}}{\partial t}$

Notice again that this equation is Ampere's law with an added term to allow changing electric flux creating magnetic fields. As one final note, the term

$$\epsilon_0 \frac{d\Phi_E}{dt} \qquad (216\text{-}10)$$

acts just like a current in a region with no current. It is called the displacement current because it is like the current charging or discharging a capacitor, but it is displaced from the real current.

We discussed earlier in the unit how the curly electric field created by a changing magnetic field differs in character from the Coulomb electric field. In contrast, the magnetic field created by a changing electric field has the same general character as that created by a current (which is why the description of displacement current works so well).

Example 216-7: AM inside capacitor.

Consider the capacitor shown in figure 216-5. At a specific time, the current used to the charge the capacitor is 0.78 A. The plates of the capacitor have a radius of 16.2 cm and are separated by 2.4 cm.

    (a) What is the magnetic field 18 cm from the center of the current carrying wire, far away from the capacitor?

    (b) In the region of the capacitor, what is the magnetic field 18 cm from the middle of the electric fields between the plates?

    (c) In the region of the capacitor, what is the magnetic field 14.1 cm from the middle of the electric fields between the plates?

Solution:

Part (a) is a direct application of ampere's law as we saw it in unit 210

$$\oint \vec{B_a} \cdot d\vec{l} = \mu_0 I_{enc} \quad \rightarrow \quad B_a = \frac{\mu_0 I}{2\pi R'}$$

which gives us

$$B_a = \frac{(4\pi x 10^{-7}\,F/m)(0.78\,A)}{2\pi(0.18\,m)} = 8.67 x 10^{-7} T.$$

(b) For the region outside the capacitor, there is no enclosed real current; however, there is a changing electric flux inside the region of interest, so

$$\oint \vec{B}_b \cdot d\vec{l} = \mu_0 \epsilon_0 \frac{d\Phi_E}{dt}.$$

This problem is done very similarly to other ampere's law problems. Specifically, the cylindrical symmetry of the capacitor plates allow us to rewrite the integral as

$$\oint \vec{B}_b \cdot d\vec{l} = B_b 2\pi r.$$

For the right side of the Ampere-Maxwell relationship, we must first remember the equation for electric field inside a parallel plate capacitor

$$E = \frac{\sigma}{\epsilon_0} = \frac{Q}{\epsilon_0 A}.$$

Therefore the flux and rate of change of flux through the area of the full capacitor is given by

$$\Phi_E = EA = \frac{Q}{\epsilon_0} \quad \rightarrow \quad \frac{d\Phi_E}{dt} = \frac{1}{\epsilon_0}\frac{d}{dt}Q = \frac{I}{\epsilon_0}.$$

These pieces can now be recombined into the Ampere-Maxwell relationship:

$$B_b 2\pi R = \mu_0 \epsilon_0 \frac{I}{\epsilon_0} \quad \rightarrow \quad B_b = \frac{\mu_0 I}{2\pi R}.$$

This is the exact same equation we had for ampere's law outside the current carrying wire, and so we can immediately write the magnetic field as $B_b = 8.67 x 10^{-7} T.$

(c) At 16.2 cm from the center of the system, we are looking for the magnetic field *inside* the capacitor. We have actually done most of the important work for this problem. We already know the electric field inside the solenoid. However, we must use the flux out to 16.2 cm and not the flux through the entire space between the capacitor plates. So

$$\Phi_E = EA_{encl} = \frac{Q}{\epsilon_0 A_{total}} A_{encl} = \frac{Q}{\epsilon_0}\frac{\pi r^2}{\pi R^2},$$

Where r is the radius we want to find the magnetic field (14.1 cm) and R is the radius of the plates (16.2 cm). Taking the derivative to the flux (similar to part b) gives

$$\frac{d\Phi_E}{dt} = \frac{d}{dt}\frac{Q}{\epsilon_0}\frac{\pi r^2}{\pi R^2} = \frac{I}{\epsilon_0}\frac{\pi r^2}{\pi R^2}.$$

We now have all the pieces to find the magnetic field from the Ampere-Maxwell relationship

$$B_c 2\pi r = \mu_0 \epsilon_0 \frac{I}{\epsilon_0}\frac{\pi r^2}{\pi R^2} \quad \rightarrow \quad B_c = \frac{\mu_0 I}{2\pi R}\frac{r}{R}.$$

Therefore

$$B_c = \frac{(4\pi x 10^{-7}\,N/A^2)(0.78\,A)(0.141\,m)}{2\pi(0.162\,m)^2},$$

or

$$B_c = 8.38 x 10^{-7}.$$

You can see that the magnetic field somewhere inside the capacitor is reduced by a factor of r/R. Just as with Ampere's law, using the Ampere-Maxwell relationship can be reduced to algebra if there is enough symmetry.

157

217 – AC Circuits

In Units 213 – 215 we studied how batteries, wires and resistors can be used to create direct current circuits. By allowing capacitors and inductors to be included in circuits, we find that time varying voltages and currents can be achieved. In this unit, we study some of the more common time-varying circuits and discuss how electricity is generated and transmitted.

Integration of Ideas

Review current and basic circuits from units 213 and 214, respectively.
Review Kirchhoff's Laws from Unit 215.

The Bare Essentials

- Complex circuits containing capacitors and inductors are solved using Kirchhoff's Voltage Law.

- Series circuits containing capacitors and inductors create oscillating currents with a characteristic time, analogous to simple harmonic motion.

Series LC Circuit

$$i(t) = \omega Q_0 sin(\omega t + \phi)$$

Description – This equation describes how a series LC circuit creates an oscillating current with characteristic frequency $\omega = \frac{1}{\sqrt{LC}}$. Q_0 is the initial charge on the capacitor.
Note: The frequency of oscillation is given by $f = \frac{\omega}{2\pi}$.

- Series circuits containing resistors, capacitors and inductors undergo damped oscillation with a characteristic frequency, analogous to damped harmonic motion.

Series RLC Circuit

$$q(t) = Q_0 e^{\frac{-Rt}{2L}} cos(\omega' t + \phi)$$

$$\text{where } \omega' = \sqrt{\frac{1}{LC} - \left(\frac{R}{2L}\right)^2}$$

Description – This equation describes how a series RLC circuit exhibits damped harmonic oscillation assuming an initial capacitor charge of Q_0.
Note: The frequency of oscillation is given by $f' = \frac{\omega'}{2\pi}$.

- AC electricity (except for photovoltaics) is generated through the use of Faraday's Law and some method for turning a turbine.

- Transformers are used to isolate sections of the power grid as well as to 'step-up' or 'step-down' the voltage to areas of the grid.

Transformer Voltage

$$V_s = V_p \frac{N_s}{N_p}$$

Description – This equation describes how the voltage in the secondary coil of a transformer (V_s) is related to the voltage in the primary coil (V_p) and the number of turns in the secondary (N_s) and primary (N_p) coils.
Note 1: If $N_s > N_p$ the voltage on the secondary side is higher than the primary side. This is called a ***step-up transformer***.
Note 2: If $N_s < N_p$ the voltage on the secondary side is lower than the primary side. This is called a ***step-down transformer***.

217-1: Extending Kirchhoff's laws

IN UNITS 214 AND 215 WE INTRODUCED and made great use of Kirchhoff's laws for both resistive circuits and RC circuits. I mentioned in those units that Kirchhoff's laws are quite general and can be used for any circuit design since they are manifestations of the conservation of charge (KCL) and the conservation of energy (KVL). In this unit, we are going to extend the use of Kirchhoff's laws to include not only the elements we've used before (resistors and capacitors), but also inductors and alternating-current sources. Where a battery is a source of constant voltage for a given circuit, and alternating-current, or AC source provides either a voltage or current that varies sinusoidally at some specific frequency.

Since KVL is all about voltage gains and drops around a closed loop, I want to first review the voltage drops across the three elements we've discussed so far

$$\text{Resistor:} \qquad \Delta V_R = IR,$$

$$\text{Capacitor:} \qquad \Delta V_C = \frac{q}{C}, \qquad (217\text{-}1)$$

$$\text{Inductor:} \qquad \Delta V_L = L\frac{dI}{dt}.$$

Using these known voltage drops, we can quickly analyze circuits that contain inductors as well as capacitors and resistors.

217-2: LC circuit – electric oscillations

Consider: *How can Kirchhoff's Laws be used for capacitors and inductors?*

Figure 217-1 shows a circuit composed of a capacitor and inductor. Assuming the capacitor is initially charged before the switch is closed, we can use KVL to write an equation governing this circuit when the switch is closed:

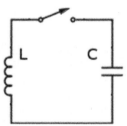

$$\Delta V_L + \Delta V_C = 0 \quad \rightarrow \quad -L\frac{dI}{dt} - \frac{q}{C} = 0, \qquad (217\text{-}2)$$

We know that, by definition, the current, I, is the rate at which the charge, q, moves, so we can rewrite this equation as

$$-L\frac{d^2q}{dt^2} - \frac{q}{C} = 0. \qquad (217\text{-}3)$$

Figure 217-1. Diagram of an LC circuit.

This is a second-order linear partial differential equation. There are two ways for us to approach the solution. First, we could go back to general principles and think about a function where the second derivative has the same form as the original function with extra constants and work from there. Fortunately, we've seen this before and so the easier way to solve equation 217-1 is to compare it with the equation for simple harmonic motion we saw in Physics I. If you remember back to the unit of oscillations of a mass, m, on a spring with spring constant, k, Newton's 2nd Law gave us

$$m\frac{d^2x}{dt^2} + kx = 0 \quad \rightarrow \quad \frac{d^2x}{dt^2} = -\frac{k}{m}x \quad \rightarrow \quad x = A\cos(\omega t + \phi), \qquad (217\text{-}4)$$

where $\omega = \sqrt{k/m}$ is the angular frequency of oscillation and ϕ is the phase angle (which allows us to start the oscillation at a place where $x \neq 0$).

We can solve equation 217-2 by analogy equation 217-3 to find

$$q(t) = -q_0\cos(\omega t + \phi) \quad \text{with} \quad \omega = \frac{1}{\sqrt{LC}}, \qquad (217\text{-}5)$$

where q_0 is the maximum charge on the capacitor. Note: the negative in this equation is arbitrary and is there to make the current positive below. The current in the system as a function of time can then be found by taking the derivative of equation 217-4 with respect to time:

$$i(t) = \omega q_0 \sin(\omega t + \phi). \qquad (217\text{-}6)$$

Therefore, a circuit consisting of an inductor and capacitor acts as an *electric oscillator*. Such oscillators are found in amplifiers, filters, tuners and mixers for their ability to oscillate at a specific frequency.

Series LC Circuit

$$i(t) = \omega q_0 sin(\omega t + \phi) \qquad (217\text{-}6)$$

Description – This equation describes how a series LC circuit creates an oscillating current with characteristic frequency $\omega = \frac{1}{\sqrt{LC}}$. q_0 is the initial charge on the capacitor.

Note: The frequency of oscillation is given by $f = \frac{\omega}{2\pi}$.

When the circuit starts off with a completely charged capacitor and no current, we would expect the total energy in the system to be

$$U = \frac{q_0^2}{2C}. \qquad (217\text{-}7)$$

Once the oscillations begin, at any time the energy stored in the circuit would be a combination of the energy stored in the electric field of the capacitor and in the magnetic field of the inductor, given by

$$U = U_C + U_L \quad \rightarrow \quad U = \frac{q(t)^2}{2C} + \frac{1}{2}Li(t)^2. \qquad (217\text{-}8)$$

Now, in order to solve this, we have to plug in what we know about $q(t)$ and $i(t)$:

$$U = \frac{q_0^2}{2C}\cos^2(\omega t + \phi) + \frac{1}{2}L\omega^2 q_0^2 sin^2(\omega t + \phi). \qquad (217\text{-}9)$$

However, if we now plug in what we know about ω, we get

$$U = \frac{q_0^2}{2C}\cos^2(\omega t + \phi) + \frac{q_0^2}{2C}sin^2(\omega t + \phi) \quad \rightarrow \quad U = \frac{q_0^2}{2C}. \qquad (217\text{-}10)$$

Therefore, conservation of energy holds in the LC circuit – no energy is lost to the environment.

Example 217-1: LC circuit

An LC circuit is constructed of a $3.20\ \mu H$ inductor and a $1.75\ nF$ capacitor. The maximum charge on the capacitor is $0.252\ \mu C$, and the capacitor is fully charged at time $t = 0$. (a) Write down the equation for current in the circuit as a function of time. (b) What is the total energy of the circuit? (c) What is the current in the circuit when half the energy is in the capacitor?

Solution:

(a) In order to find the equation for the circuit, we must first find the angular frequency:

$$\omega = \frac{1}{\sqrt{LC}} = \frac{1}{\sqrt{(3.2 \times 10^{-6}H)(1.75 \times 10^{-9}F)}} = 1.33 \times 10^7 \frac{rad}{s}.$$

The equation is then given by

$$i(t) = \omega q_0 sin(\omega t + \phi),$$

or

$$i(t) = (3.35\ A)\sin(1.33 \times 10^7 t).$$

(b) The total energy is given by

$$U = \frac{q_0^2}{2C} = 18.1\ \mu J.$$

(c) When half the energy is in the capacitor, the other half is in the current:

$$\frac{18.1\ \mu J}{2} = \frac{1}{2}Li^2 \quad \rightarrow \quad i = 2.37\ A.$$

217-2: RL circuit

> **Consider:** *How can Kirchhoff's Laws be applied to a circuit with an inductor and a resistor?*

Figure 217-2 shows a diagram of a circuit containing a resistor and an inductor. In the LC circuit of the last section, a charged capacitor acted as the initial voltage source for the circuit; however, an RL circuit needs an external source to get current going, which is why a battery is also present. When the switch is in the up position, the resistor and inductor are connected to the battery in series (charging). When the switch is in the down position, the inductor and resistor are only connected to each other (discharging).

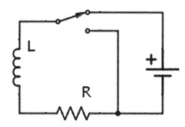

Figure 217-2. Diagram of an RL circuit.

Let's assume that the circuit has been connected to the battery for a long time, so that there is a current flowing through the circuit. If the switch is then thrown so that the resistor is connected to just the inductor, we can now use KVL to analyze the situation:

$$\Delta V_L + \Delta V_R = 0 \quad \rightarrow \quad -L\frac{di}{dt} - iR = 0. \tag{217-11}$$

Since both terms in our KVL equation are in terms of current, we can directly rearrange and integrate (similar to the RC circuit in unit 215)

$$\frac{di}{i} = -\frac{R}{L}dt \quad \rightarrow \quad i(t) = i_0 e^{-(R/L)t} \tag{217-12}$$

Note that this represents an exponential decay of the current, just as we had an exponential decay for the charge on a capacitor in an RC circuit. The equation for the increase in current when the circuit starts at $i = 0$ and is connected to the battery can now be written by analogy to what we did for charging a capacitor:

$$i(t) = \frac{\varepsilon}{R}(1 - e^{-(R/L)t}). \tag{217-13}$$

217-3: RLC circuit – damped oscillations

> **Consider:** *How can Kirchhoff's Laws be applied to a circuit with all three elements (inductor, capacitor and resistor)?*

The next step in our analysis of circuits is to include all three of the elements we've been discussing – a resistor, inductor and capacitor all in one circuit. Such a circuit can be seen in figure 217-3. Just as before, we can use KVL assuming that the capacitor is initially charged to find

$$\Delta V_L + \Delta V_C + \Delta V_R = 0 \quad \rightarrow \quad -L\frac{dI}{dt} - \frac{q}{C} - IR = 0. \tag{217-14}$$

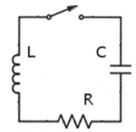

Figure 217-3. Diagram of an RLC circuit.

Just as with the LC circuit, the next step is to write this entirely in terms of charge using the fact that current is the time derivative of charge:

$$-L\frac{d^2q}{dt^2} - R\frac{dq}{dt} - \frac{1}{C}q = 0. \tag{217-15}$$

The solution to this equation is mathematically harder to obtain than that for the LC circuit. However, we can use a qualitative argument to get close: we can think of an RLC circuit as the combination of an LC circuit (oscillations) and an RL circuit (exponential decay). This is exactly what happens, and the solution is given by

$$q(t) = q_0 e^{-\frac{R}{2L}t}\cos(\omega't + \phi), \tag{217-16}$$

where q_0 is the initial charge on the capacitor and ω' is the angular frequency at which the circuit would like to oscillate:

$$\omega' = \sqrt{\frac{1}{LC} - \left(\frac{R}{2L}\right)^2}. \qquad (217\text{-}17)$$

For completeness, I do want to point out that this solution is only valid if $R < 2\sqrt{L/C}$. If the resistance is larger than this value, the circuit is known as ***overdamped*** and never oscillates – it only undergoes exponential decay. When $R = 2\sqrt{L/C}$ the system is ***critically damped*** and will also never oscillate and exponentially decay even faster.

Series RLC Circuit

$$q(t) = Q_0 e^{\frac{-R}{2L}t} cos(\omega' t + \phi) \qquad (217\text{-}16)$$

$$\text{where } \omega' = \sqrt{\frac{1}{LC} - \left(\frac{R}{2L}\right)^2} \qquad (217\text{-}17)$$

Description – This equation describes how a series RLC circuit exhibits damped harmonic oscillation assuming an initial capacitor charge of Q_0.

Note 1: The frequency of oscillation is given by $f' = \frac{\omega'}{2\pi}$.

Note 2: The solution is only valid if $R < 2\sqrt{L/C}$.

Example 217-2: RLC circuit

A series RLC circuit is constructed with a resistor of resistance 25.7 Ω, and inductor of inductance 75.2 μH, and a capacitor with capacitance 22.7 nF. You may assume Q_0 is 2.54 μC.

  (a) Will this RLC circuit maintain an oscillating current?
  (b) If so, what is the frequency of oscillation?
  (c) What is the charge on the capacitor at time $t = 2 \mu s$?
  (d) What is the current in the circuit at time $t = 2 \mu s$?

Solution:

(a) The circuit will only oscillate if $R < 2\sqrt{L/C}$, so

$$2\sqrt{\frac{L}{C}} = 2\sqrt{\frac{75.2 x 10^{-6} H}{22.7 x 10^{-9} F}} = 115 \ \Omega.$$

Since R (25.7 Ω) is less than this value, the circuit *will* oscillate.

(b) The angular frequency of oscillation is given by

$$\omega' = \sqrt{\frac{1}{LC} - \left(\frac{R}{2L}\right)^2} = \sqrt{\frac{1}{(75.2 \ \mu H)(22.7 \ nF)} - \left(\frac{25.7 \ \Omega}{2(75.2 \ \mu H)}\right)^2},$$

or

$$\omega' = 7.46 x 10^5 \frac{rad}{s}.$$

So, the frequency of oscillation is

$$f' = \frac{\omega'}{2\pi} = 1.19 x 10^5 Hz.$$

(c) The maximum charge on the capacitor at a given time is found by

$$Q_0 e^{\frac{-R}{2L}t} = (2.54 x 10^{-6} C) exp\left[\frac{-25.7 \ \Omega(2 x 10^{-6} s)}{2(75.2 \ \mu H)}\right],$$

which gives us

$$Q_0 e^{\frac{-R}{2L}t} = 1.80 \ \mu C.$$

(d) The current in the RLC circuit as a function of time is found by taking the derivative of the equation for charge (note that we must use the product rule here!):

$$i(t) = \frac{dq(t)}{dt} = -Q_0 e^{\frac{-R}{2L}t}\omega' sin(\omega' t + \phi)$$
$$- Q_0 \frac{R}{2L} e^{\frac{-R}{2L}t} cos(\omega' t + \phi).$$

From part (b), we know that

$$\omega' = 7.79 x 10^5 \frac{rad}{s},$$

and from part (c) we know

$$Q_0 e^{\frac{-R}{2L}t} = 1.80 \ \mu C.$$

Therefore, we can write the equation for current as

$$i(t) = (1.80 \ \mu C)\left(7.79x10^5 \frac{rad}{s}\right) \sin\left(7.79x10^5 \frac{rad}{s}t\right)$$
$$- (1.80 \ \mu C)\frac{25.7 \ \Omega}{2(75.2x10^{-6}H)} \cos\left(7.79x10^5 \frac{rad}{s}t\right),$$

Which at $t = 2 \ \mu s$ is

.30758

$$i(2 \ \mu s) = 1.4 \ A.$$

217-4: Alternating current and the electric transmission.

Consider: *How is electricity produced in the U.S.?*

Alternating current (ac) systems have the advantage of creation via electric generators. As we saw in unit 216, when loops of wire rotate in a magnetic field a potential difference is created in the wire, which can then drive a current. This is Faraday's law again. Many modern power plants are designed around this idea, with hydroelectric power possibly being the easiest to understand. A turbine is placed into flowing water, which causes the turbine the turn. Strong permanent magnets are attached to the turbine so that the rotational motion is converted to electric energy. This process is depicted in figure 217-4.

Many other types of power generators use this same principle. For example, a coal power plant uses thermal energy from the burning of coal to convert water to steam and the steam turns a turbine as it rises.

As the coils in the generator rotate, the potential difference induced in the coil takes on a sinusoidal form

$$\mathcal{E} = \mathcal{E}_0 \sin \omega t, \qquad (217\text{-}18)$$

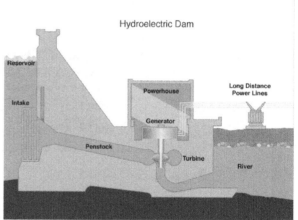

Figure 217-4. A hydroelectric power plant converting moving water to electric energy.

where $\omega = 2\pi f$ is the angular frequency of rotation of the coils. If the coils are part of a closed conducting circuit, this potential difference will drive a current in the circuit given by

$$i = i_0 \sin(\omega t - \phi). \qquad (217\text{-}19)$$

In this equation, i_0 represents the maximum current and a phase angle, ϕ, is included because the current and potential difference may not be in-phase with each other. (See unit 214 for more detail on why the current and potential may be out of phase).

Again, this type of current is called alternating, or ac, current. Much of the modern power grid in the western hemisphere is built on ac circuits because not only is the production of this type of electricity governed by Faraday's law, but also that Faraday's law can be used to deliver electricity at different places using *transformers* (see next section). In North America, the frequency of ac transmission is set at 60 Hz, while in Europe and much of Africa the frequency is standardized at 50 Hz.

So, how much do the electrons actually move in an ac current? In unit 213, we saw that the drift speed of electrons is typically on the order of 10^{-5} m/s. At 60 Hz, the electrons are changing directions 120 times in one second (since they change directions twice during one cycle). Under these conditions, the maximum displacement of an electron from equilibrium is about 8.3 x 10^{-8} m!! How can energy be transferred with such little electron motion?!? It turns out that it doesn't matter how far the electrons move – because if we choose one specific place on a wire and try to find the current at that point, it only matters how much charge *passes* that point, since current is charge per time. If the electrons are oscillating back and forth, even with a very small amplitude, they can easily pass that specific point 120 times a second – which is the current. *Remember that charge is not used up in a circuit; it is the conversion of electric potential energy, and not kinetic energy, to some other form (often thermal) that give electricity its usefulness!*

> **Connection:** Corona Discharge
>
> There are limits on how large the voltage can be in a high-voltage power line. Remember from unit 213 that current carrying wires create surface charges that are important to maintain the electric field inside the wire. If the surface charges get too large, they can create sparks via dielectric breakdown (remember - the same reason you get shocked touching a door handle in winter). This breakdown is called *corona discharge*, and it creates both an audible buzzing noise and radio-frequency interference. It is also a major source of loss in power lines when it occurs. You may have *heard* corona discharge as a transformer breaks down on a power line.

Using ac electricity for long distance transmission has advantages. First, very large power can be transmitted at relatively low current if the voltage is high. In fact, the reason that ac transmission lines use high voltage is that it reduces power loss. Consider a relatively standard high voltage power transmission at 735 kV. High power transmission lines have a resistance of about $0.22 \ \Omega/km$. If we assume we are going to transmit electricity over a distance of 500 km, the total resistance of the wires is 110 Ω.

Let's say we have a transmission line that needs to carry 375 MW of power (not an unreasonable power supply). The relationship between power transferred, current and voltage is given by

$$P = IV, \tag{217-20}$$

which suggests this transmission line would need to carry a current of 510 A. We can find the power loss in the transmission using

$$P = I^2R = (510 \ A)^2(110 \ \Omega) = 28 \ MW. \tag{217-21}$$

If we were to reduce the voltage in the line by half (therefore multiplying the current by two via $P = IV$), we get

$$P = I^2R = (1020 \ A)^2(110 \ \Omega) = 114 \ MW. \tag{217-22}$$

In the high voltage case, we lost 7.5% of the energy; however at the lower voltage, we lost 30%!!! This quick example shows why high voltage lines are used: we lose far less of the energy to thermal and other losses at high voltage. See the connection box in this section to see at least one upper limit on the voltage of a power line.

217-5: Transformers

Consider: *How is voltage transformed for different applications and transmission?*

Another reason that ac electricity is employed for electrical transmission is that Faraday's law can easily be used to control the voltage in different areas of the power grid. The devices that do this are called *electric transformers*. A simplified diagram of an ideal transformer is shown in figure 217-5. In the figure you see a coil of wire connected to an ac source is wrapped around an iron core. This is called the *primary coil*. On the other side of the iron core is a second wrapping of coil, called the *secondary coil*, which is connected to another circuit, called the load. As the ac current flows in the primary coil, it produces a magnetic field in each winding of the coil such that

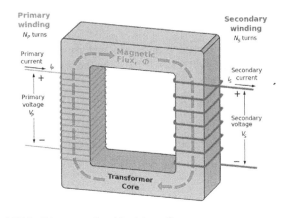

217-5. Diagram of an ideal transformer.

$$\mathcal{E}_p = -N_p \frac{d\Phi_B}{dt}, \tag{217-23}$$

where N_p is the number of turns in the primary coil. The ferromagnetic iron core will carry the magnetic field created by the primary coil through the entire core, relatively undiminished. In an *ideal transformer*, we assume that the magnetic field is carried perfectly through the entire iron core.

On the secondary side, the changing magnetic flux will induce a potential difference in each of the coils such that

$$\mathcal{E}_s = -N_s \frac{d\Phi_B}{dt}. \tag{217-24}$$

where N_s is the number of loops in the secondary coil. If the transformer is truly ideal, the magnetic flux will be the same in both the primary and secondary coils, so that we can divide our two equations and find

$$\frac{\mathcal{E}_s}{\mathcal{E}_p} = \frac{N_s}{N_p}. \tag{217-25}$$

This gives us a direct relationship between the input and output voltages in terms of the number of turns of coil on each side.

Transformer Voltage

$$V_s = V_p \frac{N_s}{N_p} \qquad (217\text{-}26)$$

Description – This equation describes how the voltage in the secondary coil of a transformer (V_s) is related to the voltage in the primary coil (V_p) and the number of turns in the secondary (N_s) and primary (N_p) coils.

Note 1: If $N_s > N_p$ the voltage on the secondary side is higher than the primary side. This is called a *step-up transformer*.

Note 2: If $N_s < N_p$ the voltage on the secondary side is lower than the primary side. This is called a *step-down transformer*.

Example 217-3: Transformer

The primary coil of an ideal transformer contains 200 turns and the secondary coil contains 115 turns. If the voltage entering the primary coil is 242 V, what is the voltage of the secondary coil?

Solution:

This is a direct application of the ideal transformer

equation:

$$V_s = V_p \frac{N_s}{N_p} = (242\ V)\frac{115}{200} = 139\ V.$$

This would be a *step-down transformer*, since the voltage in the secondary is less than the voltage in the primary.

Again, if we have an ideal transformer, conservation of energy must hold. Since the power transmitted along a power line is given by P=IV, we can write the transformer relationship in terms of currents in the primary and secondary coils as well

$$P_p = P_s \quad \rightarrow \quad I_p V_p = I_s V_s \quad \rightarrow \quad I_s = I_p \frac{V_p}{V_s}. \qquad (217\text{-}27)$$

We can now use our relationship between voltage and number of turns above to write

$$I_s = I_p \frac{N_p}{N_s}. \qquad (217\text{-}28)$$

217-6: Overview of the electric transmission grid

> **Consider:** *How does electricity get from a power plant to a home in the U.S.?*

The combination of ac currents and transformers are used to power the United States. One of the current energy drawbacks is that we do not have efficient ways to store electricity (we will explore this more in unit 226). Therefore, power plants of all types need to operate 24/7, although they can transmit different levels of power at different times due to grid demands. For example, energy demands at 3 AM when it is 55 degrees in May are generally far less than the energy demands at 3 PM on an August weekday when it is 100 degrees Fahrenheit. Regardless, energy production and transmission happens continuously.

Figure 217-6 shows a schematic of electric transmission from a generating station to users. Electricity from the power plant first goes through a set of transformers designed to step-up the voltage for high-voltage transmissions. The delivery of electricity over long distances at high voltage is known as *electric transmission*. Honestly, the U.S power grid is a mess. Long distance electric transmission is currently performed at 115, 138, 161, 230, 345, 500 or 765 kV. Each region of the

U.S. has either a Regional Transmission Organization (RTO) or Independent System Operator (ISO) which are responsible for overseeing the large-scale electric grid in that region. ISOs and RTOs came into existence in the 1990s to combat a single

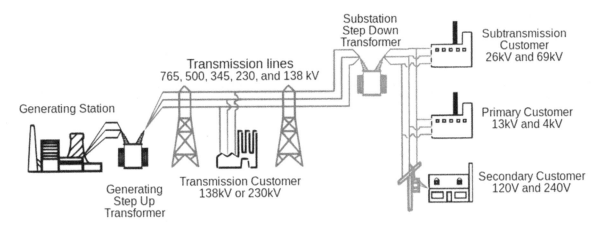

Figure 217-6. Schematic of the transmission grid in the U.S.

company owning the entire generation, transmission and distribution of electricity for a state or region. In essence, this means that the ISOs and RTOs absorbed generation and transmission lines from multiple sources. There is great concern that much of the U.S. electric grid is aging and outdated, however the details of these concerns are generally beyond the scope of this course (see the Connections box for one instance).

Regardless of what voltage electricity is transmitted long distance, once it reaches a ***distribution area***, the electricity passes through a ***substation*** where transformers are used to step-down the voltage for local distribution. Distribution that is made to residential areas then proceeds to another transformer that again steps-down the voltage to either 120 or 240 V which is then delivered to the home. Residential transformers are relatively small and can often be seen hanging from power polls. For medium-to-large scale systems, the electricity is often delivered at either 4 kV or 13 kV. At the time of writing, the Coast Guard Academy was in the process of upgrading its electric system from 4 kV to a 13 kV system.

Electric transmission in New England is overseen by ***ISO New England***. Just to be confusing, ISO New England is actually an RTO and not an ISO. Anyway, ISO New England reports that power consumption in New England ranges from around 9,000 MW (9 GW) to about 15,000 MW (15 GW) depending on the time of day and time of year.

Connection – The Great Northeast Power Failure of 2003

If any of you were in the northeast part of the U.S. on August 14, 2003, you might remember the **Great Northeast Blackout** (yes, you were young, but it was a big deal). At 1610 that day, power went out for large parts of Connecticut, Massachusetts, Rhode Island, New York, New Jersey, Maryland, Michigan, Ohio, and the **entire Canadian Province of Ontario**. In total, more than **55 million people were affected** and it was, at that time, the largest blackout in U.S. history and second only to a 1999 blackout in Brazil worldwide.

This power outage exposed a couple of failures in the U.S. transmission grid. At 1402 a 345 kV transmission line in Walton Hills, OH came into contact with trees (possibly due to a gust of wind). When the lines came into contact with the leaves, tremendous sparking due to **corona discharge** (see above box) occurred which increased the current in the line to unsafe levels. Safety equipment immediately shut down the power line as it should. Unfortunately, a software bug prevented an alarm from sounding at the power control center.

The system increased current to a number of other transmission lines to make up for the lost power. The increased current in those lines caused increased thermal energy ($P=I^2R$), which in turn caused the lines to lengthen and sag. Between 1541 and 1605, **19** transmission lines contact trees and fail as they sag below the 'right-of-way' clearance created by the power companies (this is why power companies cut trees to create large open areas around power lines).

The failure of so many lines caused the Cleveland electric grid to try and pull 2,000 MW from Michigan, which caused several power plants in Michigan to trip and pull themselves off the grid to prevent overloads.

Over the next **2 minutes**, 256 power plants pull themselves off the grid as successive failures, radiating outward, create overload situations as described above. It took up to two days to restore power to a majority of the affected area. (I personally was in graduate school in Rochester, NY at the time and vividly remember the 100 degree weather with no a/c or fans and walking up and down nine flights of pitch black stairs to my apartment for two days.)

Locally, **Millstone Nuclear Power Plant** was *the only* major power generator in Connecticut to stay online, and was, in fact, credited with stopping the migration of the blackout through eastern New England.

Here are the issues that the blackout revealed:
- The initial power line should never have been able to contact foliage. This was a failure on the part of the power company to provide adequate right-of-way clearance. (Note: This also became an issue in CT during hurricane Sandy.)
- The grid should have cut approximately 1,500 MW to Cleveland before it tried to pull electricity from Michigan, creating overload conditions. The switching mechanisms designed to limit local blackouts were overwhelmed by a relatively small number of switches, and were not able to consider global conditions, only local conditions.
- The U.S. power grid is vulnerable to terrorist attack. The lack of 'smart' grid reaction to unexpected major incidents could be exploited by those wishing to cause harm.

218 – Production and Detection of EM Radiation

Unit 218 connects our recent discussion of ac circuits with how those same circuits can be used to produce and detect em radiation. In essence, this unit is a first step towards understanding the very complex field of antenna theory. In addition, we will show how electromagnetic radiation can be derived from the fundamental relationships of electricity and magnetism.

Integration of Ideas

Review ac circuits from unit 217.

The Bare Essentials

- ac sources force current in circuits, although the phase of the current may not match the phase of the source.

- In ac circuits with sources, we strive to write the effect of each element as a *generalized Ohm's law*, $V = IX$, where X is the *reactance* of the element:

$$X_R = R$$

$$X_C = \frac{1}{\omega C}$$

$$X_L = \omega L$$

- A forced series RLC circuit has a current given by not only the voltage and resistance, but also the frequency, capacitance and inductance.

Forced Series RLC Current

$$I = \frac{V_m}{\sqrt{R^2 + \left(\omega L - \frac{1}{\omega C}\right)^2}}$$

Description – This equation defines the current in a forced series RLC circuit in terms of the source voltage, V_m, the driving frequency of the circuit, ω, and the circuit resistance, R, inductance, L, and capacitance, C.

- Resonance – the state of highest possible current - is achieved in a series RLC circuit when the denominator of the above equation is minimized:

$$\omega_r = \frac{1}{\sqrt{LC}}.$$

- Since electrons are accelerating in an ac circuit, they produce electromagnetic waves with the same frequency as the oscillating frequency of the electrons.

- Antennas are designed to efficiently radiate and detect em radiation, often using ac circuits. Common antenna designs include monopole and dipole antenna.

- The equations for em radiation can be derived from the set of Maxwell's equations studied earlier in the course: Gauss's law (electricity and magnetism), Faraday's law and the Ampere-Maxwell relationship.

218-1: ac sources and simple circuit reaction

Consider: *How does a circuit react to an ac source like those from the last unit??*

Power generation from turbines as discussed in the last unit create ac voltages. The symbol for this type of voltage source is shown in figure 218-1. We discussed in the last unit how LC and RLC circuits have certain frequencies at which they like to oscillate. What happens if that characteristic frequency does not match the frequency of the source? This will be the main question we strive to answer in this unit. It turns out that resistors, capacitors and inductors all react differently to ac sources, so we will quickly introduce each individual reaction and then put them together to investigate a driven RLC circuit. Such circuits are routinely used to produce and detect electromagnetic radiation. Let's see how this works.

Figure 218-1. Symbol used for an ac source

First, let's consider what happens when an ac source is connected to each individual element. The voltage of the ac source will be taken as

$$V_{Source} = V_0 \sin \omega t. \tag{218-1}$$

For each element, we will use KVL to determine the equation for the system and then solve for the current that develops in the circuit. Since we have used KVL many times in the last few units, I will progress rapidly through the setup so that I can report the results quickly.

In each case, we will attempt to write the current in the following form:

$$I = \frac{V}{X}, \tag{218-2}$$

where X is called the ***reactance*** of the element. We'll see later how this reactance is important when trying to combine elements in an ac circuit.

Resistor

Figure 218-2 shows an ac source connected to a resistor. KVL gives us

$$V_0 \sin \omega t - IR = 0 \quad \rightarrow \quad I_R = \frac{V_0}{R} \sin \omega t. \tag{218-3}$$

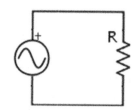

Figure 218-2. Resistive ac circuit.

As we might expect, the current that develops in this circuit looks like Ohm's law except that the current has a $\sin \omega t$, just like the voltage.

Comparing the above equation to our definition of reactance shows

$$X_R = R, \tag{218-4}$$

or that the reactance of the resistor is simply the resistance. Note that it is independent of the frequency. Graphs of the ac voltage and current are concurrently shown in figure 218-3.

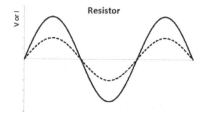

Figure 218-3. V (solid line) and I (dotted line) for a resistive ac circtuit.

Capacitor

Next, consider a capacitive ac circuit as shown in figure 218-4. Using KVL, we find

$$V_0 \sin \omega t - \frac{q}{C} = 0 \quad \rightarrow \quad q = CV_0 \sin \omega t. \tag{218-5}$$

Now, in order to find the current, we must take the derivative of the charge with respect to time

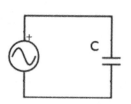

Figure 218-4. Capacitive ac circuit.

170

$$I = \omega C V_0 \cos \omega t \quad \rightarrow \quad I_C = \frac{V_0}{1/\omega C} \cos \omega t. \qquad (218\text{-}6)$$

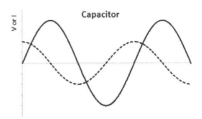

Again, comparing this with our definition of reactance above, we find

$$X_C = \frac{1}{\omega C}. \qquad (218\text{-}7)$$

Figure 218-5. V (solid line) and I (dotted line) for a capacitive ac circuit.

The other important thing to notice about the current equation above is that it has a cosine instead of a sine. This means that the current is *out of phase* with the voltage. That is, the current is a maximum when the voltage across the capacitor is zero and the current is zero when the voltage across the capacitor is maximum. This relationship is shown in figure 218-5. It is also important to note that the capacitive reactance depends inversely on angular frequency.

Inductor

Finally, consider the inductive ac circuit shown in figure 218-6. KVL gives us

$$V_0 \sin \omega t - L \frac{dI}{dt} = 0 \quad \rightarrow \quad \frac{dI}{dt} = \frac{V_0}{L} \sin \omega t. \qquad (218\text{-}8)$$

This time we need to integrate the equation in order to solve for the current through the inductor

$$I_L = -\frac{V_0}{\omega L} \cos \omega t. \qquad (218\text{-}9)$$

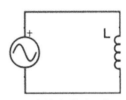

Figure 218-6. Inductive ac circuit.

Once again, we have a slightly different result. The inductive current contains a cosine, but it is also negative. Comparing again with the definition of reactance, we find

$$X_L = \omega L. \qquad (218\text{-}10)$$

The graph of V and I_L are shown in figure 218-7.

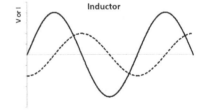

Figure 218-7. V (solid line) and I (dotted line) for an inductive ac circuit.

Summary

Each of the three basic elements we've studied in this course reacts differently when connected to an ac source. The reactance of each is summarized as

$$X_R = R,$$

$$X_C = \frac{1}{\omega C}, \qquad (218\text{-}11)$$

$$X_L = \omega L.$$

It is also important to note that we found that the phase relationship between the voltage and current was different in each case. The current in a resistive circuit is in-phase with the voltage, and the current in capacitive and inductive circuits are out of phase by 90 degrees. However, the capacitor is 90 degrees ahead of the voltage and the inductor is 90 degrees behind the voltage.

218-2: Series RLC circuit

> **Consider:** *How is an RLC circuit different if it is driven by an ac source?*

In unit 216, we found that in a series RLC circuit with an initial current but no driving source, the current oscillates at a characteristic frequency, but also decays exponentially. You might imagine that if we try to drive the circuit with an ac voltage, the energy added to the system could offset the losses of the resistor and keep the current in the circuit going as long as the source is present. This is very true; however, there is a relationship between the frequency of the source and the frequency at which the circuit wants to oscillate that determine the amplitude of the current in the circuit.

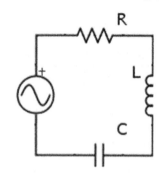

This is very similar to pushing a kid on a swing. The swing has a natural frequency of oscillation at which it will proceed if you just let the kid swing back and forth on her own. If you push the kid at the right frequency, you add energy to the system efficiently and the kid goes higher (bigger amplitude). If you try to push at the wrong times, you may be able to keep the kid going a little bit, but she will not reach the same height as if you do it at the right time. Putting this more in physics terms, if your pushes (sources of energy) are at the right frequency, the energy is transferred into the swing. If you push at incorrect times, some energy may be transferred to the swing, but not all of it.

Figure 218-8. A driven series RLC circuit

Going back to our RLC circuit, let's first apply KVL to the circuit

$$V_{source} - V_R - V_L - V_C = 0 \quad \rightarrow \quad V_{source} = V_R + V_L + V_C.$$

(218-12)

We know from our discussion of each individual element that the voltage across an inductor leads the voltage across a resistor by 90 degrees and that the voltage across a capacitor lags behind the voltage across the resistor by 90 degrees. We can visualize this with the use of a **phasor diagram**, as shown in figure 218-9. This diagram takes the periodic graph of V versus t and relates it to rotations around a circle. Moving to the right on a V versus t graph corresponds to moving counterclockwise on the phasor diagram. You can see that if we represent the voltage across the resistors as on the x-axis of the phasor diagram, then the voltage across the inductor must be on the positive y-axis because it is *ahead* of the resistor voltage. Similarly, the voltage across the capacitor is on the negative y-axis because it *lags behind* the resistor voltage.

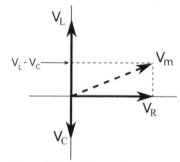

There are two important results of the phasor diagram. First, the amplitude of the net voltage can be found. Since the inductor and capacitor voltage are on the same axis, they can be added directly (or rather subtracted since they face in opposite directions). This is noted in figure 218-9 as the dotted horizontal line that helps create Vm. Then we are essentially adding two perpendicular vectors, for which we must use the Pythagorean Theorem:

Figure 218-9. Phasor diagram for the RLC circuit.

$$V_m^2 = V_R^2 + (V_L - V_C)^2.$$

(218-13)

We can now write each voltage drop using our definition of reactance ($V = IX$) as

$$V_m^2 = (IX_R)^2 + (IX_L - IX_C)^2.$$

(218-14)

Now, since all of our elements are in series, they must have the same current, so the current can be factored out of the entire right side of the equation

$$V_m^2 = I^2[(X_R)^2 + (X_L - X_C)^2].$$

(218-15)

Finally, we can solve this equation for the current in the circuit as we have for each individual element

$$I = \frac{V_m}{\sqrt{(X_R)^2 + (X_L - X_C)^2}} \quad \rightarrow \quad I = \frac{V_m}{\sqrt{R^2 + \left(\omega L - \frac{1}{\omega C}\right)^2}}.$$

(218-16)

This equation gives us the current in the circuit in terms of the forcing voltage and frequency, and the resistance, capacitance and inductance in the circuit.

Forced Series RLC Current

$$I = \frac{V_m}{\sqrt{R^2 + \left(\omega L - \frac{1}{\omega C}\right)^2}}$$

(218-16)

Description – This equation defines the current in a forced series RLC circuit in terms of the source voltage, V_m, the driving frequency of the circuit, ω, and the circuit resistance, R, inductance, L, and capacitance, C.

Example 218-1: Series RLC Circuit

A series RLC circuit is created using components with the following values:

$$Resistor - R = 2.45\ \Omega,$$
$$Inductor - L = 32.5\ \mu H,$$
$$Capacitor - C = 723\ pF.$$

If the circuit is driven by a voltage

$$V = 12.0\ V \sin(1.65x10^5 t),$$

what is the maximum current found in the circuit?

Solution:

This is a direct application of the equation for force RLC circuits.

$$I = \frac{V_m}{\sqrt{R^2 + \left(\omega L - \frac{1}{\omega C}\right)^2}}.$$

We know from the equation of oscillation that $V_m = 12\ V$ and that $\omega = 1.65x10^5\ rad/s$.

The max current can then be found by substituting all of the known values into the equation above. To simplify the calculation, I will first find a couple of the individual terms:

$$R^2 = (2.45\ \Omega)^2 = 6.00\ \Omega^2,$$

$$\omega L = \left(1.65x10^5 \frac{rad}{s}\right)(32.5\ \mu H) = 5.36\ \Omega,$$

$$\frac{1}{\omega C} = \frac{1}{\left(1.65x10^5 \frac{rad}{s}\right)(723\ pF)} = 8383\ \Omega,$$

$$\left(\omega L - \frac{1}{\omega C}\right)^2 = (5.36\ \Omega - 8383\ \Omega)^2 = 7.02x10^7\ \Omega^2.$$

Combining all terms into the main equation gives us

$$I = \frac{12.0\ V}{\sqrt{6.00\ \Omega^2 + 7.02x10^7\ \Omega^2}} = 0.0014\ A$$

218-3: Resonance

Consider: *How can we maximize the current in an RLC circuit?*

Equation 218-16 shows that the current in a forced series RLC circuit not only depends on the voltage of the source and the circuit resistance, but also on the frequency of the source, the inductance and the capacitance. The current in the circuit is maximized when the denominator of equation 218-16 is minimized. In order to do this, we need

$$\omega L - \frac{1}{\omega C} = 0.$$

(218-17)

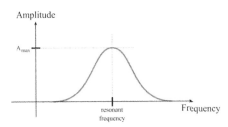

Figure 218-10. Current amplitude versus driving frequency for a series RLC circuit.

Solving this for the angular driving frequency we find

$$\omega_r = \frac{1}{\sqrt{LC}}.$$ (218-18)

A circuit where this condition is met is said to be in ***resonance*** and ω_r is called the ***resonant angular frequency***. It is important to note that a circuit that does not meet this condition will still conduct a current, just that the maximum value of that current will be less than at resonance. Figure 218-10 shows a plot of current versus ω for this circuit. Note, again, that at resonance, the maximum current in the circuit is given by I = V/R as we saw in a DC circuit, but drops off quickly when the resonance condition is not met.

Example 218-2: Resonance

Consider an RLC circuit for which $R = 11\,\Omega, L = 0.21\,mH, C = 5.2\,\mu F$ and the applied voltage is $Vm = 25.0\,V$. What is the maximum current in the circuit if $f = 3.00\,kHz$? What would be the maximum current at resonance?

Solution:

This problem is very similar to example 218-1 with an added question about resonance. The angular driving frequency at resonance of this circuit is

$$\omega_r = \frac{1}{\sqrt{LC}} = \frac{1}{\sqrt{(0.21\,mH)(5.2\,\mu F)}} = 3.03x10^4 \frac{rad}{s}.$$

This corresponds to a frequency of

$$f_r = \frac{\omega_r}{2\pi} = \frac{3.03x10^4\,rad/s}{2\pi} = 4.82\,kHz.$$

So, you can see that a driving frequency of 3 kHz is within a factor of two of resonant frequency.

As in example 218-1, let's first find some of the individual terms of the RLC equation before putting it all together:

$$\omega = 2\pi f = 2\pi(3.00\,kHz) = 1.88x10^4 \frac{rad}{s},$$

$$R^2 = (11\,\Omega)^2 = 121\,\Omega^2,$$

$$\omega L = \left(1.88x10^4 \frac{rad}{s}\right)(0.21\,mH) = 3.95\,\Omega,$$

$$\frac{1}{\omega C} = \frac{1}{\left(1.88x10^4 \frac{rad}{s}\right)(5.2\,\mu F)} = 10.2\,\Omega,$$

$$\left(\omega L - \frac{1}{\omega C}\right)^2 = (3.95\,\Omega - 10.2\,\Omega)^2 = 39.1\,\Omega^2.$$

Combining all terms into the main equation gives us

$$I = \frac{25.0\,V}{\sqrt{121\,\Omega^2 + 39.1\,\Omega^2}} = 1.99\,A.$$

At resonance, the current equation reduces to

$$I = \frac{V_m}{R} = \frac{25.0\,V}{11\,\Omega} = 2.27\,A.$$

So, we can see the difference between the current at 3.00 kHz and at resonance (4.82 kHz), although both currents are reasonable.

218-4: ac circuits create and detect em waves.

Consider: *Why do ac circuits produce em waves?*

We've discussed a number of times how accelerating charged particles create electromagnetic (em) waves. Well, electrons in an alternating current are, by definition, accelerating. Therefore, circuits such as the forced series RLC circuit will produce em waves at the same frequency as the oscillation of the electrons in the circuit. The resonance condition we just discussed is very important in producing such waves. The current amplitude in the ac circuit is directly related to the power of the em wave emitted. Therefore, a forced RLC circuit will produce a strong wave at its resonance frequency and less strong waves off resonance.

Also, since the resonant angular frequency of the circuit,

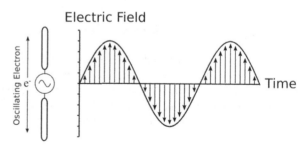

Figure 218-11. Description of how an em wave is produced by oscillating electrons (and therefore oscillating electric fields).

$$\omega_r = \frac{1}{\sqrt{LC}},$$ (218-19)

depends on the inductance and capacitance, changing either of these values will change the frequency of the emitted em wave. At a very basic level, this is how transmitters – such as radio towers and your cell phones – send signals. Standard power transmission lines and small circuits are very inefficient at transmitting and receiving em signals; however, since sections of wire facing opposite directs often lead cancellation of the incoming or outgoing signal. In order to efficiently couple the power radiated by oscillating electric charges to em waves, an *antenna* can be added to the circuit. An antenna is a conducting part of the circuit that is shaped very specifically to emit em radiation in a well-defined pattern.

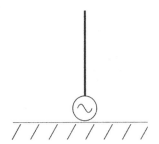

Figure 218-12. Schematic diagram of a monopole antenna.

We will consider two of the most common types of antenna used today – monopole and dipole antennas. A **monopole antenna** consists of a single length of wire connected to the source (in the case of transmission) or the receiver (in the case of reception). The schematic of a monopole antenna can be seen in Figure 218-12. A **whip antenna** is a monopole antenna that is designed to be flexible – they get their name from the whip-like motion when disturbed. Whip antenna can be found on some cell phones, walkie-talkies and on police cars, fire engines and military vehicles for GPS and radio reception. Examples of whip antennas can be seen in figure 218-13.

Figure 218-13. Examples of whip antennas: A retractable whip antenna on an old portable stereo (left), a tethered whip antenna on a military vehicle (middle) and whip antennas on various walkie-talkies (right).

Monopole antennas are considered omnidirectional, meaning that they have no preferred direction for radiation. Consider a vertically situated monopole antenna (similar to Figure 218-12). Electromagnetic waves propagate away from the antenna in all directions along the horizontal plane. There is, however, generally a decrease in the power radiated as the vertical angle increases from parallel to the ground (horizontal) to vertically upward. Very little radiation is directed upward from the monopole antenna situated as described.

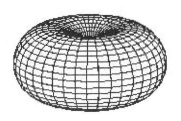

Figure 218-14. Radiation pattern of a ¼ wave monopole antenna.

Although the description above is generally true for monopole antennas, the exact pattern depends on the antenna length in relation to the wavelength of the signal being created. Commonly, monopole antennas are designed as ¼-wave antennas, meaning that the length of the antenna is designed to be ¼ of the wavelength of the carrier wave. The radiation pattern of such an antenna can be seen in Figure 218-14.

The standard FM radio broadcast range is from 87.5 MHz to 108.0 MHz. To think about what length of monopole antenna would work as an FM receiver, consider an em wave at 100 MHz. What wavelength does this correspond to? In order to figure this out, we use the relationship between frequency, wavelength and the speed of light discussed in earlier units - $\lambda f = c$. For 100 MHz, we find

$$\lambda = \frac{c}{f} = \frac{3 \times 10^8 \ m/s}{100 \times 10^6 \ Hz} = 3 \ m.$$ (218-20)

This suggests that an ideal size for an FM antenna is approximately, 0.75 m (one-quarter of three meters). If you think back to cars with external whip radio antennas, this is a good approximation of their length.

Again, the radiation pattern of a monopole antenna depends on the length of the antenna in relation to the wavelength. Figure 218-15 shows the radiation pattern of a 3/2-wavelength antenna. Notice in this case how there are two lobes – one lower and one higher – where em waves are radiated, but there is a gap at mid-angles where very little radiation is emitted. The desired pattern of radiation is an important factor in even simple antenna design.

Although monopole antennas are the easiest to understand, they are not the most common antenna type used – that honor goes to the *dipole antenna*. In its simplest form, a dipole antenna is made by two monopole antennas. The antenna shown in Figure 218-11 as producing the electromagnetic wave is one example of a dipole antenna – the ac source is at the center of two monopole antennas pointing in opposite directions. "Rabbit ear" antennas (Figure 218-16) found on old TV sets are another example of dipole antennas.

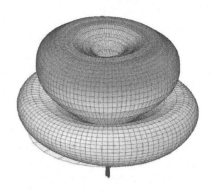

Figure 218-15. Radiation pattern of a 3/2-wave monopole antenna.

Figure 218-16. Rabbit ear antenna with a circular "UHF" antenna.

Dipole antennas have a number of advantages over their simpler monopole cousins. First, as is seen with the rabbit ear dipole, each pole of the dipole antenna can be oriented separately, allowing for maximum alignment with an incoming em wave. This can be especially powerful if two signals, emanating from different directions, are carried on similar frequencies – it is very likely that a given configuration will receive one of the signals more readily than the other. In addition, the size of dipole antennas fits very well with modern communications.

The modern 4G LTE signals recently rolled out by the major wireless communications companies (AT&T, Verizon, etc.) carry signals at 700 and 1700 MHz. For illustrative purposes, let's consider the 1700 MHz carrier frequency. The most common design for a dipole antenna is a $\frac{1}{2} - \lambda$ setup, which is essentially two $1/4 - wave$ monopole antennas combined. At this size, a dipole antenna has a radiation pattern that is almost exactly like Figure 218-14, essentially guaranteeing that a signal generated by the antenna (your phone in this case) will reach a cell tower. What is the optimal size for an antenna at the larger LTE frequency? To determine this, we use the same process as we did with the monopole antenna – we use the frequency-wavelength relationship to find the wavelength of an em wave at 1700 MHz and then find the ideal antenna size. In this case, the wavelength is

$$\lambda = \frac{c}{f} = \frac{3 \times 10^8 \; m/s}{1.7 \times 10^9 \; Hz} = 0.17 \; m = 17 \; cm. \tag{218-21}$$

Therefore, the entire length of a $\frac{1}{2} - \lambda$ dipole antenna for the LTE signal is only 8.5 cm. This antenna can easily fit inside the case of a modern smartphone.

Modern antenna design goes far beyond the two configurations we just discussed. Depending on the application, antennas can look like long spindles of crossed wires (directional dipole antennas), parabolic dishes (satellite dishes), or even snowflakes (fractal antennas) among other designs. However, one thing is true for all of these antenna systems – they produce the em waves they radiate through the acceleration of electrons. The specific design of the antenna only supports the direction, power and efficiency of the radiation.

218-5: Maxwell's equations and em waves (optional).

In Units 210 and 216, we worked with three of the fundamental equations of electromagnetism – Gauss's law (for electric fields), Faraday's law and the Ampere-Maxwell relationship. It turns out that there is a fourth fundamental equation that is important for symmetry between electric and magnetic fields – Gauss's law for magnetism. From Unit 210, Gauss's law can be written

$$\oint \vec{E} \cdot d\vec{a} = \frac{q_{encl}}{\epsilon_0}. \tag{213-2}$$

The right side of this equation represents the net charge enclosed by a Gaussian surface – a fictitious surface that allows us to count enclosed charge to relate to the electric flux. However, magnetic fields do not have single magnetic charges – north poles always come with south poles which cannot be separated. If you break a magnet into a billion small pieces – you still have a billion small magnets right down to the individual magnetic dipole moments of the atoms that make up a magnet. Therefore, there is no Gaussian surface for magnetic fields that can ever contain a net 'magnetic charge' – the north poles and south poles will always cancel each other out. To put this another way, physicists say there are no magnetic monopoles. Therefore, ***Gauss's law for magnetism*** can be written

$$\oint \vec{B} \cdot d\vec{a} = 0, \tag{218-22}$$

with a corresponding differential form

$$\vec{\nabla} \cdot \vec{B} = 0. \tag{218-23}$$

We now have four fundamental equations for electricity and magnetism, which combined are called Maxwell's equations. These equations are summarized in Table 218-1.

Table 218-1. Maxwell's equations.

Equation	Integral Form	Differential Form	Description
Gauss's law (electricity)	$\oint \vec{E} \cdot d\vec{a} = \dfrac{q_{encl}}{\epsilon_0}$	$\vec{\nabla} \cdot \vec{E} = \dfrac{\rho}{\epsilon_0}$	Relates electric charge to electric field.
Gauss's law (magnetism)	$\oint \vec{B} \cdot d\vec{a} = 0$	$\vec{\nabla} \cdot \vec{B} = 0$	No magnetic monopoles exist
Faraday's law	$\oint \vec{E}_{ind} \cdot d\vec{l} = -\dfrac{d\Phi_B}{dt}$	$\vec{\nabla} \times \vec{E} = -\dfrac{\partial \vec{B}}{\partial t}$	A changing magnetic field creates an electric field
Ampere-Maxwell relationship	$\oint \vec{B} \cdot d\vec{l} = \mu_0 I_{enc} + \mu_0 \epsilon_0 \dfrac{d\Phi_E}{dt}$	$\vec{\nabla} \times \vec{B} = \mu_0 \vec{J} + \mu_0 \epsilon_0 \dfrac{\partial \vec{E}}{\partial t}$	Magnetic fields are created by electric currents and changing electric fields.

Taken together, Maxwell's equations describe all of classical electrodynamics – including electromagnetic waves as we will see below. In addition, note that these four equations are quite symmetric except for the fact that no magnetic monopoles exist. If such monopoles are ever discovered, Gauss's law for magnetism would be modified to have a magnetic charge term, q_m, and Faraday's law would need an additional term for magnetic current, I_m. Physicists love symmetry, so finding monopoles and symmetrizing Maxwell's equations would be a major success for the field. However, at present, we have no reason to believe magnetic monopoles exist.

In order to show how Maxwell's equations lead to electromagnetic waves, it is best to use the differential form of each equation. We start out with a set of perpendicular electric and magnetic fields that are allowed to vary in space and time. For completeness, let's say that the electric field is along the y-axis and the magnetic field is along the z-axis:

$$\vec{E}(x,t) = \{0, E(x,t), 0\} \qquad \vec{B}(x,t) = \{0, 0, B(x,t)\}. \tag{218-24}$$

We now assume that both fields are situated in free space, so that there are no electric charges or currents. Therefore, $\rho = 0$ and $\vec{J} = 0$, With this simplification, Maxwell's equations are

$$\vec{\nabla} \cdot \vec{E} = 0 \qquad\qquad \vec{\nabla} \times \vec{E} = -\frac{\partial \vec{B}}{\partial t}$$
$$\vec{\nabla} \cdot \vec{B} = 0 \qquad\qquad \vec{\nabla} \times \vec{B} = \mu_0 \epsilon_0 \frac{\partial \vec{E}}{\partial t} \tag{218-25}$$

Note that in free space, Maxwell's equations do take on the symmetry we just described above. In order to simplify notation, I will drop the functional form of each field; however, do remember that both fields are a function of x and t.

If we first want to consider Faraday's law, we need to find the curl of the electric field:

$$\vec{\nabla} \times \vec{E} = -\frac{\partial \vec{B}}{\partial t} \quad \rightarrow \quad \begin{vmatrix} \hat{\imath} & \hat{\jmath} & \hat{k} \\ \frac{\partial}{\partial x} & \frac{\partial}{\partial y} & \frac{\partial}{\partial z} \\ 0 & E & 0 \end{vmatrix} = \begin{bmatrix} 0 \\ 0 \\ \frac{\partial E}{\partial x} \end{bmatrix} = \begin{bmatrix} 0 \\ 0 \\ -\frac{\partial B}{\partial t} \end{bmatrix}. \tag{218-26}$$

Since each field only has z-components, we can write

$$\frac{\partial E}{\partial x} = -\frac{\partial B}{\partial t}.$$

(218-27)

This means that a time variation in the magnetic field gives rise to a spatial variation in the electric field and vice versa. We can now apply our fields to the Ampere-Maxwell relationship:

$$\vec{\nabla} \times \vec{B} = \mu_0 \epsilon_0 \frac{\partial \vec{E}}{\partial t} \quad \rightarrow \quad \begin{vmatrix} \hat{i} & \hat{j} & \hat{k} \\ \frac{\partial}{\partial x} & \frac{\partial}{\partial y} & \frac{\partial}{\partial z} \\ 0 & 0 & B \end{vmatrix} = \begin{bmatrix} 0 \\ -\frac{\partial B}{\partial x} \\ 0 \end{bmatrix} = \begin{bmatrix} 0 \\ \mu_0 \epsilon_0 \frac{\partial \vec{E}}{\partial t} \\ 0 \end{bmatrix}.$$

(218-28)

Writing out only the y-component, we find

$$\frac{\partial B}{\partial x} = -\mu_0 \epsilon_0 \frac{\partial E}{\partial t}.$$

(218-29)

Similar to our equation from Faraday's law above, this equation tells us that a time-varying electric field gives rise to a spatially varying magnetic field and vice versa.

We can now relate these equations by taking a second derivative of Faraday's law and using some calculus (the order of differentiation does not matter):

$$\frac{\partial}{\partial x}\frac{\partial E}{\partial x} = -\frac{\partial}{\partial x}\frac{\partial B}{\partial t} = -\frac{\partial}{\partial t}\frac{\partial B}{\partial x}.$$

(218-30)

Finally, we can substitute in our Ampere-Maxwell relationship

$$\frac{\partial^2 E}{\partial x^2} = -\frac{\partial}{\partial t}\frac{\partial B}{\partial x} = -\frac{\partial}{\partial t}\left(-\mu_0 \epsilon_0 \frac{\partial E}{\partial t}\right) \quad \rightarrow \quad \frac{\partial^2 E}{\partial x^2} = \mu_0 \epsilon_0 \frac{\partial^2 E}{\partial t^2}.$$

(218-31)

If you remember back to Unit 201 (equation 201-2), this is the equation of a wave for the electric field. So, the most simple solution is

$$\vec{E} = E_0 sin(\omega t - kx)\hat{e},$$

(218-32)

which is our equation for an em wave (equation 204-1)!

In a similar fashion, Maxwell's equations can be used to derive a wave equation for the magnetic field, leading to our basic magnetic wave formula (equation 204-2):

$$\vec{B} = B_0 sin(kx - \omega t)\hat{b}.$$

(218-33)

This is really quite a remarkable result – by simply working with the equations for electricity and magnetism, we found electromagnetic waves. This result can even be traced back to the statement

accelerated charges create electromagnetic waves.

If you consider the wave equation for the electric field, you should be able to convince yourself that the term $\partial^2 E / \partial t^2$ **_is_** an acceleration term since the electric field is caused by a charge density.

219 – Geometrical Optics and Rays

In this unit, we begin to explore the practical uses of electromagnetic radiation in the visible and near visible part of the spectrum. Specifically, we will explore the three models of optics – the study the light. This unit will introduce the three main models of optics, and begins a two-unit exploration of geometrical optics: the model of how light interacts with objects of everyday size.

Integration of Ideas

Review electromagnetic waves from unit 204.
Review the index of refraction from unit 204.

The Bare Essentials

- Depending on the part of the EM spectrum and intended use, EM radiation can be viewed as either a ray (geometrical optics), wave (physical optics) or particle (quantum optics).

- When all objects in a system are much larger than the wavelength of the EM radiation, the radiation can be treated as a *ray* – that is as if the radiation travels in a straight line between points.

- We refer to EM radiation that follows this property as *light*, even though it may not be in the optical part of the spectrum.

- When light reaches an interface with differing indices of refraction, some of the light is reflected from the surface and (usually) some of the light is transmitted into the new material.

Reflectance and Transmittance

$$1 = R + T$$

Description – This equation describes how the percentage of light hitting a surface that is reflected (R) and transmitted (T) must equal one, stating that no light is lost.
Note: R and T are determined by the indices of refraction of the two materials and the angle of incidence through the *Fresnel equations*.

- The law of reflection states that the angle of the reflected ray and incoming ray are the same relative to the normal.

Law of Reflection

$$\theta_i = \theta_r$$

Description – The law of reflection states that the angle of incidence of a ray is equal to the angle of reflection for the outgoing ray.
Note: Both angles *must* be measured relative to the normal – a line perpendicular to the surface at the point of reflection.

- The law of refraction (Snell's Law) relates the angle of the incoming wave (relative to the normal) to the angle of the ray (relative to the normal) in the new material.

Law of Refraction (Snell's Law)

$$n_1 sin(\theta_1) = n_2 sin(\theta_2)$$

Description – The law of refraction relates the angle of incident of a ray from a material with index of refraction n_1 to the angle of refraction of the light in the new material with index of n_2.
Note: Both angles *must* be measured relative to the normal – a line perpendicular to the surface at the point of reflection.

- The Fresnel equations describe how much light is reflected and transmitted at an optical interface.

Fresnel Reflectivity

$$R = \frac{1}{2}\left[\left|\frac{n_1 \cos\theta_1 - n_2 \cos\theta_2}{n_1 \cos\theta_1 + n_2 \cos\theta_2}\right|^2 + \left|\frac{n_1 \cos\theta_2 - n_2 \cos\theta_1}{n_1 \cos\theta_2 + n_2 \cos\theta_1}\right|^2\right]$$

Description – This equation describes the percentage of light reflected from a surface, where n_1 and θ_1 are the index and angle in the initial medium and n_2 and θ_2 are the index and angle in the final medium.
Note: You must usually use Snell's law to get θ_2.

219-1: Three models of light

In this unit, we begin a discussion of *light* and *optics*. So, what exactly do we mean by these terms? Technically, light is the common agent that stimulates our eyes. To put a more physics spin on light, it is the part of the electromagnetic spectrum that our eyes are designed to detect – generally wavelengths from 400 nm to 750 nm. Optics is the scientific study of light. Today, both the terms light and optics are used more broadly. For example, it is not uncommon to hear of *infrared optics* or *UV optics* for those areas of the spectrum that we cannot directly see with our eyes.

In physics, a ***model*** is a set of rules that can be used under certain conditions. For example, when we approach a projectile motion problem, we can immediately assume that there is the acceleration of gravity pointing vertically down, and no acceleration in the horizontal directions. This is the *model* we are using for those problems. Depending on the wavelength of light and the size of the materials with which the light is interacting, there are three main models for optics:

Geometrical Optics – light is treated as a ***ray***, a straight line that can bend only at interfaces (in homogeneous materials). This model works well when all objects and interfaces are much larger than the wavelength of light.

Physical Optics – light is treated as a wave. This model works well when light is interacting with objects and interfaces that are on the same order of magnitude as the light's wavelength.

Quantum Optics – light is treated as a particle (photon). This model works well when we are considering the interaction of light with atoms and molecules.

In this and the next unit, we will focus on geometrical optics. Units 221 and 222 will deal with physical optics and quantum optics, respectively. *Be very careful*, the rules that are used for one model of optics do not generally carry over into the other models.

219-2: Geometrical optics and rays

Consider: *What happens when light, treated as a ray, hits a boundary?*

Geometrical optics is the model we are most familiar with in everyday life. Anything dealing with regular lenses – from eyeglasses, to binoculars, to telescopes and even your bathroom mirror – deals with geometrical optics. In all of these cases, the objects are so much bigger than the sub-micrometer wavelength of light, that the light can be treated as moving in a straight line.

For each ray, we simply use a straight arrow:

$$\longrightarrow$$

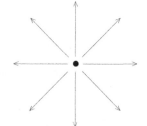

Figure 219-1. Rays emanating from a point source.

Extending this idea, the rays that come off of a points source of light would extend spherically away from the source as shown in figure 219-1.

As discussed in Unit 204, the speed at which light moves in a medium is determined by the ***index of refraction*** of the medium. I'm now going to take this one step further and say that any interface between two mediums in optics is defined by a change in index of refraction of those mediums. For example, there is an optical interface between air and water because air has an index of refraction very close to one, whereas water has an index of refraction 1.33. On the flip side, if two materials have the same index of refraction, there will be essentially no optical interface between them. Certain types of gels have been developed (called index matching gels) that have the same index of refraction as glass, so that when they are placed in contact with glass there is no discernible optical interface.

Now that we have clear definitions of optical rays and interfaces, we can talk about what happens when a ray hits an optical interface. At almost all such surfaces, some of the light reflects off the surface and some transmits through the surface and into the new medium. The percentage of light that reflects back is called the reflectance, R, and the percentage of light that is transmitted is called the transmittance, T. For the time being, we are going to assume that optical materials do not absorb light. In that case, since all of the light is either transmitted of reflected, the sum of our percentages (R and T) must be 1.

> **Reflectance and Transmittance**
>
> $$1 = R + T \qquad (219\text{-}1)$$
>
> **Description** – This equation describes how the
> percentage of light hitting a surface that is reflected (R)
> and transmitted (T) must equal one, stating that no light
> is lost.
> **Note:** R and T are determined by the indices of refraction
> of the two materials and the angle of incidence through
> the *Fresnel equations*.

219-3: Reflection

Consider: *What happens to light that bounces off of (reflects from) a surface? Does it matter if the surface is rough or smooth?*

Light that is reflected off of a surface reflects at the same angle that it came into the surface as can be seen in figure 219-2. Surfaces that are highly efficient at reflecting light are called **mirrors**. There are a few important points to note in figure 219-2. First, note that the incident angle θ_i and the reflected angle θ_f are measured relative to the **normal** – a line that is perpendicular to the surface. **In geometrical optics, all angles are measured relative to the normal.**

The fact that light reflects off of a surface at the same angle it comes in at (incident angle) is called the **law of reflection**. The law of reflection is simple and universal as long as we remember that the angles are measured at the specific point a ray hits the surface.

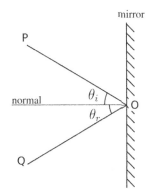

Figure 219-2. (Specular) Reflection of light from a surface.

> **Law of Reflection**
>
> $$\theta_i = \theta_r \qquad (219\text{-}2)$$
>
> **Description** – The law of reflection states that the angle
> of incidence of a ray is equal to the angle of reflection
> for the outgoing ray.
> **Note:** Both angles *must* be measured relative to the
> normal – a line perpendicular to the surface at the point
> of reflection.

There are three important points to remember when considering reflection at a point

1) The incident ray, the reflected ray and the normal all intercept at one point and lie in the same plane,
2) All angles are relative to the normal,
3) θ_i and θ_r are on opposite sides of the normal.

If the surface of the mirror is perfectly flat, all rays incident at a certain angle will reflect at the same angle. This is known as **specular reflection**. However, most surfaces are not perfectly flat, leading to rays at different locations reflecting at different angles, a process known as **diffuse reflection**. Figure 219-3 shows a diagram of diffuse reflection.

We would not be able to see without reflection. What we see as surfaces all around us comes from light reflected off the surface. The only exception to this rule is light that comes directly from sources, such as a light bulb, or the sun (of

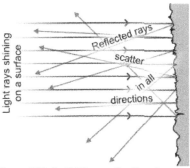

Figure 219-3. Diffuse reflection from an uneven surface.

course, do not look directly at the sun). Given this, you might wonder why different materials have different colors? White light (as from the sun or from a light bulb) contains all of the colors of the spectrum. A given material will reflect and transmit each wavelength making up the white light differently. For example, a material might be a good reflector for colors in the red portion of the spectrum, but absorb many others. We would see this object as red. This is the general idea behind **spectroscopy**, the area of science that deals with identifying materials based on their reaction to different wavelengths of light.

Example 219-1: Corner reflector

The mirror shown below (called a corner reflector) consists of two perfectly flat mirrors connected at a right angle to each other. If a ray of light is incident at 45° relative to one side of the mirrors, in which direction will the light reflected from the second mirror be traveling?

Solution:

We must use the law of reflection twice in order to answer this question.

First, when the ray reflects off of the first surface, the angle of reflection will be the same as the angle of incidence: 45°.

The ray will now be incident upon the second mirror at 45°, and, again, reflect at that same angle.

Therefore, in the end, the ray will be directed in the exact opposite direction relative to the incoming ray after leaving the second mirror.

219-4: Refraction

Consider: *What happens to light that transfers through a boundary?*

When light moves between two mediums with different indices of refraction, optical rays bend at the surface. The only exception to this rule is when the incident light is perpendicular to the surface (so that $\theta_i = 0$). To fully understand why this happens, we have to employ the wave nature of light and either Huygens's principle or Fermat's principle of least time. If you are interested, the derivation can be found at the end of unit 220 (physical optics).

Conceptually, you can understand why light rays bend at an interface between surfaces with a different index of refraction by remembering that the speed at which light moves in a medium is directly related to the index:

$$v = \frac{c}{n}. \qquad (219\text{-}3)$$

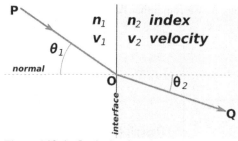

Figure 219-4. Optical refraction at an interface.

Consider figure 219-4 which shows a ray moving from a region of low index of refraction (n_1) to one with high index (n_2). Think of this as a car approaching a very rough surface at an angle. As the first wheel touches the rough surface, it will tend to slow down. However, the other front wheel is still on the smooth initial surface and continues fast. With one wheel slow and the other fast, this tends to cause the car to turn into the slow surface. Although this is not really a good analogy, it does relay why light bends to everyday phenomena. Again, for a full derivation, see unit 221.

The crux of the matter is that when light moves to a material with a higher index of refraction, its angle relative to the normal must decrease. Considering figure 219-4 again, the angles must be between 0 and 90 degrees, which suggests using either the sine or cosine of the angle. Since a small angle should correspond to being close to the normal ($\theta = 0$), we choose sine.

Law of Refraction (Snell's Law)

$$n_1 sin(\theta_1) = n_2 sin(\theta_2) \qquad (219\text{-}4)$$

Description – The law of refraction relates the angle of incident of a ray from a material with index of refraction n_1 to the angle of refraction of the light in the new material with index of n_2.

Note: Both angles *must* be measured relative to the normal – a line perpendicular to the surface at the point of reflection.

A couple of important points:

1) *Always* measure angles relative to the normal – I cannot stress this enough!
2) When moving to higher index, the angle should decrease; when moving to lower index, the angle should increase,
3) See table 207-1 for some common indices of refraction.

Example 219-2: Picking up the rock.

Have you ever tried to pick up something that is submerged underwater only to find it is not where you expect it to be? This is due to the refraction of light. Let's say that you are looking down at the surface of a pond at an angle of 30° (relative to the water) and want to pick up a rock that is 20 cm below the surface. How far is the rock from where it appears to be on the bottom?

Solution:

This is a direct application of the law of refraction and trigonometry. Here is a figure of the problem:

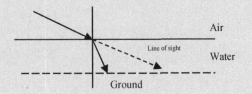

Air

Line of sight

Water

Ground

The very first thing to notice is that the angle we must use for Snell's law (relative to the normal) is 60 degrees, since what we were given (30°) is relative to the water. Now, the line of sight says that the rock should be

$$x_{sight} = \frac{20\ cm}{\cot 60°} = 34.6\ cm$$

from where the initial ray hits the water.

However, the angle of the ray in the water is

$$n_1 \sin\theta_1 = n_2 \sin\theta_2 \quad \rightarrow \quad 1(\sin 60°) = 1.33(\sin\theta)$$

or $\theta = 40.6°$. Therefore,

$$x_{real} = \frac{20\ cm}{\cot 40.6°} = 17.1\ cm.$$

The rock is $34.6\ cm - 17.1\ cm = 17.5\ cm$ from where we think it should be!

Connection: Mirages

There are certain situations where the index of refraction of a material is not constant, such as when there are wide temperature variations. One such situation is air near the surface of a hot road. The temperature of the air near the road is very hot, but drops off rapidly and continuously as we move above it. There is a corresponding change in the index of refraction of the air as well. This causes a continuous bending of light near the road and can lead to an *inferior mirage* – a wavy double image of an object (such as a car) on the road. The wavyness of these mirages often makes them look like water even though no water is there at all.

Figure 219-5. A mirage formed by a car on a hot road.

219-5: Total internal reflection

Consider: *If you are underwater and look up, why can you see outside the water above you, but not when you look at much of an angle?*

We've discussed how if you are moving from a high index of refraction to a low index of refraction, the ray bends away from the normal. As can be seen in figure 219-6, there is a specific angle, called the critical angle (θ_c) at which the refracted ray becomes parallel to the interface – meaning that it is no longer refracted *into* the new medium. At this and any greater angle, there is only reflection at the surface and not refraction. This situation is called ***total internal reflection*** because all of the light is reflected back into the initial medium.

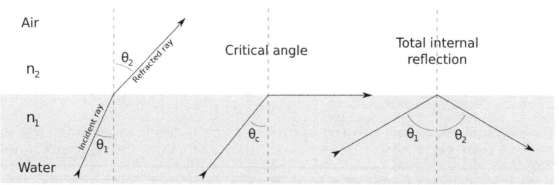

Figure 219-6. The development of total internal reflection. At the critical angle, the refracted ray would be parallel to the interface. At angles above θ_c, there is no refracted light, only reflected light.

We can find the critical angle by using the law of refraction and noting that the refracted angle goes to 90°:

$$n_1 \sin \theta_1 = n_2 \sin \theta_2 \quad \rightarrow \quad n_1(\sin \theta_c) = n_2(\sin 90°), \tag{219-5}$$

which leads to

$$\sin \theta_c = \frac{n_2}{n_1}. \tag{219-6}$$

Example 219-3: Optical Fibers

An optical fiber (or fiber optics) is a hair-thin piece of pure glass designed to carry light over vast distances. Light entering an optical fiber is directed so that its angle is greater than the critical angle. Total internal reflection then keeps the light entirely inside the fiber with very little loss. If the inside of an optical fiber is pure glass with n = 1.66, and the exterior of the fiber is air (n = 1), what is the critical angle for this fiber?

Solution:

This is a direct application of the critical angle:

$$\sin \theta_c = \frac{n_2}{n_1} = \frac{1}{1.66}.$$

Therefore $\theta_c = 37.0°$.

In reality, optical fibers are made as cylinders with two types of glass, the outer glass of index 1.5 (called the cladding) and the inner with index 1.66 (called the core). This allows for a much greater critical angle:

$$\theta_c = \sin^{-1}\frac{1.5}{1.66} = 64.6°.$$

219-6: Fresnel Reflection

There is a formula that can be used to predict the percentage of incoming light that will be reflected when it hits an interface between materials of two different indices of refraction. The set of equations that predict this are known as **the Fresnel equations** after the physicist who discovered them, *Augustin-Jean Fresnel* (said Fre'-nel). The full set of Fresnel equations can take into account more about electromagnetic waves than we have discussed to this point. Therefore, I present the equation for reflectivity, *R*, for *unpolarized* light:

Fresnel Reflectivity

$$R = \frac{1}{2}\left[\left|\frac{n_1 \cos\theta_1 - n_2 \cos\theta_2}{n_1 \cos\theta_1 + n_2 \cos\theta_2}\right|^2 + \left|\frac{n_1 \cos\theta_2 - n_2 \cos\theta_1}{n_1 \cos\theta_2 + n_2 \cos\theta_1}\right|^2\right] \qquad (219\text{-}7)$$

Description – This equation describes the percentage of light reflected from a surface, R, where n_1 and θ_1 are the index and angle in the initial medium and n_2 and θ_2 are the index and angle in the final medium, respectively.
Note: You must usually use Snell's law to get θ_2.

This is not the easiest looking equation, but it is one of the most plug-and-chug equations you will find in all of our introductory physics. If you look closely, you'll see that all you need are the index of refraction of the two mediums, n_1 and n_2, and the two angles, θ_1 and θ_2. Keep in mind you must *always* use Snell's law to get θ_2, the angle of refraction. The equation is quite powerful. In essence, it can tell you whether a given set of materials will be a good mirror, or good at transmitting light. It will even tell you the angle for total internal reflection as can be seen in figure 219-7.

Also, if you remember back to the beginning of the unit, if we know the reflectivity, R, we also know the transmissivity, T, by the conservation of energy:

$$1 = R + T. \qquad (219\text{-}1)$$

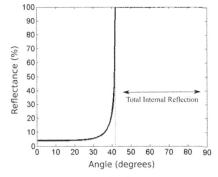

Figure 219-7. Plot of reflectance versus incident angle for a glass to air interface.

Example 219-4: Air-water reflection

What percentage of light incident upon water ($n_2 = 1.33$) from air perpendicular to the surface ($\theta_i = 0$) is reflected?

Solution:

This is a direct application of the Fresnel reflectivity equation:

$$R = \frac{1}{2}\left[\left|\frac{n_1 \cos\theta_1 - n_2 \cos\theta_2}{n_1 \cos\theta_1 + n_2 \cos\theta_2}\right|^2 + \left|\frac{n_1 \cos\theta_2 - n_2 \cos\theta_1}{n_1 \cos\theta_2 + n_2 \cos\theta_1}\right|^2\right]$$

Note that since both the angle of incidence and refraction are 0° in this case, all $\cos\theta = 1$, so

$$R = \frac{1}{2}\left[\left|\frac{n_1 - n_2}{n_1 + n_2}\right|^2 + \left|\frac{n_1 - n_2}{n_1 + n_2}\right|^2\right] = \frac{1}{2}\left[\left|\frac{1 - 1.33}{1 + 1.33}\right|^2 + \left|\frac{1 - 1.33}{1 + 1.33}\right|^2\right] = 0.02$$

So, 2% of the light is reflected, suggesting that 98% of the light is transmitted into the water.

185

Example 219-5: Air and water again

What percentage of light incident upon water ($n_2 = 1.33$) from air is reflected when the light is incident on the water at 32° relative to the normal?

Solution:

This is a direct application of the Fresnel reflectivity equation:

$$R = \frac{1}{2}\left[\left|\frac{n_1 \cos\theta_1 - n_2 \cos\theta_2}{n_1 \cos\theta_1 + n_2 \cos\theta_2}\right|^2 + \left|\frac{n_1 \cos\theta_2 - n_2 \cos\theta_1}{n_1 \cos\theta_2 + n_2 \cos\theta_1}\right|^2\right]$$

First, we must use Snell's law to find the angle of refraction:

$$n_1 sin(\theta_1) = n_2 sin(\theta_2) \quad \rightarrow \quad \theta_2 = \sin^{-1}\left(\frac{n_1}{n_2}\sin\theta_1\right) = 23.5°.$$

Then we use our known information in the reflectivity equation

$$R = \frac{1}{2}\left[\left|\frac{1(\cos 32) - 1.33(\cos 23.5)}{1(\cos 32) + 1.33(\cos 23.5)}\right|^2 + \left|\frac{1(\cos 23.5) - 1.33(\cos 32)}{1(\cos 23.5) + 1.33(\cos 32)}\right|^2\right] = 0.04$$

So, only 4% of the light is reflected, suggesting that 96% of the light is transmitted into the water.

220 – Mirrors and Lenses

Mirrors and lenses are the two most important optical instruments discussed for geometrical optics. Mirrors and thin lenses can be described by a relatively simple equation that can be used to quickly determine where the image of even complex systems is formed. In this unit, we explore mirrors and lenses in general and then apply the concepts to the optical instrument you use every day – your eyes.

Integration of Ideas

Review optical rays, reflection and refraction from the previous unit.

The Bare Essentials

- Mirrors are designed so that most of the incident light is reflected from the surface.

- Plane mirrors produce an image that is the same distance behind the mirror as the object is in front of the mirror (a virtual image).

- Spherical mirrors produce an image that can either be in front of the mirror (real image) or behind the mirror (virtual image).

Spherical Mirrors

$$\frac{1}{d_o} + \frac{1}{d_i} = \frac{1}{f}$$

Description – This equation relates the distance of an object from a mirror (d_o) to the distance of the image it produces (d_i) and the mirrors focal length (f)

Note 1: If d_i is negative, the image is virtual (behind the mirror).

Note 2: f is positive for concave mirrors and negative to convex mirrors. The concavity is determined by the direction the ray hits the mirror from.

Note 3: The magnification of the image is given by M = -d_i/d_o. Note that a negative magnification means that the image is inverted.

- The image of mirrors and other simple optical systems can be found by creating *ray diagrams*.

- Objects designed to create images of objects by transmitting light are called lenses.

- Thin lenses are an approximation where the lens is thin enough that a single number can capture their effects: the focal length.

Thin Lens Equation

$$\frac{1}{d_o} + \frac{1}{d_i} = \frac{1}{f}$$

Description – This equation relates the distance of an object from a lens (d_o) to the distance of the image it produces (d_i) and the lens's focal length (f).

Note 1: If d_i is negative, the image is virtual (in front of the lens).

Note 2: f is positive for converging lenses and negative for diverging lenses.

Note 3: The magnification of the image is given by M = -d_i/d_o. Note that a negative magnification means that the image is inverted.

- A lens that is thicker in the center than at its edge is a converging lens; one that is thicker at its edge compared to its center is a diverging lens.

- A system of thin lenses and mirrors can be solved by sequentially making the image of one element the object of the next element in line.

- The human eye uses a combination of liquid refraction and refraction at the lens of the eye to produce sharp images.

220-1: The design of lens and mirror systems.

Consider: *What are some of the important terms related to lenses and mirrors?*

W
E USE OPTICAL DEVICES EVERY SINGLE TIME WE SEE SOMETHING. Even if there are no overt mirrors or lenses between you and what you are viewing, there is a lens in your eye that allows for a crisp clear image. If your eye does not allow for a sharp image on its own, you probably wear either contact lenses or glasses, which include optical elements designed to help your eye produce clarity. This unit is going to extend our discussion of geometrical optics to give a very practical introduction to two important optical devices – lenses and mirrors.

First though, a couple of definitions:

Optical Image – also just called an image. A reproduction of an object by reflection, refraction or diffraction. There are two types of images:

Real image – an image that can be displayed on a screen. These are often found in front of mirrors or behind lenses

Virtual image – an image that cannot be displayed on a screen. These are often found behind mirrors or in front of lenses.

Lens – An optical element that refracts and transmits light. Often lenses are used to create an optical image, whether that image be directly for viewing or for use by another optical element.

Mirror – An optical element that reflects light. Often mirrors are used to create an optical image, whether that image be directly for viewing or for use by another optical element.

For both mirrors and lenses, we will create diagrams, known as **ray diagrams**, which will help us visualize what happens to light when it interacts with optical elements. Here are some important definitions for ray diagrams:

Optical or Principal axis – axis which is perpendicular to the center of the optical element passing through the optical elements. When we have multiple elements, we will try to line them up so they share a common axis.

Object distance (d_o) – the distance of an object from the center of an optical element, along the optical axis.

Object height (h_o) – the distance from the top of the object to the optical axis along a line that is perpendicular to the optical axis.

Image distance (d_i) - the distance of an image from the center of an optical element, along the optical axis.

Image height (h_i) – the distance from the top of the object to the optical axis along a line that is perpendicular to the optical axis.

Focal point (f) – A measure of an optical element's ability to bend light. Light rays that are initially parallel to the principle axis will pass through the optical axis at the focal point of the element after striking the element. The *focal length (f)* is the length along the optical axis from the center of the optical element to the focal point.

Many of these features can be seen in the lens diagram of figure 220-1. In this figure, the optical axis is represented by the dotted line in the center. You can see that there are two focal points, one in front of the lens and one behind it, both with focal lengths, f. The vertical arrows represent the object height (left) and image height (right). Finally, both the object distance and image distance are explicitly noted.

You may notice in figure 220-1 that there are also three **rays** that connect the object to the image through the lens. These rays are called the principle rays, and allow us to pinpoint the image of a mirror or lens regardless of whether the image is real or virtual.

All three of the principle rays are formed starting at the tip of the object.

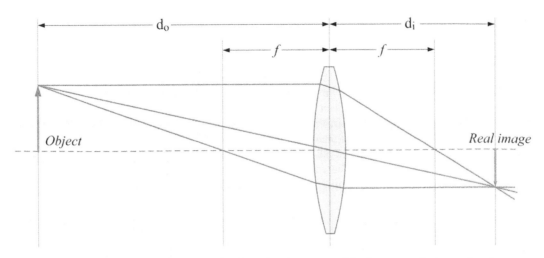

Figure 220-1. Ray diagram for a converging lens showing many of the important features of such a diagram.

1) The first ray moves parallel to the optical axis until it strikes the element. The ray then proceeds through the focal point of the element.
2) The second ray goes directly to the center of the element on the optical axis and continues without changing direction for lenses or reflects for mirrors.
3) The third ray proceeds through the focal point to the optical element and then proceeds parallel to the optical axis.

In principle, you only need to do two of the three principle rays in order to find an image. Keep in mind that rays will reflect off of mirrors and traverse through lenses. Note that figure 220-1 shows the three principle rays for a converging lens forming a real image. Figure 220-2 shows the three principle rays for a converging mirror forming a real image.

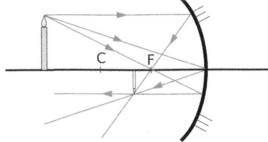

Figure 220-2. Example of principle rays for a concave mirror.

Magnification

The magnification of an image is the ratio of the image size to the object size. A magnification less than one means the image is smaller than the object and a magnification greater than one means that the image is greater than the object. In addition, the sign of the magnification tells us whether the image is upright (positive magnification) or inverted (negative magnification).

$$M = \frac{h_i}{h_o} = -\frac{d_i}{d_o} \tag{220-1}$$

220-2: Mirrors

Consider: *How do I determine where the image of a mirror is located?*

Mirrors are optical elements that change the way light rays move by reflection. In this section, we are going to consider three mirror types – plane mirrors, concave mirrors and convex mirrors.

Plane Mirrors

We all have experience with plane mirrors, also known as flat mirrors.– such as the mirrors we use in bathrooms and restrooms. A plane mirror is made by a very flat reflective surface. A ray diagram for a plane mirror is shown in figure 220-3. The direction of each ray after hitting the mirror is determined by the law of reflection as studied in unit 218. Note from the figure that rays from each part of the object diverge after reflecting from the mirror. In this case, they will never converge in front of the mirror forming an image. In order to find the image we must **backtrack** each ray until they converge at one point. The point the rays converge to is the image. Since the image for a flat mirror is behind the mirror (where none of the light actually goes), this is a **virtual image**. A couple of important points for plane mirror images

189

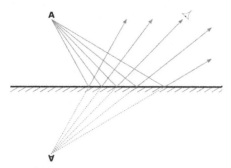

Figure 220-3. Ray diagram for a flat mirror.

mind for plane mirrors

- $d_i = -d_o$
- $h_i = h_o$

(220-2)

meaning that the image is the same distance behind the mirror (negative) as the object is in front of the mirror, and that the height of the object and the image are the same.

There are a couple of very important points to keep in

Connection: How can we see a virtual image?

We just saw that plane mirrors produce a virtual image where there is no light. How is it that we can see a nice crisp image in a plane mirror then? The answer is that your eye is also an optical element containing a converging lens that produces a real image. We will study how this works later in the unit.

1) The image formed by a plane mirror is behind the mirror, and therefore a virtual image,
2) A plane mirror produces an image that *is reversed* back to front, and *not reversed* side to side!

Spherical Aberration

Many introductory physics textbooks talk about spherical mirrors – mirrors that are made as part of the arc of a sphere. These mirrors can be made with the bulge towards the incoming light (a convex mirror) or with the bulge facing away from the incoming light (a concave mirror). As a reminder, figure 220-2 shows a concave mirror. There is a substantial problem with spherical mirrors – light from far away (light parallel to the optical axis) does not actually focus to a point, as can be seen in figure 220-4. This effect is known as **spherical aberration** and is present for both spherical mirrors and spherical lenses. Now, it is best to use a parabolic shape, known as a parabolic reflector; however, parabolic reflectors are very hard and expensive to produce. Luckily, as long as a very small part of the arclength of the sphere is used in creating a spherical mirror or lens, the light does converge very close to a focus, so they are used as a good approximation.

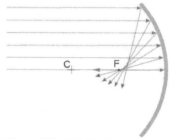

Figure 220-4. Spherical aberration and the lack of a focal point for a spherical mirror.

Concave and Convex Mirrors

For spherical mirrors using a small percentage of the spherical arc, there is a direct relationship between the object distance, image distance and focal length, known as the mirror equation. Note the radius of curvature of the mirror is related to the focal length ($r = 2f$).

Spherical Mirrors

$$\frac{1}{d_o} + \frac{1}{d_i} = \frac{1}{f}$$

(220-3)

Description – This equation relates the distance of an object from a mirror (d_o) to the distance of the image it produces (d_i) and the mirrors focal length (f)

Note 1: If d_i is negative, the image is virtual (behind the mirror).

Note 2: f is positive for concave mirrors and negative to convex mirrors. The concavity is determined by the direction the ray hits the mirror from.

Note 3: The magnification of the image is given by $M = -d_i/d_o$. Note that a negative magnification means that the image is inverted.

For convex mirrors, the focal length is negative (diverging mirror), and the mirror produces a virtual image, no matter where the object is relative to the mirror. Table 224-1 shows important features of convex mirrors.

Table 220-1. Convex mirrors and images. Note: f is negative for convex mirrors

Object distance	Image Details	Ray Diagram
All distances	• Virtual • Upright • Smaller than object	

For concave mirrors, the type of image depends not only on the focal length, but also on the distance of the object from the mirror in relation to the focal point. Table 224-2 summarized concave mirrors.

Table 220-2. Concave mirrors and images. Note: f is positive for concave mirrors.

Object distance	Image Details	Ray Diagram
$d_0 < f$	• Virtual (behind mirror) • Upright • Larger than object	
$d_0 = f$	• No image is formed (outgoing rays are parallel)	

$d_0 > f$	• Real
	• Inverted
	• Magnification
	  - Larger if $f < d_0 < 2f$
	  - Same if $d_0 = 2f$
	  - Smaller if $d_0 > 2f$	

191

Example 220-1: Mirror example 1.

An object of height 2.7 cm is placed 4.7 cm to the left of a concave mirror with a focal length of 3.8 cm.
- (a) Where is the image of this mirror formed?
- (b) Is the image real or virtual
- (c) What is the magnification and size of the image? Is it upright or inverted?

Solution:

For this problem, we will use the mirror equation, because the problem specifies that we have an object a specified distance from a mirror.

Conceptually, it is important to note that the object is located farther away from the mirror than the focal point. This suggests that we have the situation shown in the last row of Table 220-2, and that we should have a real, inverted image that is larger than the object ($f < d_0 < 2f$).

(a) To find the image position, we use the mirror equation:

$$\frac{1}{d_o} + \frac{1}{d_i} = \frac{1}{f} \quad \rightarrow \quad \frac{1}{4.7 \ cm} + \frac{1}{d_i} = \frac{1}{3.8 \ cm}.$$

Solving this equation for d_i, we find

$$d_i = 19.8 \ cm.$$

(b) The result of part (a) suggests that we have a real image that is 19.8 cm to the left of the mirror (on the same side as the object).

(c) For the magnification:

$$M = -\frac{d_i}{d_0} = -\frac{19.8 \ cm}{4.7 \ cm} = -4.21.$$

Therefore the size of the image is

$$h_i = Mh_o = (-4.21)(2.7 \ cm) = -11.4 \ cm.$$

Since the magnification is negative, the image is inverted.

The ray diagram for this problem would look very similar to the ray diagram in the last row of Table 224-2. Also, note that all of our predictions were found to be true mathematically – real, inverted, larger image.

Example 220-2: Mirror example 2.

Let's say we have the same problem as example 220-1, except that the mirror is convex, so that its focal length is -3.8 cm.
- (a) Where is the image of this mirror formed?
- (b) Is the image real or virtual
- (c) What is the magnification and size of the image? Is it upright or inverted?

Solution:

We must approach this problem very similarly to example 220-1. Conceptually, we should expect a virtual, upright, smaller image, as can be seen in Table 220-1.

(a) To find the image position, we use the mirror equation:

$$\frac{1}{d_o} + \frac{1}{d_i} = \frac{1}{f} \quad \rightarrow \quad \frac{1}{4.7 \ cm} + \frac{1}{d_i} = \frac{1}{-3.8 \ cm}.$$

Solving this equation for d_i, we find $d_i = -2.10 \ cm$.

(b) The result of part (a) suggests that we have a virtual image that is 2.10 cm behind the mirror (where the light does not actually go).

(c) For the magnification:

$$M = -\frac{d_i}{d_0} = -\frac{-2.10 \ cm}{4.7 \ cm} = 0.477.$$

Therefore the size of the image is

$$h_i = Mh_o = (0.447)(2.7 \ cm) = 1.21 \ cm.$$

Since the magnification is positive, the image is upright.

The ray diagram for this problem would look very similar Table 220-1. Also, note that all of our predictions were found to be true mathematically – virtual, upright, smaller image.
.

220-3: Thin lenses

Consider: *How do I determine where the image of a lens is located?*

Thin lenses act very similarly to spherical mirrors, except that light refracts through the lens as opposed to reflecting off of the mirror surface. Lenses can be made of many shapes, some of which are shown in figure 220-5. For our purposes, the most important thing about a thin lens is how the thickness of the lens compares at its center to its edge. If the lens is thicker in the middle than it is on its edge, it is a converging lens and will have a positive focal length. If the lens is thicker on its edge than it is at its center, then it is a diverging lens with a negative focal length. In figure 220-5 the first three lenses would be converging lenses and the last three would be diverging lenses.

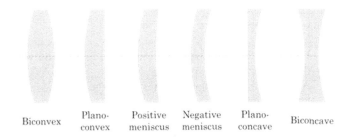

Biconvex Plano-convex Positive meniscus Negative meniscus Plano-concave Biconcave

Figure 220-5. Different shapes for thin lenses.

As with mirrors, incoming rays that are parallel to the optical axis will converge to the focal point. Table 224-3 shows this for both a converging and diverging lens.

Table 220-3. The converging of light to a focal point for a converging and diverging lens.

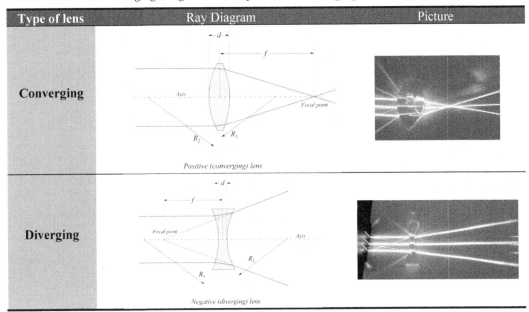

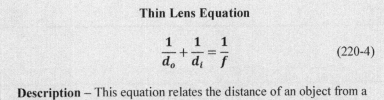

The equation used to determine the image distance from the lens is very similar to that of a spherical mirror

---

Thin Lens Equation

$$\frac{1}{d_o} + \frac{1}{d_i} = \frac{1}{f} \qquad (220\text{-}4)$$

Description – This equation relates the distance of an object from a lens (d_o) to the distance of the image it produces (d_i) and the lens's focal length (f).

Note 1: If d_i is negative, the image is virtual (in front of the lens).

Note 2: f is positive for converging lenses and negative for diverging lenses.

Note 3: The magnification of the image is given by $M = -d_i/d_o$. Note that a negative magnification means that the image is inverted.

The only situation in which a lens forms a real image is shown back in figure 220-1, namely that we have a converging lens and the object is farther from the lens than the focal length. All other lens situations create a virtual image: converging lenses with object closer than the focal length and all diverging lenses.

Some conventions to keep in mind for lenses:

1) Focal length is positive for converging lenses and negative for diverging lenses
2) A positive image distance means the image is on the far side of the lens (real image).

Figure 220-6 shows the ray diagram for a converging lens with the object closer to the lens than the focal point, causing a virtual image (image on the same side of the lens as the object).

Table 220-4 summarized the images formed by thin lenses.

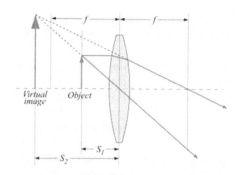

Figure 220-6. Ray diagram for a converging lens with an object closer than the focal point.

Table 220-4. Images formed by thin lenses

Object distance	Image Details	Ray Diagram
Converging Lens $d_0 < f$	• Virtual (in front of lens) • Upright • Larger than object	

Converging Lens $d_0 > f$	• Real
	 • Inverted
	 • Magnification
	   - Larger if $f < d_0 < 2f$
	   - Same if $d_0 = 2f$
	   - Smaller if $d_0 > 2f$	

Diverging Lens	• Virtual • Upright • Smaller than object	

Example 220-3: Focal length, focal length.

Light from a distant star strikes a converging lens with focal length 1.2 m. Where does the light focus?

Solution:

This is a conceptual problem that is important enough to go over one more time. Light from very far away (as from a star) would be parallel to the optic axis. When this light strikes a converging lens, it will converge at the focal point of the lens – at 1.2 meters on the far side of the lens in this case. This is the definition of focal length for a thin lens!

Example 220-4: Virtual times.

A candle, standing 32.4 cm tall, is placed 2.50 cm to the left of a diverging lens with a focal length of -11.2 cm. Describe the image formed by this lens.

Solution:

Since we have a diverging lens, we would expect the image to be virtual, upright and smaller than the object. It is true that the object is well inside the focal point of the lens, but that should not change the quality of the image for this diverging lens.

In order to describe the image, we must use the thin lens equation and the magnification equation

$$\frac{1}{d_o}+\frac{1}{d_i}=\frac{1}{f} \rightarrow \frac{1}{2.50\ cm}+\frac{1}{d_i}=\frac{1}{-11.2\ cm}$$

$$\rightarrow d_i = -2.04\ cm,$$

$$M = -\frac{d_i}{d_0} = -\frac{-2.04\ cm}{2.50\ cm} = 0.816.$$

Therefore, we do indeed have a virtual image (2.04 cm to the left of the lens) that is upright (positive magnification) and smaller than the object ($|M| < 1$).

220-4: Complex Systems

Consider: *What do I do if there is more than one mirror/lens?*

When a system contains multiple mirrors and lenses, we sequentially make the images of one element the object of the next element. When doing this, take great care with the distances as all distances must be measured relative to the element you are using for a specific step. Also, when working with one optical element, make sure you treat the problem as if the current element is the only element in the system at that time.

The magnification of an optical system is simply the product of the magnification of the individual pieces:

$$M_{total} = M_1 M_2 M_3 ... \tag{220-5}$$

Example 220-5: Two lenses

An object 23.2 cm tall is placed 114 cm to the left of a diverging lens with focal length $f_1 = -43\ cm$. A second converging lens with focal length $f_2 = 11.2\ cm$ is placed 22.3 cm to the right of the first lens. Describe the final image of the system in terms of where it is, its size and whether it is real or virtual and upright or inverted.

Solution:

As described above, this is really two one-lens problems, where we take the image of the first lens and make it the object of the second lens.

Lens 1
For the first lens, we have $d_o = 114\ cm$ and $f_1 = -43\ cm$. Using the thin lens equation, we can find the position and magnification of the image:

$$\frac{1}{d_o}+\frac{1}{d_i}=\frac{1}{f} \rightarrow \frac{1}{114\ cm}+\frac{1}{d_i}=\frac{1}{-43\ cm}$$

$$\rightarrow d_{i_1} = -31.2\ cm,$$

$$M_1 = -\frac{d_i}{d_0} = -\frac{-31.2\ cm}{114\ cm} = 0.274.$$

Lens 2
Since the image of lens 1 is 31.2 cm to the left of the lens and the distance between the lenses is 22.3 cm, the image of the first lens becomes the object of the second lens, 31.2 + 22.3 = 53.5 cm to the left of the second lens. We can again use our equations to find the position of the image and magnification of the second lens:

$$\frac{1}{d_o}+\frac{1}{d_i}=\frac{1}{f} \rightarrow \frac{1}{53.5\ cm}+\frac{1}{d_i}=\frac{1}{11.2\ cm}$$

$$\rightarrow d_{i_2} = 14.2\ cm,$$

$$M_2 = -\frac{d_i}{d_o} = -\frac{14.2\ cm}{53.5\ cm} = -0.265.$$

Therefore the final image is 14.2 cm to the right of the second lens (real image). Also, the overall magnification

$$M_{total} = M_1 M_2 = (0.274)(-0.265) = -0.073$$

tells us that the image is much smaller than the original object (1.7 cm tall).

Example 220-6: Lens and mirror

Imagine if we replace the second lens in example 220-5 with a mirror of focal length 11.2 cm (the same focal length of the second lens in the example). Where is the final image of the system?

Solution:

Since the first lens and the object distance from the first lens is the same as in the previous example, we can immediately use the results of that problem here:

$$d_{i_1} = -31.2 \ cm \qquad M_1 = 0.274.$$

We now just have the problem of the mirror. Since the image of the first lens is 31.2 cm to the left of the lens, it is $31.2 + 22.3 = 53.5$ cm to the left of the mirror. This becomes our object distance for the mirror.

We can now use the mirror equation, since we treat the mirror as an isolated system now that we know its object distance (53.5 cm):

$$\frac{1}{d_o} + \frac{1}{d_i} = \frac{1}{f} \quad \rightarrow \quad \frac{1}{53.5 \ cm} + \frac{1}{d_i} = \frac{1}{11.2 \ cm}$$

$$\rightarrow d_{i_2} = 14.2 \ cm,$$

$$M_2 = -\frac{d_i}{d_0} = -\frac{14.2 \ cm}{53.5 \ cm} = -0.265.$$

Note the similarities to the math in the previous example; however, the interpretation is slightly different. Since light is reflected from the surface of the mirror, a positive image is on the same side of the mirror as the object. Therefore, the image is 14.2 cm *to the left* of the mirror.

The overall orientation and magnification are the same as example 220-6, however:

$$M_{total} = M_1 M_2 = (0.274)(-0.265) = -0.073$$

Example 220-7: Refracting telescope.

A basic refracting telescope consists of two converging lenses, one called the objective lens (with focal length f_o) and one called the eyepiece (with focal length f_e). The two lenses are placed a distance, $L = f_o + f_e$, apart from each other. Imagine that we want to construct a telescope to view a star such that the telescope has an objective lens with $f_o = 2.70 \ m$, and an eyepiece with $f_e = 2.12 \ cm$. Describe the magnification of this system and how the image of the star can be seen.

Solution:

Since this is a multiple lens system, we will treat each lens separately and then combine magnifications at the end.

Objective lens

Light from a star striking the objective lens of a telescope is coming from very far away. Therefore, we can treat the light rays as if they are all parallel to the optic axis (i.e., the object is essentially at $d_o \approx \infty$. As can be seen in the converging lens of Table 224-3, the light will form an image at the focal point of the lens ($d_i = f_o$).

The magnification of the object lens is then given by

$$M_o = -\frac{d_i}{d_o} = -\frac{f_o}{\infty},$$

which I will not try to simplify for the time being.

Eyepiece

Now remember, in the setup for the telescope, the focal points of the two lenses lie on top of each other ($L = f_o + f_e$), so that the image of the objective becomes the object of the eyepiece at the focal point of the eyepiece. Since rays emanating from the focal point of a converging lens image at ∞, we can say d_o for the eyepiece is f_e, and d_i for the eyepiece is ∞. This gives an eyepiece magnification of

$$M_e = -\frac{d_i}{d_o} = -\frac{\infty}{f_e}.$$

Although the magnification of each of our lenses could not be well defined, let's look at what happens to the magnification of the system:

$$M_{total} = M_o M_e = \left(-\frac{f_o}{\infty}\right)\left(-\frac{\infty}{f_e}\right) = \frac{f_o}{f_e}.$$

Therefore, our telescope would give a magnification of

$$M_{total} = \frac{f_o}{f_e} = \frac{2.70 \ m}{0.0212 \ m} = 127.$$

Therefore, this telescope would have a magnification of 127.

The ray diagram for the simple telescope we just described is shown to the right The light leaving the eyepiece (on the right of the image) is parallel to the optic axis ($d_i = \infty$). So, how can we *see* the star? It turns out we need another optical instrument – your eye. The human eye acts as a converging lens and is most relaxed when looking at very distance objects (parallel light rays). We will study more on the human eye in section 5 below.

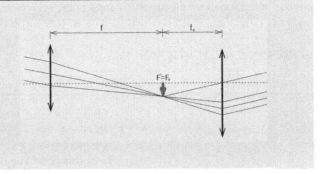

220-5: The human eye and vision

Consider: *How does the human eye work as an optical system?*

The human eye is a remarkable optical instrument. Start off by taking a look at the basic anatomy of the eye in figure 226-7. Light first enters the eye through the **cornea** (the front surface of the eye). Just behind the cornea is a fluid-filled chamber known as the anterior chamber, which is filled with **aqueous humor**. Light then passes through the **pupil** (a hole) to the **lens**. The combination of the shape of the cornea, the aqueous humor and the shape of the lens act together as a converging optical element. Although the cornea and aqueous humor are relatively stable, we can change the shape of the lens, and therefore the focal length of the system, by using the **ciliary muscles** attached to the lens. Although the process is often automatic, when you purposefully focus on one distance, you are using your ciliary muscles to affect the shape of your lens.

When everything is working correctly, light that passes through the cornea and lens then proceeds through the **vitreous humor**, the fluid filling most of the eye, and is focused on the back inside surface of the eye, called the **retina**. When the eye is most relaxed, it is designed to focus light from very far away – even from infinity. When we try to focus on close objects, the ciliary muscles cause the lens to bulge in the middle. The ideal human eye can focus on an object as close as 25 cm.

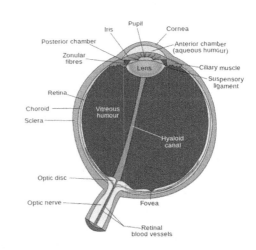

Figure 220-7. Basic anatomy of the human eye.

Light detection happens by two structures embedded in the cornea. Color vision is detected by **cones**, a set of three very similar cells, one of each corresponds roughly to red, green and blue color detection. Black and white vision (low light vision) is detected by **rods**. There is a small area at the back of the retina, almost across from the lens, called the fovea, which contains a very high concentration of cones. This is the area that is responsible for very precise color vision. Rods are spread much more evenly across the retina. An interesting point: the area where the optic nerve attaches to the retina is called the optic disc, and contains no rods or cones. Therefore, we all have a bind spot where this connection occurs, which is approximately 20° off axis.

Normal vision

Figure 220-8 shows roughly how rays entering the eye are focused to the back of the eye. Although it is difficult to show on a diagram such as figure 220-8, approximately 80% of the refraction of the cornea-lens system happens at the front of the cornea, and *not* at the lens. (This is a major misconception about the eye). The lens is responsible for approximately 20% of the refraction, and use of the ciliary muscles can change the focal length of the lens by 7-8% in a healthy eye. To see why this is, consider table 224-5, which gives

Table 220-5. Index of refraction (n) for materials in the eye.

Material	n
Air	~1
Cornea	1.38
Aqueous humor	1.34
Lens	~1.41 (varies)
Vitreous humor	1.34

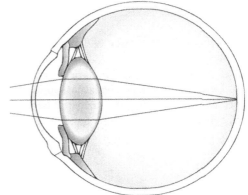

Figure 220-8. Rays converging on the back of the retina in a healthy eye.

the index of refraction for the important areas of the eye. Note that the transition from air to cornea is by far the largest change in index of refraction, which is why refraction is so strong at this interface.

Nearsightedness and farsightedness

Two of the most common vision problems are nearsightedness (myopia) and farsightedness (hyperopia or hypermetropia). Myopia is a very common condition where the eye is too long along its optical axis. This causes rays from a very distance

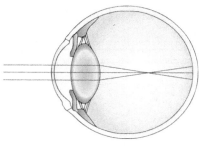
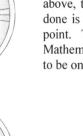

object the focus too far in front of the retina for the lens to correct. In order to fix this condition, a diverging lens is placed in front of the eye with either eye glasses or contact lenses. The focal length of the needed lens can be determined using the thin lens equation. The farthest point one can clearly see is called the **far point**. As noted above, the far point should be at infinity for normal vision. So, what needs to be done is to take an object at infinity and use the lens to place its image at your far point. That image then becomes the object for your eye, and corrects the vision. Mathematically, we use the this lens equation, noting that since we need the image to be on the same side as the object, it is a virtual image (negative):

$$\frac{1}{d_o} + \frac{1}{d_i} = \frac{1}{f} \quad \rightarrow \quad \frac{1}{\infty} + \frac{-1}{d_{fp}} = \frac{1}{f}, \tag{220-6}$$

where I have called the distance to the your far point d_{fp}. Solving, this for the focal length gives

$$f = -d_{fp}. \tag{220-7}$$

Note that the focal length is negative, meaning that we have a diverging lens (as we knew from above conceptually). One problem, corrective lenses are not denoted by their focal length, but rather by their **power**, which is simply the inverse of the focal length (with units of m^{-1}):

$$P = \frac{1}{f}. \tag{220-8}$$

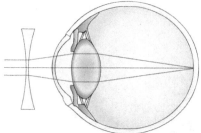

Figure 220-9. Myopia and its correction.

Therefore, if you are near sighted, you could use this process to get a lens power that is probably quite close to your prescription.

Hyperopia, on the other hand, occurs when the eye is too short along its optical axis and the rays therefore converge behind the retina. In order to correct this, a converging lens must be used. Figure 220-10 shows this arrangement. I noted before that the closest distance that the ideal human eye can focus on is 25 cm. The closest point that a person can focus is called the **near point**. We can use a very similar process to what we did for myopia to figure out the focal length and power of a lens that is needed to correct vision. In this case, we want to take 25 cm (call this the object) and place it at the near point for a person's eye (call this the image) so that this image becomes the object for the eye itself.

Using the thin lens equations, we find

$$\frac{1}{d_o} + \frac{1}{d_i} = \frac{1}{f} \quad \rightarrow \quad \frac{1}{25 \, cm} + \frac{-1}{d_{np}} = \frac{1}{f}, \tag{220-9}$$

where d_{np} is the near point of the person's uncorrected eye. Again, since we need the image to be on the same side of the lens as the object, it must be a virtual image, which is why the second term is negative.

This hyperopia equation is most easily solved for the power of the lens needed, since $P = 1/f$. Also, the standard units of lens power is the inverse meter, so I will convert 25 cm to 0.25 m. Solving, we find

$$P = \frac{d_{np} - 0.25m}{0.25 \, d_{np}}, \tag{220-10}$$

which can be simplified slightly to

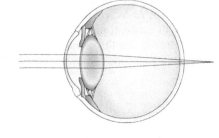
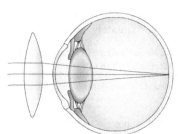

Figure 220-10. Hyperopia and its correction.

$$P = \frac{4d_{np} - 1\,m}{d_{np}}.$$ (220-11)

Two other vision conditions

I wanted to make quick note of two other very common vision problems:

Astigmatism – This condition is caused by irregularities in the curvature of the cornea or lens. These irregularities allow for light to focus at some points on the back of the retina along some axes, but not on others, leading to a blurry image. Astigmatism is often correctable these days with glasses, contact lenses or laser eye surgery. Figure 220-11 is a simple test for astigmatism. With one eye at a time (cover the other eye lightly), look at the very center of the image. Do all the radiating lines look the same or are some lighter or more blurry? If you see differences in the lines, this is a sign that you may have astigmatism.

Presbyopia – This very common condition comes with age. The lens of the eye becomes less pliable as we get older. Therefore, over time the ciliary muscles are less effective at focusing on near objects – remember, the resting state of the eye is to focus on far away objects. The inability to correct for close focus is called presbyopia. This is why many older citizens need to use reading glasses.

Figure 220-11. An astigmatism test. Focus on the center of the diagram. If the radial lines do not all appear to be the same darkness to you, you may have astigmatism.

> **Connection: Near point and myopia**
>
> It turns out that the near point of a myopic eye is often much closer than 25 cm. Although less precise, an interesting little exercise to do (if you are myopic) is to measure how closely to your face you can focus on the page of a book and then correct this 'too close' near point to the 'correct' 25 cm. For many with myopia, this result is very close to the lens power you get for doing it the correct way above. Try it!!
>
> $$P = \frac{4d_{np} - 1\,m}{d_{np}}.$$
>
> Note: the power will be negative in this case, since d_{np} will be less than 25 cm!

Color blindness

I mentioned earlier that color vision is created using three types of cone cells in the retina of the eye. Each type of cone responds to different color light as shown in figure 220-12. Earlier, I called the cones red, green and blue. More technically, they are denoted by their wavelength relationship: Short (S – blue), medium (M – green) and long (L – red). In normal color vision, the brain takes input from the various cones to create a composite color that we perceive. Notice that all of the cones taken together cover the entire visual part of the spectrum from 400 to 700 nm.

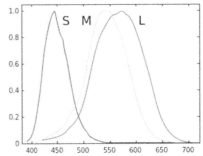

Figure 220-12. Response of each type of cone to color.

Colorblindness is actually a catch-all for any condition where one or more of the cones do not act the way they should, or where one or more of the cones is simply missing. The genes that regulate the colorblindness are carried on the X chromosome, so males are much more likely to suffer from these conditions than females. Since males only have one X chromosome, the genetic anomaly need only be present on that chromosome. For females to present colorblindness, the anomaly must be present on **both** X chromosomes, which is quite rare. Worldwide, about 8% of the male population and less than 1% of the female population suffer from colorblindness.

Colorblindness can be broken into two general categories:

1) **Dichromacy**, where one of the three cones is completely missing (missing the red cone being the most common with missing the green cone a close second).
2) **Anomalous Trichromacy**, where one of the three cones is present, but not functioning correctly. 73% of all colorblindness is this type.

I (the author) am colorblind. The specific type of colorblindness that I have is called **Deuteranomaly**, which affects about 5% of males and 0.4% of females. I have all three types of cones; however, my green cone is inefficient and has a slightly shifted peak color when compared to normal. I am often asked what things look like to me, and my first reaction is that things look the way they always have!! It is a very hard question since I have always seen colors the way I do. However, after many years of talking with people, I can now give some comparisons. Red and green are rather dull colors to me – more like grey than the bright hues of normal vision. I *can* often tell between the two because, to me, green is a much lighter 'shade' than red. The red and green in traffic lights are hard to tell apart; luckily green is either always on the bottom or on the right. Also, the new LED traffic lights (the ones that look pixilated) are much easier to tell apart. Christmas colors are kind of boring for me; although it is great to have all the lights around.

If you are interested, go to http://www.color-blindness.com/coblis-color-blindness-simulator/ where you can upload a picture (or use one of theirs) and it will show you what it looks like to people with different types of colorblindness. If you do try this, keep in mind that when you click from normal color to deuteranomaly, I see the *exact* same picture.

Connection: Four cones?

Recent studies suggest that some women (possibly as high as 50% of the population) have four types of color receptors, a condition called tetrachromacy. It is unclear whether this is truly a fourth type of cone, or a separate type of pigment living on one of the three common cones. Another study suggested that 2-3% of women do have a fourth cone, but it was unclear whether it is functional or not. Regardless, an extra cone or pigment greatly increases hue discrimination, and there is substantial evidence that some people fall into this category.

221 – Physical Optics

When the wavelength of light is on the same order of magnitude as the objects that it interacts with, we cannot use the ray model of light. In these cases, we must treat light fully as a wave, and include many of the wave properties discussed earlier in the course. The model of light used on this scale is called physical optics. Because physical optics effects are highly dependent on wave frequency, they are not readily seen in everyday life; however, with the use of a laser, we can explore the wave properties of light in detail.

Integration of Ideas

Review the basic idea of mechanical waves.
Review the idea of combining waves.

The Bare Essentials

- When the wavelength of light is close to the size of objects it is interacting with, we must consider the light to be a wave.

- Light (actually all EM radiation) has a property called polarization, which describes the plane(s) over which the electric field oscillates.

- The Law of Malus describes the intensity that emerges from a linear polarizer.

Law of Malus

$$I = I_0 cos^2\theta$$

Description – The Law of Malus gives the intensity of light, I, that is transmitted through a linear polarizer. I_0 is the intensity of the incoming light and θ is the angle between the plane of the electric field of the incoming light and the polarization direction of the polarizer.

- Light passing through two slits will produce an *interference pattern* on a screen that depends on the wavelength of light and the spacing between the two slits.

Double Slit Interference

$$d sin\,\theta_{int} = m\lambda \qquad m = 0, 1, 2 \ldots$$

Description – This equation describes the set of bright fringes where light passing through two slits constructively interfere to form an interference pattern. θ_{int} is the angle from the midpoint of the slits, m is the order of the fringe, λ is the wavelength of light and d is the distance between the slits.

- Light passing through a single slit comparable in size to the wavelength of the light will be *diffracted* through the slit.

Single Slit Diffraction

$$a\,sin\,\theta_{diff} = n\lambda \qquad n = 1, 2, 3 \ldots$$

Description – This equation describes the set of dark fringes caused by light diffracting through a single slit of width a. n is the order of the fringe and λ is the wavelength of light.

- *Diffraction* places limits on how closely two objects can be separated and still be *resolved*.

**Optical Resolution
(Circular Aperture)**

$$\theta_{res} = sin^{-1}\left(\frac{1.22\lambda}{a}\right)$$

Description – This equation describes the angle to the first dark fringe produced by diffraction through a circular aperture of diameter a.
Note 1: If two objects are separated by less than θ_{res}, they cannot be resolved at that wavelength.
Note 2: The factor of 1.22 comes from the circular nature of the aperture and is approximately the first zero of the Bessel Function of the first kind.

- *Dispersion* describes how light of different colors moves with different speeds in materials (different indeces of refraction). This effect leads to rainbows.

221-1: The wave nature of light

BACK IN UNIT 204, WE DISCUSSED electromagnetic waves and noted that the light we see with is a small part of the electromagnetic spectrum. Therefore, it stands to reason that one of the ways we could analyze light is by considering it as a wave. This is the model of *physical optics*. Geometrical optics, as studied in the last unit, simplifies the analysis and removes many of the wave-like properties of light when light interacts with objects sized much greater than the wavelength of light. However, when light interacts with objects that have dimensions on the order of the size of the wavelength, we must consider all of the wavelike features.

It is also important to note that although this chapter is primarily about light waves, many of the phenomena we discuss are applicable to all waves, including water waves, sound waves, etc. It is more common to see the effects of interference and diffraction (as described below) with light, which is why these effects are often discussed at this point in introductory physics. I will remind you at various places along the way of how other types of waves exhibit the properties described; however, our main focus is on physical optics.

221-2: Polarization

> **Consider:** *Why is the electric field direction so important in an em wave?*

Consider the following equations for an electromagnetic wave:

$$\vec{E} = E_0 sin(\omega t - kx)\hat{e}, \tag{221-1}$$

$$\vec{B} = B_0 sin(\omega t - kx)\hat{b}. \tag{221-2}$$

As a reminder, these equations describe how the electric and magnetic fields of the wave are oscillating with an angular frequency ω and angular wave number k. We know the wave is propagating in the $+x$ direction because x is in the argument of the sine function and the sign between the ωt and kx terms is negative. The amplitudes of the electric and magnetic fields are E_0 and B_0, respectively. Remember that $E_0 = cB_0$, so that the magnitude of the electric field is *much* larger than the magnitude of the magnetic field. Finally, remember that the direction of $\vec{E}$, $\vec{B}$, and the propagation of the wave must all be perpendicular to each other. If you are having trouble remembering these features, I strongly recommend reviewing unit 204.

Since the direction of the fields and direction of propagation are all perpendicular to each other, to characterize the wave, we only need two of these values. As noted above, we know the direction of propagation from the sine function. Since the magnitude of the electric field is so much larger than the magnitude of the magnetic field, we will choose the direction of the electric field as the other important value. We call this direction the *polarization* of the wave, and is given by $\hat{e}$.

Many sources emit light that has a random polarization leading to an *unpolarized wave*. This means that if we were to measure the direction of the electric field at points along the wave, we would get random, unpredictable results. Light from a light bulb, fluorescent lights and the sun are all unpolarized before hitting any surfaces.

However, if the electric field oscillates along predictable axes as the wave propagates, the wave is said to be *polarized*. If the electric field oscillates along only one axis, it is said to be *linearly polarized*. If the tip of the maximum electric field traces out a circle as the wave propagates, the wave is *circularly polarized*. If the shape is an oval, it is *elliptically polarized*, etc. Some of the common polarization states are shown in figure 221-1.

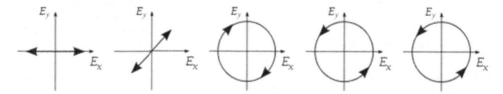

Figure 221-1. Some common polarization states of light propogating along the +z-direction. From left to right: linear polarization along an axis, linear polarization at 45°, right circular polarization, left circular polarization and elliptical polarization.

The polarization state of light can be altered by sending the light through a *linear polarizer* or a *wave plate*, A linear polarizer is a material that will only allow one polarization of light to pass through. A waveplate is a material that has a different index of refraction for different linearly polarized light (called a birefringent material). Waveplates are often used to change light between linear and circular polarizations. The effect of these polarizers can be seen in figure 221-2.

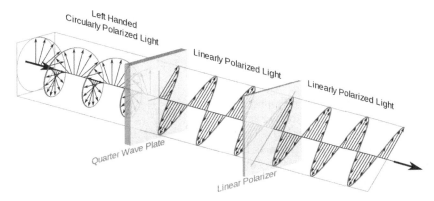

Figure 221-2. The effects of waveplates and polarizers on light.

Both unpolarized light and circularly polarized light can be thought of as made up of polarizations along the two axes that are perpendicular to the direction of propagation. Therefore, **both unpolarized light and circularly polarized light lose half of their intensity when they pass through a linear polarizer**. When linear polarized light is incident upon a linear polarizer, the percentage of light that is transmitted is given by the law of Malus:

Connection: Polarizing sunglasses

Polarizing sunglasses contain linear polarizers in their lenses, with the transmission axis oriented vertically. Sunlight coming from the sky is relatively unpolarized, so half of the light passes through the lenses. However, light reflected off of flat horizontal surfaces (such as water) is polarized along the horizontal axis and is therefore mostly blocked by the sunglasses. Put another way, polarizing sunglasses greatly reduce glare from light reflected off of water as well as horizontal metal surfaces.

Law of Malus

$$I = I_0 cos^2\theta \qquad (221\text{-}3)$$

Description – The Law of Malus gives the intensity of light, I, that is transmitted through a linear polarizer. I_0 is the intensity of the incoming light and θ is the angle between the plane of the electric field of the incoming light and the polarization direction of the polarizer.

The light that emerges from a polarizer is aligned with the polarization axis of the polarizer.

Example 221-1: Law of Malus

Unpolarized light with an intensity of $22.1\ W/m^2$ is incident upon a pair of linear polarizers. The first polarizer has its transmission axis aligned $25°$ from the vertical. The second polarizer has its transmission axis aligned at $42°$ from the vertical. What is the intensity of light that emerges from the second polarizer?

Solution:

This problem contains both a conceptual interpretation of polarization and polarizers and a direct application of the Law of Malus. Key terms are polarization and angles.

When unpolarized light passes through a linear polarizer,

half of the incident intensity passes through, therefore light striking the second polarizer has intensity $\frac{1}{2}(22.1\ W/m^2) = 11.05\ W/m^2$.

Now as the light passes through the second polarizer, the intensity will go drop in accordance with the Law of Malus, keeping in mind that the angle is the angle between the two axes - $\theta = 42° - 25° = 17°$:

$$I = I_0 cos^2\theta = \left(11.05\frac{W}{m^2}\right)cos^2 17° = 10.1\frac{W}{m^2}.$$

Also note that the net intensity drop due to the polarizers is the product of the drop of each.

221-3: Double slit interference

Consider: *Does light interfere like mechanical waves do?*

When a wave is incident upon a small hole, the wave will **diffract**, or bend around the edges, as it passes through the hole. Another way of saying this is that the hole acts as a secondary source of a spherical wave emanating from the hole. This effect can be seen on the left side of figure 221-3.

As we will see later in the unit, the way light diffracts through a hole is dependent on the size of the hole and the wavelength of the light used. This second point, the wavelength is very important – we don't notice diffraction in our everyday life because white light is made up of all the colors of the rainbow. In order to see diffraction effects, we must use a monochromatic (one-color) source. Also, we must use a **coherent source**. A coherent source is one which produces waves that are in-phase with each other, meaning that all of the peaks of the sine waves are on top of each other. An **incoherent source** (such as a light bulb or the sun) does not satisfy this requirement as the light waves have a random phases relative to each other. Luckily, lasers produce relatively monochromatic, coherent light that can be used for such as experiment.

At S2 in figure 221-3, the two small slits again act as light sources, sending out spherical waves. Then, at a given point on a screen (at F),

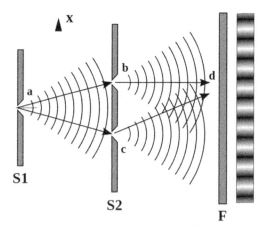

Figure 221-3. Setup for Young's double slit experiment. Slit 1 (S1) acts as a single source. The two slits at S2 act as secondary sources. The detector (F) shows the overall interference pattern.

the waves from each source combine, the only problem is that, except for the exact center between the two slits, light from each slit has traveled a different distance to make it to the screen. The superposition principle says that to find the overall amplitude of the light on the screen, we simply add the amplitudes of each waves; however, since they have traveled different distances, the amplitude at the screen could be different for each wave. As a couple of examples, take a position where the difference in pathlength is exactly one wavelength. Such a situation, known as **constructive interference**, is shown in figure 221-4(a). In this case, the combined wave has the highest possible amplitude and we would expect the spot at that point to be very bright. Contrast this with a point where the difference between the two paths is

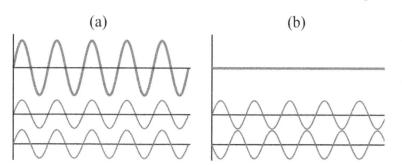

(a)   (b)

Figure 221-4. (a) Constructive interferences when the pathlength differs by a multiple of λ. (b) Destructive interference when the pathlengths are half a wavelength different.

exactly one-half of one wavelength, as shown in figure 221-4(b). In this case, the waves cancel at everywhere and the amplitude is zero, a situation known as **destructive interference**. We would expect to have a dark spot at this place on the screen. This pattern should repeat for all possible pathlength differences of one wavelength and one-half wavelength as can be seen on the very right side of figure 221-3.

Mathematically, we can say that a bright spot appears where the difference in path length is any integer multiple of the wavelength, which I will call mλ. How do we determine the pathlength difference in our situation though? Consider figure 221-5. If the distance between the two slits is d and the angle from the slits to the point on the screen we care about is θ, then the difference in pathlength between the two waves is

$$\Delta l = d sin\ \theta. \qquad (221\text{-}4)$$

Since we want our difference in pathlength to be mλ, we can then write

$$m\lambda = d sin\ \theta. \qquad (221\text{-}5)$$

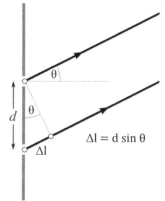

Figure 221-5. Diagram of double-slit experiment.

This is the angular condition to find bright spots on a far screen due to a double slit setup. This experiment has been conducted innumerable times and is called Young's double slit experiment after its original designer, Thomas Young.

Double Slit Interference

$$d\sin\theta_{int} = m\lambda \qquad m = 0, 1, 2 \ldots \qquad (221\text{-}6)$$

Description – This equation describes the set of bright fringes where light passing through two slits constructively interfere to form an interference pattern. θ_{int} is the angle from the midpoint of the slits, m is the order of the fringe, λ is the wavelength of light and d is the distance between the slits.

Note: The subscript *int* stands for interference to help you separate this from other angles discussed in this unit.

Note, there is always a bright spot in the middle of the screen (straight out from the middle of the two slits). This is known as the zeroth order maximum ($\theta = 0$). In order, the next set of maxima are called the 1st order maximum, 2nd order maximum, etc. The order of the maxima corresponds to the value of m in the double slit equation.

Now that we know the angles to the bright spots on a far screen, we can use a little bit of trigonometry to find where the bright spots are on the screen, relative to the midpoint of the experiment. If the screen is a distance, D, along the axis of the experiment from the two slits, and the angle to the maximum is known from θ_{int}, then the distance along the screen from center to maximum, y, is given by

$$y = D\tan\theta_{int}. \qquad (221\text{-}7)$$

I should also note that we could run the exact same process for the dark fringes in the interference pattern. The only difference is that we need the destructive interference that happens when the two waves are shifted by half a wavelength. Therefore, the equation for the dark fringes is

$$d\sin\theta_{int,dark} = (m + 1/2)\lambda \qquad m = 0,1,2 \ldots \qquad (221\text{-}8)$$

Note: This is one of those situations where if you understand where the equation for bright spots came from, it is very easy to figure out the equation for dark spots.

Example 221-2: Distance between spots

Green light (543 nm) is incident upon two small rectangular, vertical slits separated by 23.1 μm. What is the spacing between the zeroth order and first order maxima on a screen 2.12 m from the slits?

Solution:

This is a light interference problem. Key terms are light, maxima and two slits.

The first step is to find the angle of separation between the given maxima. By definition, the zeroth order maxima is at $\theta = 0$. The first order maxima ($m = 1$) is found using the interference equation:

$$d\sin\theta_{int} = m\lambda.$$

Rearranging and substituting our known values, we find

$$\theta_{int} = \sin^{-1}\frac{m\lambda}{d} = \sin^{-1}\frac{(1)(543\ nm)}{(23.1\ \mu m)} = 1.3°.$$

We now have the angle for both the zeroth order (0°) and first order (1.3°) maxima. We can use each of these angles in the equation relating distance on a screen to the angle:

$$y = D\tan\theta_{int}.$$

Since the zeroth order maxima has an angle of zero, we can immediately write

$$y_0 = 0.$$

For the first order maximum, we find

$$y_1 = (2.12\ m)\tan 1.3° = 0.048m = 4.8\ cm.$$

Therefore the separation between the two bright spots is $4.8\ cm - 0 = 4.8\ cm$.

221-4: Single-slit diffraction

Consider: *What happens when light goes through a very small hole?*

I noted in the last section that light will bend around a corner due to diffraction. In this section, we are going to look more closely at how diffraction works and what sorts of patterns it produces. Consider a single slit with a width, a. As discussed before, when light impinges on the slit, the slit acts as a secondary source for spherical waves – which we used in determining the pattern for Young's double slit experiment above. However, not only does the overall slit act as a secondary source, but each point within the slit acts as a source, a principle first put forth by Christiaan Huygens in 1678:

> **Huygens' Principle**
>
> Every point on a wavefront may be considered the source for secondary spherical wavelets.

Huygens' Principle is very powerful and used extensively in advanced optics. This principle holds even for light waves traveling through empty space; however, in that case, it tends not to be very interesting, since applying Huygens' principle in free space to a spherical wave gives you a spherical wave and applying it to a linear wave gives you a linear wave. Exciting, eh? However, when we include boundaries, using Huygens' principle solves many problems in optics (including the bending of light at an interface – see the optional section at the end of this unit).

Consider figure 221-6. Although the figure resembles that from our double slit experiment in section 221-3, note that there is only one opening. Let's say the opening has a width a. The figure not only shows waves traveling from each side of the opening, but also one from the middle of the opening. This wave in the middle is one of the wavelets we will use to apply Huygens's principle to the opening.

We now use a very similar reasoning to that of our double slit. Namely, I am going to look for an angle from the opening at which the wave from one end of the opening and the wave from the center of the opening destructively

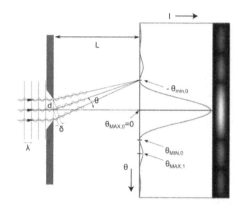

Figure 221-6. Setup for describing single-slit diffraction.

interfere. That is to say, the angle at which their waves will be out of phase by one-half of a wavelength. The geometry is exactly the same as the geometry for the double-slit experiment, which led to

$$d\sin\theta = m\lambda. \tag{221-9}$$

However, we have to make a couple of modifications. First, whereas d was the distance between the double slits, we must use the distance between the side of the opening and the center of the opening, $a/2$. Also, for the double slit, we were looking for bright spots (where the waves were in-phase) and now we are looking for dark spots (where the waves are out of phase by $\lambda/2$. With these substitutions, we find:

$$\frac{a}{2}\sin\theta = m\frac{\lambda}{2}. \tag{221-10}$$

The equation can be simplified further by canceling the 2s. Also, I will differentiate the angle in the equations from the one for interference by using the subscript "diff" for diffraction:

> **Single Slit Diffraction**
>
> $$a \sin \theta_{diff} = n\lambda \qquad n = 1, 2, 3 \dots \tag{221-11}$$
>
> **Description** – This equation describes the set of dark fringes caused by light diffracting through a single slit of width a. n is the order of the fringe and λ is the wavelength of light.
> **Note:** The diffraction equation gives you **dark spots**.

The right side of figure 221-6 shows what a diffraction pattern looks like in both graphical and picture forms. Note that the center maximum (zeroth order maximum) in a diffraction pattern is much brighter than the maxima farther out (higher order maxima).

We found the diffraction equation by dividing the single hole into two pieces. We could continue our exploration of diffraction by dividing the hole up into 4, 8, 16, etc., sections and doing the analysis again. I'll ask that you trust me that you get the same results as boxed above and leave the work as an exercise for those interested.

Note: The smaller the slit, the larger the diffraction pattern! (a and $sin\,\theta_{diff}$ are inversely proportional.)

Example 221-3: Diffraction effects

Red light ($\lambda = 632\ nm$) passes through a single slit with width 2.36 μm. What are the angles of the 1st, 2nd, 3rd and 4th dark fringes in the diffraction pattern?

$$\theta_1 = \sin^{-1}\frac{(1)(632\ nm)}{2.36\ \mu m} = 15.5°,$$

Solution:

$$\theta_2 = \sin^{-1}\frac{(2)(632\ nm)}{2.36\ \mu m} = 32.4°,$$

This is a direct application of our equation for diffraction. Key terms include single slit and dark fringes.

$$\theta_3 = \sin^{-1}\frac{(3)(632\ nm)}{2.36\ \mu m} = 53.4°.$$

Since we know this is a diffraction problem, we start with our equation for single-slit diffraction:

$$a\ sin\ \theta_{diff} = n\lambda.$$

For the 4th dark fringe, we have a problem – the argument of the inverse sine function,

Also, since we were asked for the angles, I will rearrange this equation to solve for θ_{diff}:

$$\frac{n\lambda}{a} = \frac{(4)(632\ nm)}{2.36\ \mu m} = 1.07,$$

$$\theta_{diff} = \sin^{-1}\frac{n\lambda}{a}.$$

Is greater than one, which makes the inverse sine undefined. This suggests that the angle of diffraction would be greater than 90° which is not possible – light would have to go through the slit in one direction and then travel back the other direction through the material the slit is cut in. Therefore, there are only three dark fringes in this setup with angles given by θ_1, θ_1 and θ_1 above.

We can now find each angle directly by substituting the known values into our equation:

221-4: Diffraction and interference can happen at the same time

Consider: *Does the light forming an interference pattern also have a diffraction pattern since the light went through small holes?*

Both diffraction and interference occur as light moves through one or more slits. Since, conceptually, the setups are very similar, you might guess that both effects happen at the same time. In fact, when doing a Young's double slit experiment, in addition to the nice, regular interference pattern we find from two slits, a diffraction pattern is superimposed over the interference pattern causing the intensity of higher-order maxima to decrease rapidly.

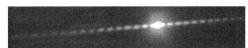

Figure 221-7. Interference pattern with diffraction.

This effect can be seen in figure 221-7, where you can see the even spacing of the interference pattern drop off in intensity as you move away from the central maximum. If you need to calculate the combined effects of interference and diffraction, you do each separately and then combine the effects later.

221-5: Optical Resolution

Consider: *Does diffraction place a limit on the imaging of objects?*

Diffraction of light has stark consequences for optical instruments, including binoculars, cameras, telescopes and the human eye. This is especially true for high-resolution applications such as those used in surveillance and anti-terror activities. Think of it this way – if we want to image just a tiny pin prick in a piece of paper, what would a camera pointed at the pin prick see?

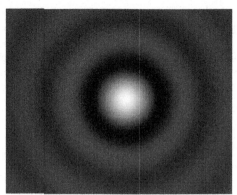

Figure 221-8. Two dimensional diffraction pattern of a circular hole.

It would be nice if the light from behind the pin hole flowed in straight lines to the camera creating a perfect image of the hole. Unfortunately, we now know that the light will diffract as it passes through the hole and therefore spread out in a diffraction pattern. The two dimensional diffraction pattern of a circular hole is shown in figure 221-8. This type of diffraction pattern is known as an airy pattern. Also, remember, that the smaller the hole, the larger the diffraction pattern!

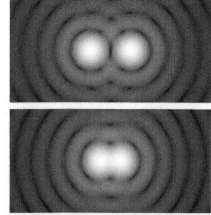

Now, imagine that we want to make two pinholes near each other in a piece of paper. Since light will diffract through both pinholes, each image will be smeared out. This means that the two pinholes can only be so close together on the paper before their diffraction patterns start to overlap and we can no longer tell their images apart.

This effect is known as ***diffraction limited imaging***. We can relatively easily get an equation describing the diffraction limit. What we need is for the peak of each diffraction pattern to be separated so that their first minima coincide. Any closer than this and the two peaks will overlap and we lose our ability to resolve them separately. Therefore, the two pinholes must be separated such that their diffraction patterns are more than the distance to the first diffraction minima apart. This effect can be seen in figure 221-9.

There is one complication – since our diffraction pattern is two dimensional, the circular nature of the opening must be taken into account. The mathematics behind this is quite complicated, so we will skip over the details here. The crux of the situation is that it adds a 1.22 to the equation:

Figure 221-9. Overlapping diffraction patterns of two pinholes as they approach each other.

Optical Resolution
(Circular Aperture)

$$\theta_{res} = sin^{-1}\left(\frac{1.22\lambda}{a}\right) \qquad (221\text{-}12)$$

Description – This equation describes the angle to the first dark fringe produced by diffraction through a circular aperture of diameter a.
Note 1: If two objects are separated by less than θ_{res}, they cannot be resolved at that wavelength.
Note 2: The factor of 1.22 comes from the circular nature of the aperture and is approximately the first zero of the Bessel Function of the first kind.

The equation for resolution above is also called the ***Rayleigh criterion*** after Lord Rayleigh. Do keep in mind that this criterion is only for *circular apertures*. For any other shapes, the equation must be derived from basic physics principles. However, for our purposes, you can apply this equation to any two-dimensional hole.

Example 221-4: Human eye resolution

The human pupil is a circular aperture that allows light to enter the eye. Our bodies can vary the diameter of the pupil from about 2 mm in bright light to about 8 mm in dark conditions. What are the maximum and minimum angular resolutions for the human eye at 543 nm? What object separation does this correspond to at a distance of 100 meters?

Solution:

The angular resolutions can be directly calculated from the information given:

$$\theta_{2mm} = sin^{-1}\left(\frac{1.22\lambda}{a}\right) = sin^{-1}\left(\frac{(1.22)(543\ nm)}{2\ mm}\right) = 0.019°,$$

$$\theta_{8m} = sin^{-1}\left(\frac{(1.22)(543\ nm)}{8\ mm}\right) = 0.0047°.$$

For object separation at 100m, we can use the separation of a screen equation from the interference section:

$$y = Dtan\ \theta_{int},$$

which gives us

$$y_{2mm} = (100m)tan\ (0.019°) = 3.3cm,$$

$$y_{8mm} = (100m)tan\ (0.0047°) = 8.2mm.$$

As you can see, the 'perfect' human eye would be able to make out very fine details at 100m. Most of our eyes are not capable of this diffraction limited viewing.

221-6: Dispersion and rainbows

Consider: *What causes rainbows?*

Dispersion is a phenomenon where the index of refraction of a material depends on the frequency of light traveling through it. What this does is cause the light of different colors to spread out. As we will see, this is how prisms work and how rainbows form. We have been able to ignore this effect up until now because of two reasons

1) In geometrical optics, we dealt with thin lenses and so there was not enough distance for dispersion to become important,
2) In physical optics, we are dealing with monochromatic (one color) light in order to see the wave effect.

Figure 221-10 shows how the index of refraction of a number of types of glass depends on the wavelength of the light traversing the material. Remember that although the index of refraction of a material is most directly related to the speed of light in that material (n = c/v), the index of refraction is also important in how terms of how much light bends at an interface through the law of refraction

$$n_1 \sin \theta_1 = n_2 \sin \theta_2. \qquad (221\text{-}13)$$

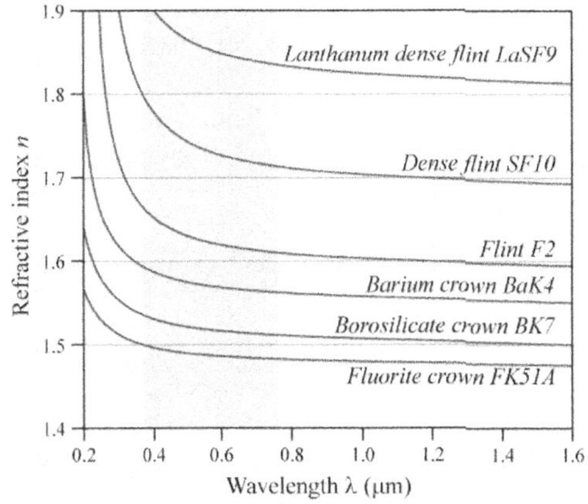

Figure 221-10. Index of refraction versus wavelength for a number of glasses showing dispersion.

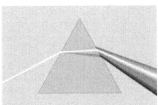
Figure 221-11. How a prism works based on dispersion.

Therefore, if different colors of light are bending at different angles at an interface, the different colors move at separate angles through the material. This effect can be seen for a prism in figure 221-11. In some materials red light moves faster than blue light (called **normally dispersive materials**) and in others blue light moves faster than red light (called **anomalously dispersive materials**). It was originally thought that most materials exhibited normal dispersion in the optical region of the spectrum; however, we now have many examples of both normal and anomalous dispersion in normal transparent media.

Rainbows are one of the most common visual displays of dispersion that we see. White light enters water droplets and is dispersed in the water. The light must then reflect off of the back side of the droplet (through total internal reflection) and

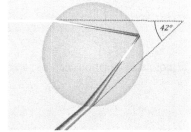
Figure 221-12. How a rainbow forms through dispersion.

209

continue back towards the front side of the raindrop. When the light leaves the drop, the colors of light have been separated through dispersion. A couple of important features: 1) since the light must undergo total internal reflection on the back of the water drop, this limits the angles at which rainbows can form; 2) Since we can only see rainbows via reflection, the sun must be behind you or to the side of you and rain in front of you. These criteria are why we do not always see a rainbow when the sun is out and it is raining – there are more stringent conditions.

221-7: Huygens' Principle and refraction (optional)

Consider: *Where does the law of refraction (Snell's law) really come from?*

In unit 219, I gave a (bad) analogy of a car moving from a smooth surface to a rough surface to describe why light bends at an optical interface. Now that we have learned a little about Huygens' principle, we can use this to better describe why light bends the way it does.

Figure 221-13 shows spherical wavelets (al la Huygen's principle) forming as a linear light wave hits an optical interface. In this diagram the index of refraction of the lower medium is higher than that of the upper medium. Since the index in the lower medium is higher, the speed of light is lower. Note that the spherical wavelets are smaller on the right side because those areas of the wave hit the interface later than the wave on the left side of the diagram; therefore the waves on the right side had less time to propagate in the new medium

By tangentially connecting the spherical wavelets in figure 221-13, you can see how the wavefronts have changed directions. We can also include some mathematical detail to show exactly how the wave changed direction. Let's call the time in figure 221-13, which is the time that wavelets begin to form on the right side of the wave, t_2. Now, let's backtrack in time to when the wavelets on the right side of the wave began to form and call this time t_1.

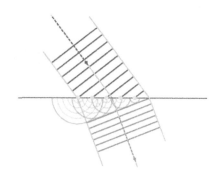

Figure 221-13. Wavelets forming at an optical interface.

During the time from t_1 to t_2, the right side of the wave was moving in the upper material at a speed I'll call v_1, and the left side of the wave was moving in the lower material at a speed v_2. So, over the course of the time $t_2 - t_1$ each wave moved a distance

Right

$$d_{right} = v_1(t_2 - t_1) = v_1\Delta t \qquad (221\text{-}14)$$

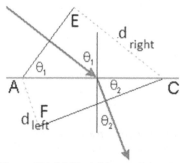

Figure 221-14. Variables used for the refraction problem.

Now we need to use a little trigonometry to relate the angles. Figure 221-14 shows the same setup as figure 221-13, but with our distances over Δt labeled as well as introducing angles to the problem. θ_1 is the angle of the incoming ray relative to the normal and θ_2 is the angle of the outgoing ray with respect to the normal. By definition d_{right} is parallel to the incoming ray and d_{left} is parallel to the outgoing ray. Line segment AE was formed perpendicular to the incoming ray so that the angle between AC and AE is the same as the angle of incidence (θ_1). Likewise, line segment AF was formed perpendicular to the outgoing ray so that the angle between AC and AF is the same as the angle of refraction (θ_2).

Now, we can use trigonometry to write down the known angles

$$\sin\theta_1 = \frac{d_{right}}{AC} = \frac{v_1\Delta t}{AC} \qquad \sin\theta_2 = \frac{d_{left}}{AC} = \frac{v_2\Delta t}{AC}. \qquad (221\text{-}15)$$

Each of these equations can be solved for $\frac{\Delta t}{AC}$, giving us

$$\frac{\Delta t}{AC} = \frac{\sin\theta_1}{v_1} = \frac{\sin\theta_2}{v_2}. \qquad (221\text{-}16)$$

Recall that the index of refraction of a material is defined as $v = c/n$. We can now substitute this into the equation above:

$$\frac{n_1 \sin \theta_1}{c} = \frac{n_2 \sin \theta_2}{c}.$$

(221-17)

Finally, we cancel the speed on light on both sides of the equation, and we are left with the law of refraction:

$$n_1 \sin \theta_1 = n_2 \sin \theta_2.$$

(221-18)

Although, I used a number of paragraphs to explain every step, the actual derivation of the law of refraction from Huygens' Principle only took four lines of math. Most calculations starting primarily with Huygens' Principle consume considerably more space; however, I hope this shows how powerful the principle can be for even a simple situation.

222 – Quantum Optics

When scientists began to look at how light interacts with individual particles, such as electrons, they discovered a startling result – light was also acting as a particle itself. The model of light used when discussing interactions on the very small scale of subatomic particles is called quantum optics. In this unit, we explore some of the important results of this model.

Integration of Ideas

Review the ideas of light as rays and waves.
Review blackbody radiation from Unit 204.

The Bare Essentials

- In certain situations, light also acts as a large conglomeration of particles known as photons. This model of light is known as *quantum optics*

- The wave model of light was consistent with almost all experiments up until approximately 1900; however, after this time, some experiments were not consistent with the wave model and suggested that light acts as a particle.

- An important energy unit in quantum optics is the energy that one electron would gain being accelerated through a one-volt potential difference – the electron-volt (1 eV = 1.6 x 10^{-1} J).

- Individual photons contain energy that is related to their wavelength.

Photon Energy

$$E = hf = \frac{hc}{\lambda}$$

Description – This equation describes the energy contained in an individual photon in terms of Planck's constant, h, the speed of light, c, the frequency, f, and/or the wavelength of light, λ.
Note 1: The quantity hc will be encountered quite often, its value is $hc = 1240$ $eV \cdot nm$.
Note 2: $h = 6.626 \times 10^{-34} J \cdot s = 4.136 \times 10^{-1}$ $eV \cdot s$
Note 3: $hc = 1240$ $eV \cdot nm$.

- Lasers are a ubiquitous application of quantum optics in modern technology.

- The Photoelectric Effect is a process by which light of a given energy can knock an electron from the surface of a metal so that is then free to move.

The Photoelectric Effect

$$K_{max} = \frac{hc}{\lambda} - W$$

Description – This equation describes how an electron liberated by a photon of energy $\frac{hc}{\lambda}$ will have a maximum kinetic energy (K_{max}) depending on the work function (W) of the material from which the electron came.
Note 1: W is the *work function* of the metal, and is an intrinsic property of the material, representing the potential energy the electron must overcome to be liberated.
Note 2: This equation is essentially the conservation of energy assuming the electron has its maximum kinetic energy when released.

222-1: The particle nature of light

Consider: *Wait, so light can act like a particle?*

FOR MOST OF THE 19TH CENTURY, scientists were confident in the wave nature of light. Maxwell's equations showed that light was a combination of electric and magnetic fields (electromagnetic waves), and light produced interference and diffraction pattern as would be expected from light as a wave. However, near the end of the 19th centuries, experiments started to emerge that directly contradicted this wave model.

Consider light hitting metal with many free electrons. Although the electrons are free to move within the metal, they are bound to the metal by a potential energy. When light waves hit the metalic surface, the energy of the wave should be transferred to particles in the metal. This can happen in two main ways; 1) energy is transferred into the vibrational states of bound atoms becoming thermal energy, or 2) energy could be transferred to free electrons, increasing their kinetic energy. We would expect both of these effects to happen routinely – and they do! Now, consider a very intense beam of light hitting the surface of a metal. Continuing the two processes above, we would expect the metal surface to warm up, but also the electrons to eventually have enough kinetic energy to overcome the potential energy holding them to the metal – and they would escape the metals surface; a process known as ejection.

Physicists performed this type of experiment many times, and when using ultraviolet light, found exactly what I just described. However, as they decreased the frequency of the light, eventually they found a color where no electrons were ejected at all – no matter how bright or intense they made the light. This made no sense – if the light were simply a wave coming in, no matter what the frequency, it should only be a matter of time before electrons are ejected. This is simply physics – energy is power multiplied by time. If the power is lower, but you increase the time, the total energy transferred should be the same, and the electron should eject.

So, what was happening here? Well, in 1901, Max Planck proposed a theoretical model, where light could be treated as discrete bundles of energy as opposed to electromagnetic waves. Originally, this was considered an odd mathematical trick that allows the correct **blackbody radiation curves** (see unit 204) and ejection of electrons to be modeled correctly. However, by the 1920's these discrete bundles of energy, now called **photons**, became an important part of the theoretical background of **quantum mechanics** – a theory of the very small that treats every bound interaction in small energy packets known as quanta.

How do we know if light is acting as a particle or a wave? Is it really one of these models, but just acts like the other in certain situations? The answers to these two questions are: it depends, and no. **Light is both a particle and wave at the same time.** This effect is known as **particle-wave duality** and is at the heart of quantum mechanics. We'll see in unit 223 that particle-wave duality not only holds for photons, but for all subatomic particles such as electrons. In fact, in very special situations, even things that can be seen with the naked eye can exhibit this weird phenomenon.

For now, let us get back to photons. Just as electromagnetic waves, photons carry energy, and we now know that the energy of an individual photon depends on its frequency.

Photon Energy

$$E_p = hf = \frac{hc}{\lambda} \qquad (222\text{-}1)$$

Description – This equation describes the energy contained in an individual photon in terms of Planck's constant, h, the speed of light, c, the frequency, f, and/or the wavelength of light, λ.

Note 1: The quantity hc will be encountered quite often, its value is $hc = 1240 \, eV \cdot nm$.

Note 2: $h = 6.626 \times 10^{-34} J \cdot s = 4.136 \times 10^{-15} eV \cdot s$

Note 3: $hc = 1240 \, eV \cdot nm$.

Needless to say, the energy of an individual photon is very small; a photon in the optical area of the spectrum has energy around 10^{-19} J. In order to counter this, we use a new unit of energy, called the **electron-volt (eV)** to describe energies on this scale. One eV is the energy gained or lost by an electron as it travels through a potential difference of one volt:

$$1 \, eV = (1.6 \; x \; 10^{-19} \, C)(1 \, V) = 1.6 \; x \; 10^{-19} J. \qquad (222\text{-}2)$$

One of the nice things about the eV is that if you remember the charge on an electron, you automatically know the conversion factor between eV and Joules – because it is numerically the same as the charge on the electron.

The h in the photon energy equation is the proportionality constant between frequency and energy and is called ***Planck's constant*** after the same Max Planck described above. The value of Planck's constant is given in the box above in both $J \cdot s$, and $eV \cdot s$, so that you have access to it in whichever units are best suited.

Also, if you notice *Note 3* in the photon energy box, the value of the combined hc (Planck's constant multiplied by the speed of light) is $1240\ eV \cdot nm$. Using this value can make calculations of photon energy very easy if you have the wavelength of light in nanometers – which of course makes sense since the wavelength of visible light is approximately $400 - 750$ nm. Since this is an important value for these calculations, I mention it again:

$$hc = 1240\ eV \cdot nm. \qquad (222\text{-}3)$$

> **Connection**: Your amazing eyes
>
> The rods in your eyes respond to a single optical photon. This, in itself, is not too surprising. However, it only takes about nine photons for a completely dark-adjusted eye to visually respond and see a pin-prick of light. Compare this number to the number of photons given off by a typical light bulb (see example below), and you can get a feel for just how amazing this is. Also, to put in another way, these nine photons only carry about 3×10^{-18} Joules of energy, but your eye and brain can respond to such small energy.

Example 222-1: How many photons?

How many photons per second does an 8 watt LED light bulb emit, assuming that all 8 watts go into light. Note that an 8 watt LED gives off the same light equivalent as a standard 60 watt incandescent bulb.

Solution:

Remembering that the units of watts are joules per second, we can directly relate the power of the light bulb to the photons released per second using the photon energy equations.

LED light bulbs give off light across the optical spectrum. So, for this problem, I will first assume an average wavelength for the light emitted – 550nm.

At 550 nm, energy of one photon is

$$E_p = \frac{hc}{\lambda} = \frac{1240\ eV \cdot nm}{550\ nm} = 2.25\ eV \text{ per photon.}$$

Also, I will now convert our 8 watts into eV/s:

$$8\ W = \frac{8\ J}{s} \cdot \frac{1\ eV}{1.6 \times 10^{-1}\ J} = 5.0 \times 10^{19} eV/s.$$

Therefore, our number of photons per second is

$$n = \frac{5.0 \times 10^{19} eV/s}{2.25\ eV \text{ per photon}} = 2.2 \times 10^{19} photons/s.$$

As you can see, this is an immense number of photons!

Example 222-2: Photon energy we can see

To give a feel for the energy range of optical photons, calculate the energy per photon (in eV) of a 400 nm photon (violet light) and a 700 nm photon (red light).

Solution:

This is a direct application of our photon energy equation:

$$E_p(400\ nm) = \frac{hc}{\lambda} = \frac{1240\ eV \cdot nm}{400\ nm} = 3.1\ eV,$$

$$E_p(700\ nm) = \frac{hc}{\lambda} = \frac{1240\ eV \cdot nm}{700\ nm} = 1.8\ eV.$$

You can see here why the eV is such a useful unit for photons – optical photons are just a few eVs.

222-2: Quantum optics and lasers

Consider: *How does a laser (such as a laser pointer) work?*

The term ***quantum optics*** is used to describe optics where light is treated as a particle. Remember that geometrical optics treated light as a ray and worked when light was interacting with large objects. Wave optics directly dealt with light as a wave and works when light interacts with objects, openings, masks, etc., which are about the same size as the wavelength of light. Quantum optics tends to be used when we care about the interaction of light with a single atom or molecule.

One example of this interaction is in your eye. As noted in the connection box above, rods in your eye respond to a single photon by exciting. The photoreceptors in your eye work backwards from what you might expect. When not receiving light, the rods and cones in your eyes create a constant current, known as a dark current. When a photon strikes a large molecule within the rods and cones known as opsin, electron transduction causes proteins to change shape and close an ion channel, stopping the dark current. Although chemically very complex, the process starts with a single photon interacting with a single molecule – a hallmark of quantum optics.

Another of the great successes of quantum optics is the **laser**. Laser was originally an acronym for light amplification by the stimulated emission of radiation, although it is now considered by most to be a separate term on its own. Being a feature of quantum optics, you might now guess that lasers work by the interaction of individual photons with electrons in atoms – which is exactly how they work.

Lasers make use of quantum energy levels in atoms and molecules. Although we will study these in greater depth in the next unit, for now it is just important to know that electrons bound to atoms can only exist in very specific energy levels (one of the hallmarks of quantum mechanics as noted earlier). When a photon strikes an electron in an atom, if the photon carries the correct energy for the next energy level, the electron will jump to that level. If the photon does not contain the correct energy, the photon will pass by the electron and leave it unaffected. The same is true for electrons that are in an excited state. If there is a lower energy level in an atom than where they reside, they can drop down to the lower level and emit a photon whose energy is exactly the difference in energy between the two levels. This process is shown in Figure 222-1.

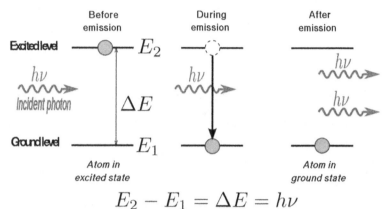

$$E_2 - E_1 = \Delta E = h\nu$$

Figure 222-1. The energy levels of a hypothetical laser and how an incident photon creates a second photon by stimulated emission.

In this generalized laser, electroncs are excited to a higher level, usually by heat or electric activity. As noted above, electrons in excited states will decay to the ground state emitting a photon. For the laser process to happen, this must happen spontaneously (a process known as spontaneous emission). However, once a photon has been emitted spontaneous, it will then stimulate the emission of photons in other nearby atoms with excited electrons. An interesting property of these stimulated photons is that they are **in-phase** with the photon that stimulated them (remember back to constructive and destructive interference). As more photons fill the laser cavity, the stimulated emission can get very bright. One end of the laser is engineered to let a small percentage of this stimulated emission out (called the output coupler), and the light that emerges is very much one color (since it came from one atomic transition) and very bright (because all of the photons are in-phase).

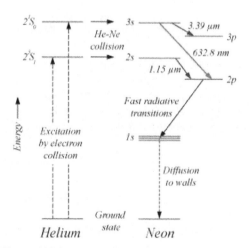

Figure 222-2. Energy level diagram for a helium-neon laser.

A very common type of laser is the Helium-Neon laser, also known as a HeNe. A HeNe laser contains a cavity filled with helium and neon gas (with about 5-10 times as much helium as neon). High energy electrons created by a heated filament, collide with helium atoms to excite helium electrons to an excited state (the 2S state using notation you should be familiar with from chemistry). As helium atoms collide with neon atoms in the gas, energy is transferred exciting neon electrons to their 3S state. The 3S state cannot relax to the 1S state because of quantum mechanical restrictions that are beyond the scope of this course, so the 3S electrons relax to the 2p energy level, which has an energy difference that corresponds to a 633 nm photon. As this process unfolds, the cavity becomes filled with these photons. An output coupler at the end of the cavity then allows the red photons to exit as a beam of red laser light. This process is summarized in Figure 222-2, which shown an energy-level diagram for the situation. As can be seen in the Figure, there are a few other processes that go on inside a helium-neon laser cavity; however, the process I described is the main one for creating the classic red light of a HeNe laser.

Lasers are a ubiquitous quantum technology in our society. Lasers are used in

- Laser eye surgery and soft tissue surgery
- Telecommunications and television technology

- Laser cutting and welding
- Laser sights for firearms and missiles
- Laser light displays
- Holography
- CD, DVD and Blue Ray players
- Microscopy

In chemistry and physics, lasers are also used for ***spectroscopy***, the science of using light to determine material properties. Since the color (frequency) of laser light is very well known and controlled, that light can be used to probe materials, and determine their optical properties. We will discuss this in greater detail in the next unit.

Example 222-3:

Consider Figure 222-2 showing the energy level diagram of a helium-neon laser. There are three optical transitions (in the upper right hand corner). Characterize the three wavelengths given off by this type of laser. How many of these wavelengths are visible? If any are not visible, what part of the em spectrum are they?

Solution:

Inspecting the Figure, we can see that the three wavelengths emitted by the laser are $632.8\ nm$, $1.15\ \mu m$ and $3.39\ \mu m$. Since the human eye is capable of detecting wavelengths from $400 - 750$ nm, only the $632.8\ nm$ transition gives off

visible light. Comparing this to Table 204-3 in unit 204, we see that this is red light.

The other two transitions are outside of the visible range of humans. Both are in the infrared region of the spectrum and would not contribute to anything we could see coming from the laser.

As a side note, there is also a transition not shown in Figure 222-2 from the 3s level to the 2p level. Thermal broadening of this level allows another transition at 543 nm – in the green part of the spectrum. Lasers that exploit this transition are called green HeNe's.

222-3: The photoelectric effect

Consider: *How do we <u>know</u> light acts as a particle?*

I noted earlier that experiments around the turn of the century refuted the wave-only model of light. One of those experiments, which I alluded to earlier, is known as the photoelectric effect. In this effect, photons impinging on a metal surface cause electrons to be ejected from the metal. In a pure wave model of light, we would expect

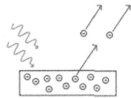

Figure 222-3. Diagram of the photoelectric effect.

1) The rate at which electrons are ejected should have to do with the intensity of light. That is, if we turn up the energy per second on a specific area of metal, the number of ejected electrons should increase for all wavelengths.
2) At low intensities, there should be a delay between when the light is turned on and when the electrons are ejected since it takes a while for the electrons to accumulate enough energy to be ejected from the material.
3) The rate of ejection of electrons should be material dependent, because the potential energy holding the electrons to the metal will be different for different metals.
4) The maximum kinetic energy of the ejected electrons should depend on the intensity of light – that is, more energy in equals more energy out.

However, when the experiment is performed in the laboratory, here is what happens

1) The rate of electron ejection does, indeed, depend on the intensity of light; however, there is a frequency below which there are no electrons ejected even if the intensity is turned up incredibly high. This frequency is called the ***cutoff frequency***.
2) There is no delay between when the light is turned on and electrons are ejected if the frequency is above the cutoff frequency.

3) If the frequency is above the cutoff frequency, the rate of electron ejection does not depend on the material. However, the cutoff frequency *is* material dependent.
4) The maximum kinetic energy of ejected electrons *does not* depend on the intensity of light. However, the *number* of ejected electrons does depend on the intensity.

Each of these points makes sense if we treat light as a single particle interacting with a single electron. Below the cutoff frequency, the photon simply does not have enough energy to liberate an electron – thus why there is a cutoff frequency. Above the cutoff frequency, there is a one-to-one correspondence between photons and electrons, so increasing the number of photons per second does not increase the kinetic energy of each electron, but it does increase *the number* of electrons ejected.

Overall, the photoelectric effect is governed by the conservation of energy. If a photon of energy, E_p, hits an electron near the surface of a metal, that energy will be transferred to the electron. Each material has a specific energy with which it holds these free electrons called the **work function, W**. Now, if the photon energy is less than the work function, the electron does not have enough energy to leave the surface of the metal. However, if the photon has more energy than the work function, then the electron accepts all of the energy from the photon, loses the amount of energy characterized by the work function as potential energy to leave the surface of the metal and then any remaining energy is left as electron kinetic energy. Writing this out mathematically, we see

$$E_p - W = K_{max}, \qquad (222\text{-}4)$$

> **Connection:** Solar panels and streetlights
>
> Both solar panels and streetlights make use of the photoelectric effect to light the world. Solar panels are based solely on the photoelectric effect. Light from the sun causes electron ejection from a metal and those electrons are used to create electricity. Another application is streetlights. Streetlights have a small solar panel on their top. When sunlight hits the panel, creating a small current, the light knows to be off, but when the light goes down, the current stops and the streetlight knows to turn on. Ever notice the streetlights turn off sometimes in lightning storms? This is because the bright flash of light is confused by the streetlight as the sun, so it shuts off and must reset.

where E_p is the energy of the photon, W is the work function of the material and K is the maximum kinetic energy that the electron can have based on this interaction. Now, remembering that the photon energy can be written

$$E_p = hf = \frac{hc}{\lambda}, \qquad (222\text{-}5)$$

we can write the equation for the photoelectric effect as in the box below.

The Photoelectric Effect

$$K_{max} = \frac{hc}{\lambda} - W \qquad (222\text{-}6)$$

Description – This equation describes how an electron liberated by a photon of energy $\frac{hc}{\lambda}$ will have a maximum kinetic energy (K_{max}) depending on the work function (W) of the material from which the electron came.

Note 1: W is the *work function* of the metal, and is an intrinsic property of the material, representing the potential energy the electron must overcome to be liberated.

Note 2: This equation is essentially the conservation of energy assuming the electron has its maximum kinetic energy when released.

Table 222-1. Work function for various materials.

Material	W (eV)
Silver	4.26
Aluminum	4.06
Gold	5.10
Boron	4.45
Carbon	5.00
Copper	5.00
Iron	4.67
Mercury	4.48
Potassium	2.29
Magnesium	3.66
Sodium	2.36
Lead	4.25
Silicon	4.60
Uranium	3.63
Tungsten	4.32
Zinc	3.63

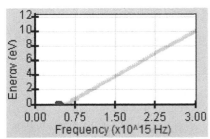

Figure 222-4. Plot of electron energy versus frequency for Sodium.

Table 222-1 gives the work function for a few common metals. Please note that many metals have a range for their work function based on atomic structure, temperature, etc., and that the values given in the Table are only illustrative for each material.

A common way to characterize the photoelectric effect is to graph the maximum kinetic energy of ejected electrons versus the frequency of light as shown in Figure 222-4. This plot is essentially a graph of the photoelectric effect equation. The slope of all such graphs should be Planck's constant, and the x-intercept the cutoff frequency for that material. I leave it as exercise to determine the meaning of the y-intercept of such a graph.

Example 222-4:

The work function for Cesium is $2.14 \, eV$. What is the maximum wavelength of light that will allow electrons to be ejected by Cesium in a photoelectric effect experiment?

Solution:

This is a direct application of the equation for photoelectric effect:

$$K_{max} = \frac{hc}{\lambda} - W.$$

By inspection, you can see that (for a certain material), the only variable on the right side of the equation is the wavelength. Since λ is in the denominator of a term, a larger wavelength will lead to a smaller K_{max}.

Therefore, the maximum wavelength is found when $K_{max} = 0$:

$$0 = \frac{hc}{\lambda_{max}} - W$$

Rearranging the equation for λ_{max} and substituting known values gives

$$\lambda_m = \frac{hc}{W} = \frac{1240 \, eV \cdot nm}{2.14 \, eV} = 579 \, nm.$$

Longer wavelengths will not cause electrons to be emitted at all and shorter wavelengths will cause electron ejection with a larger K_{max}.

Example 222-5:

During a photoelectric experiment with lithium the maximum potential difference between the plates of the experiment is $2.7 \, V$, when illuminated with ultraviolet light at $223 \, nm$. What is the maximum wavelength that will allow electrons to be ejected for lithium?

Solution:

This problem will use the equation for the photoelectric effect twice. Key phrases include photoelectric experiment and maximum wavelength for ejected electrons.

First, we must start with the equation for the photoelectric effect:

$$K_{max} = \frac{hc}{\lambda} - W.$$

If a maximum potential difference of 2.7 V is formed in the experiment, that means that the maximum kinetic energy of the ejected electrons is 2.7 eV.

With this information, we can solve for the work function of lithium:

$$2.7 \, eV = \frac{1240 \, eV \cdot nm}{223 \, nm} - W \quad \rightarrow \quad W = 2.9 \, eV.$$

Now that we know the work function, we can go back and find the maximum wavelength for ejected photons, remembering that this occurs when the maximum KE is zero ($K_{max} = 0$):

$$0 = \frac{1240 \, eV \cdot nm}{\lambda_{max}} - 2.9 \, eV.$$

Solving for the maximum wavelength, we find

$$\lambda_{max} = 428 \, nm.$$

This light is in the visible part of the spectrum (blue).

223 – Matter Waves

In the last few units, we have seen that light can act as both a wave and as a particle. It turns out that matter we usually think of as particles (such as electrons) can also act as waves. In this unit, we continue our exploration of this particle-wave duality.

Integration of Ideas

Review the basic idea of mechanical waves.
Review the idea interference of light waves from unit 221.
Review standing waves from unit 202

The Bare Essentials

- Just as light can act as both a particle and a wave, so can objects that we normally think of as just particles (electrons, protons, you, etc.). This is an effect known as particle-wave duality.

- The effective wavelength of a massive particle is given by the deBroglie wavelength.

---

deBroglie Wavelength

$$\lambda = \frac{h}{p} = \frac{hc}{\sqrt{2Kmc^2}}$$

Description – This equation defines the deBroglie wavelength of a massive particle in terms of Planck's constant and either its momentum or kinetic energy and mass.
Note: The deBroglie wavelength for macroscopic objects is exceedingly small, which explains why they act like particles.

- Two-slit interference experiments can now be done on massive particles as well as other wave phenomena.

Double Slit Interference

$$d \sin \theta_{int} = n\lambda$$

Description – This equation describes the set of bright fringes where a massive particle with de Broglie wavelength λ passing through two slits constructively interfere to form an interference pattern. θ_{int} is the angle from the midpoint of the slits, and n is the order of the fringe.
Note: for massive particles, λ must be given by the de Broglie wavelength.

- Massive particles trapped between two walls exhibit quantization of their energy levels (wavelengths) just as standing waves do with their frequencies.

Energy of a Particle in a Box

$$E_n = \frac{h^2 n^2}{8mL^2} \qquad n = 1, 2, 3, ...$$

Description – This equation describes the possible energy levels of a massive particle in a box of length L, analogous to a standing wave we encountered before.
Note: The "wavefunction" of this particle in a box is given by
$\Psi(x) = A sin\left(\frac{n\pi x}{L}\right)$ for x between 0 and L and zero everywhere else.

- An electron in a hydrogen atom is also bound to the atom, causing its energy levels to be quantized.

Energy in the Bohr Model

$$E_n = \frac{-ke^2}{2a_0 n^2} = \frac{-13.6\ eV}{n^2} \qquad n = 1, 2, 3, ...$$

Description – This equation describes the possible energy levels of an electron in the Bohr model of the hydrogen atom. a_0 is known as the Bohr radius ($a_0 = 0.053\ nm$).

- Vibration levels in molecules are also quantized and resemble harmonic oscillations.

Vibrational Harmonic Oscillations

$$E_n = \frac{h\omega}{2\pi}(n + 1/2) \qquad n = 0, 1, 2, 3, ...$$

Description – This equation describes the possible energy levels for atomic vibrations which resemble simple harmonic motion.

223-1: Wave properties of matter

Consider: *What does it mean for a particle to act like a wave?*

THE QUANTUM REVOLUTION CHANGED THE WAY HUMANS view the world of the very small. We saw in the last unit that light can be viewed as both a particle and a wave, and in fact, always has properties of both. Although our macroscopic experiences have taught us to treat particles and waves differently, in quantum mechanics they are really two parts of a whole – a particle-wave duality.

This same duality can be extended to *all* subatomic particles. Although subatomic particles were always thought of as particles, it wasn't until 1952 that physicists were using bubble chambers to detect the tiny particles moving through a supercritical fluids. On the flip side, back in 1923, Louis de Broglie posited that any subatomic particle will display wavelike properties, and just two years later C. Davisson and H. Germer found evidence of electron interference. Over the next 90 years, physicists kept trying to find wave-like features in larger and larger systems. At the time I'm writing this, the largest object used in these wave experiments was done with molecules containing 810 atoms (with a mass over 10,000 amu).

In his original work, de Broglie used Einstein's photon energy to extend wavelike features to massive particles. Remember from the last unit that the energy of a (massless) photon is

$$E_p = \frac{hc}{\lambda}. \tag{223-1}$$

Now, if we go back to our definitions of momentum and (total) energy from the relativity sections, we find

$$p = \gamma m v \quad \text{and} \quad E = \gamma m c^2. \tag{223-2}$$

Using these relations, we can find the ratio of a particle's momentum to its total energy:

$$\frac{p}{E} = \frac{\gamma m v}{\gamma m c^2} = \frac{v}{c^2}. \tag{223-3}$$

Now, if we apply this relationship to a photon moving at $v = c$, we find the rather simple relationship

$$p = \frac{E}{c}. \tag{223-4}$$

Finally, if we apply the photon energy from above:

$$p = \frac{1}{c}\frac{hc}{\lambda} = \frac{h}{\lambda}. \tag{223-5}$$

de Broglie brilliantly then suggested that since special relativity places no special difference between matter and energy ($E = mc2$), then this relationship should hold for *all* particles. Many experiments have subsequently supported the de Broglie hypothesis, and the wavelength associated with this equation is named in his honor.

The de Broglie wavelength can also be written in terms of the kinetic energy of the particle using the classical relationship

$$K = \frac{p^2}{2m} \quad \rightarrow \quad p = \sqrt{2Km}. \tag{223-6}$$

The de Broglie relationship is then

$$\lambda = \frac{h}{p} = \frac{h}{\sqrt{2Km}} = \frac{hc}{\sqrt{2Kmc^2}}. \tag{223-7}$$

The final version of this equation is helpful because the numerical value of $hc = 1240\ eV \cdot nm$ can be used, and mc^2 in the denominator has units of energy ($E = mc^2$).

deBroglie Wavelentgh

$$\lambda = \frac{h}{p} = \frac{hc}{\sqrt{2Kmc^2}} \qquad (223\text{-}7)$$

Description – This equation defines the deBroglie wavelength of a massive particle in terms of Planck's constant and either its momentum or kinetic energy and mass.

Note: The deBroglie wavelength for macroscopic objects is exceedingly small, which explains why they act like particles.

Example 223-1: Wavelength of a baseball

What is the de Broglie wavelength of a 150 g baseball moving at 40 m/s (100 mph)?

Solution:

This is a direct application of the de Broglie wavelength

$$\lambda = \frac{h}{p} = \frac{h}{mv} = \frac{6.626 \times 10^{-34} J \cdot s}{(0.155\ kg)(40\frac{m}{s})}$$

$$\lambda = 1.07\ x\ 10^{-34} m.$$

Note that this is an incredibly small wavelength, which is why we do not notice the wave nature of the baseball.

The wave nature of macroscopic items is so incredible small that they do not resemble waves. Think of it this way, as the wavelength of a wave gets smaller and smaller, the peaks and troughs get closer and closer together until you cannot distinguish between the individual parts of the wave and they start to look like a straight line (in fact, this *is* the ray of geometrical optics).

Since subatomic particles act as waves with a given wavelength, we can do the equivalent of Young's double-slit experiment with such particles. The math and physics of such an experiment are exactly the same as for the double-slit experiment in unit 220, so I give the result without further derivation:

Double Slit Interference

$$d\ sin\ \theta_{int} = n\lambda \qquad (223\text{-}8)$$

Description – This equation describes the set of bright fringes where a massive particle with de Broglie wavelength λ passing through two slits constructively interfere to form an interference pattern. θ_{int} is the angle from the midpoint of the slits, and n is the order of the fringe.

Note: for massive particles, λ must be given by the de Broglie wavelength.

Example 223-2: Electron interference

Imagine you want to do the double-slit experiment with electrons. What must be the spacing between the two slits for electrons moving at 200 m/s to produce a first bright fringe 2.1° from the central axis?

Solution:

Although this is a direct application of the double-slit interference equation, we must first replace the wavelength

with the de Broglie wavelength equation

$$d\ sin\ \theta_{int} = n\frac{h}{p}.$$

Now, we know that we have the angle to the first-order maxima, so $n = 1$ for $\theta_{int} = 2.1°$. We can also replace the electron momentum with its mass and velocity.

$$d \sin \theta_{int} = \frac{h}{mv}.$$

We can then solve for d:

$$d = \frac{h}{mv \sin \theta_{int}} = \frac{6.626 \times 10^{-34} J \cdot s}{(9.1 \times 10^{-31} kg)(200 \, m/s)(\sin 2.1°)}$$

$$d = 9.9 \times 10^{-5} m.$$

This separation is about 100 microns or 0.1 mm. I should note that 200 m/s is incredibly slow for an electron. In real (especially older) experiments, electron speeds of 10^7 m/s were more realistic, which means that the separation would need to be on the order of the size of an atom (10^{-9} m). This is why direct two-slit interference was not noted until recently.

223-2: Energy quantization and the Schrödinger equation

Consider: Is there a general wave equation for particles? What does this equation tell us about important systems?

The de Broglie wavelength works very well for **free particles** – particles that do not have a potential energy associated with them. Let's now look at a way to generalize the de Broglie wavelength for when particles are under some form of potential energy. First, I'm going to start with our kinetic energy form of the de Broglie wavelength,

$$\lambda = \frac{hc}{\sqrt{2Kmc^2}}, \tag{223-9}$$

and solve for the kinetic energy:

$$K = \frac{h^2}{2m\lambda^2}. \tag{223-10}$$

The conservation of energy tells us that an object's kinetic energy, K, plus its potential energy, U, is its total energy, E. Therefore, the kinetic energy can be written as the total energy minus the potential energy, leaving us with

$$E - U = \frac{h^2}{2m\lambda^2}. \tag{223-11}$$

Now, in an isolated system, the total energy, E, should be constant, but the potential energy, U, could change with position (a simple example being gravitational potential energy changing with height). Since the only term that is not constant on the right side is the wavelength, this means that as the potential energy of a particle changes, so will its wavelength.

Conceptually, I would argue that the particle's wavelength has to do with how large the curvature is relative to the entire wave. A wave with a large maximum curvature will curve very quickly, reducing the overall wavelength, and a wave with a small maximum curvature will take a long time to change and therefore have a long wavelength. Short and long relative to what though? Relative to the wave itself. Mathematically, we can write this as

$$\lambda^2 = \frac{measure \ of \ the \ wave}{curvature \ of \ the \ wave} = 4\pi^2 \frac{-\Psi(x)}{\Psi''(x)}, \tag{223-12}$$

where $\Psi(x)$ is called the wavefunction of the wave, $\Psi''(x)$ represents the second derivative of the wavefunction, since the second derivative to the function gives us its curvature, and $4\pi^2$ is a constant required to make this wavelength work for sinusoidal waves. The negative is there because the wavefunction curves upward most at its minimum point (or curves downward most at its maximum point) This function fits our general description – small curvature is a large wavelength and large curvature is a small wavelength. So, why is the wavelength squared in this expression? The answer is simple: we need to match units. Since we are taking spatial derivatives (the wavefunction is only a function of x), the second derivative of the wavefunction has the same units of the wavefunction divided by x^2.

Now, we can plug this expression back into our generalized de Broglie relation to find

$$E - U = \frac{h^2}{2m}\frac{1}{\lambda^2} = \frac{-h^2}{2m}\frac{\Psi''(x)}{4\pi^2\Psi(x)}. \tag{223-13}$$

This equation can further be simplified to

$$\frac{-\hbar^2}{2m}\frac{d^2\Psi}{dx^2} = (E - U)\Psi. \qquad (223\text{-}14)$$

where $\hbar$ is a constant known as "h-bar" and is equal to $h/2\pi$. This equation is called Schrödinger's equation and can be used (in principle) to find the wavefunction of any particle if we know its total energy and potential energy. Although easy to pass by, this is a very powerful statement. Schrödinger's equation plays the same role in quantum mechanics that Newton's laws play in classical mechanics. With Newton's laws, if we know the initial state and forces on an object, we can find the way it behaves. In quantum mechanics, if we know the potential energy and total energy for a particle, we can find how it behaves.

As a second-order differential equation, Schrödinger's equation can be very hard to solve in all but the simplest situations. However, there are a couple of features I want to mention and then we will apply it to a couple of very common and tractable geometries.

1) The solution of Schrödinger's equation gives the wavefunction for a particle in its given potential landscape. Although this is incredibly important in quantum mechanics, we will not go into the details of the wave function much here.
2) Solutions to the Schrödinger's equation in the presence of a potential function are **_quantized_**, meaning that only certain specific energies are allowed. In classical mechanics, a system can have any energy between a minimum and maximum (which can also be infinite). However, this is not true in quantum mechanics. We do not notice quantum energy levels in our everyday life because they are very close together on our scale (as we will see below).

We are going to focus on the second point above. What are the possible energies for a couple of important geometries? Below, we are going to look at the particle in a box, electron in a hydrogen atom, and thermal vibrations of atoms and molecules.

Particle in a box

Consider a region of space as shown in figure 223-1. In the region between x = 0 and x = L, there is no potential energy; however, outside of this region, there is an infinite potential. The way to view this is an area of flat ground between two very steep and very high cliff walls. Although you are free to move between the walls (classically), there is no way you can get over the top.

To find a solution to the Schrödinger equation in the region of zero potential, we simply put zero in for U in the equation:

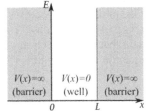

Figure 223-1. Potential energy for the particle in a box.

$$\frac{-\hbar^2}{2m}\frac{d^2\Psi}{dx^2} = (E - 0)\Psi = E\Psi. \qquad (223\text{-}15)$$

But we know the solution to this equation! It asks for a function whose second derivative is negative constants multiplied by the original function. This is either sine or cosine. Since we know we cannot go beyond the very high walls, I will say that the function must be zero at both x = 0 and x = L. Of our two choices, it is sin(x) which is zero at zero, so I make that choice:

$$\Psi = A\sin kx, \qquad (223\text{-}16)$$

where I also imposed an arbitrary amplitude, A, and added a constant to the phase.

Believe it or not, that is the solution to this quantum mechanical problem. OK, let's look at the energy in the box:

$$E = K + U = K + 0 = K = \frac{p^2}{2m}. \qquad (223\text{-}17)$$

In this equation, I included our known relationship between kinetic energy and momentum since kinetic energy is the total energy in this case. Now, we can use the de Broglie wavelength since we are in a region of zero potential to find

$$E = K = \frac{p^2}{2m} = \frac{1}{2m}\left(\frac{h}{\lambda}\right)^2. \qquad (223\text{-}18)$$

Finally, we actually know what possible wavelengths this function can have. This is simply a standing wave between two fixed points. We need each end to be a node, meaning that we can fix any multiple of a half wavelength in the box, meaning the wavelength can be any multiple of 1/2L:

$$E = \frac{h^2}{2m}\left(\frac{n}{2L}\right)^2 = \frac{h^2 n^2}{8mL^2}. \qquad (223\text{-}19)$$

This is the energy condition for the particle in a box. Since n must be an integer, you can see that the energies are not continuous, but come at discrete values, as promised above.

Energy of a Particle in a Box

$$E_n = \frac{h^2 n^2}{8mL^2} \quad n = 1, 2, 3, ...\qquad (223\text{-}19)$$

Description – This equation describes the possible energy levels of a massive particle in a box of length L, analogous to a standing wave we encountered before.

Note: The "wavefunction" of this particle in a box is given by $\Psi(x) = A\sin\left(\frac{n\pi x}{L}\right)$ for x between 0 and L and zero everywhere else.

Note: if you'd like to use the $hc = 1240 \ eV \cdot nm$ trick, you can multiply top and bottom of the particle in a box energy equation to get

$$E_n = \frac{(hc)^2 n^2}{8(mc^2)L^2}. \qquad (223\text{-}20)$$

The hydrogen atom

One of the early successes of "wave mechanics" as quantum mechanics is sometimes called, was to assume that the electron in a hydrogen atom must be a wave and see what happens. More specifically, Rutherford (working off of earlier work of Bohr) said that for an electron to orbit a hydrogen nucleus as a wave, every trip around must be equal to an integer multiple of the electron's wavelength. That is to say, the circumference of the orbit must be an integer times the electron wavelength:

$$2\pi r = n\lambda. \qquad (223\text{-}21)$$

We also know that it is the coulomb potential that holds the electron near the nucleus and that it is moving in a circle (centripetal acceleration), so

$$\frac{mv^2}{r} = \frac{ke^2}{r^2} \quad \rightarrow \quad v = \sqrt{\frac{ke^2}{mr}}. \qquad (223\text{-}22)$$

We can use this to find the momentum and relationship to the de Broglie wavelength\

$$p = mv = \sqrt{\frac{mke^2}{r}} = \frac{h}{\lambda}. \qquad (223\text{-}23)$$

Finally, we can combine equations 223-21 and 223-23 to eliminate the wavelength from this equation and find

$$r_n = \frac{n^2 h^2}{4\pi^2 mke^2}. \qquad (223\text{-}24)$$

This equation for radius has a lot of constants and the level of the energy, n. Note that I added the subscript n to remind you that the radii are also quantized in this case. Substituting known values in for the constants gives us

$$r_n = n^2 a_0 = (0.053 \ nm)n^2, \qquad (223\text{-}25)$$

where a_0 is called the bohr radius and represents the radius of the n = 1 level. r_n are the possible radii for the electron in the Bohr atom. We can find the associated energies again by using conservation of energy

$$E = K + U = \frac{1}{2}mv^2 - \frac{ke^2}{r} = \frac{1}{2}m\left(\sqrt{\frac{ke^2}{mr}}\right)^2 - \frac{ke^2}{r} = -\frac{ke^2}{2r}. \tag{223-26}$$

Plugging in the equation we just found for the r_n, gives us the energy levels of an electron in the hydrogen atom.

Energy in the Bohr Model

$$E_n = \frac{-ke^2}{2a_0 n^2} = \frac{-13.6\ eV}{n^2} \qquad n = 1, 2, 3, \dots \tag{223-27}$$

Description – This equation describes the possible energy levels of an electron in the Bohr model of the hydrogen atom. a_0 is known as the Bohr radius ($a_0 = 0.053\ nm$).

Atomic vibrations (harmonic oscillator)

Atomic bonds act very much like classical springs. Take, for example, an O_2 molecule consisting of two oxygen atoms. The covalent bond has a characteristic length that it would like to maintain. If the oxygen atoms are too close together, the bond pushes them apart; if they are too far apart, the bond pulls them together. This is exactly how a classical spring works. So, it is not to much of a stretch to suggest a harmonic potential energy function:

$$U(x) = \frac{1}{2}kx^2, \tag{223-28}$$

where x would represent how far an oxygen atom is from its equilibrium position.

Now, imagine putting this function in for U(x) in the Schrödinger equation. This extra x^2 makes for a non-linear differential equation, which is extremely hard to solve. Therefore, I will present the results for the energy of the harmonic potential below. However, I want to stress again, that this function represents the ***vibrational energy levels of atoms in molecules***. It is easy to lose track of the physics with the math, which is why I'm reminding you what's going on here.

Vibrational Harmonic Oscillations

$$E_n = \frac{\hbar\omega}{2\pi}(n + 1/2) \qquad n = 0, 1, 2, 3, \dots \tag{223-29}$$

Description – This equation describes the possible energy levels for atomic vibrations which resemble simple harmonic motion.

223-3: Putting it all together (examples)

Consider: *How do I actually use the information from this unit?*

So, what is the takeaway for these three systems? For now, you are really just trying to recognize which system is being used and apply the equation to find energy levels. At this point, this is very plug and chug. In the next unit, we will use these three systems to find important ***spectra*** – energy levels of light that are either absorbed or emitted from such systems. For now though, consider the following examples.

Example 223-3: Particle in a box

Determine the lowest three energy levels for an electron trapped in a box with length, $L = 13.7\ nm$.

Solution:

This is a direct application of our equation for the particle in a box. Key terms are energy levels and box.

$$E_n = \frac{h^2 n^2}{8mL^2} = \frac{(hc)^2 n^2}{8(mc^2)L} = \frac{(1240 eV \cdot nm)^2}{8(511{,}000\ eV)(13.7 nm)} n^2,$$

or

$$E_n = (0.027\ eV)n^2$$

The lowest three values of n may not be substituted to find the lowest three energy levels:

$$E_1 = (0.027\ eV)(1)^2 = 0.027\ eV,$$

$$E_2 = (0.027\ eV)(2)^2 = 0.110\ eV,$$

$$E_3 = (0.027\ eV)(3)^2 = 0.247\ eV.$$

Example 223-4: Hydrogen atom

The electron in a hydrogen atom is excited to the n = 11 state. What is the energy and electron radius associated with this energy level? How many times large is an a hydrogen atom at this energy level relative to its ground state?

Solution:

This problem is a direct application of the electron bound energy equation and hydrogen atom radius equations. Key phrases include hydrogen atom, energy level and radius.

First, we start with the energy level equation for the hydrogen atom

$$E_n = \frac{ke^2}{2a_0 n^2} = \frac{-13.6\ eV}{n^2},$$

which for n=11 is

$$E_{11} = \frac{-13.6\ eV}{11^2} = -0.112\ eV.$$

As for the radius of a hydrogen atom at $n = 11$, we can apply this value to the equation for hydrogen radius:

$$r_{11} = n^2 a_0 = (0.053\ nm)(11)^2,$$

or

$$r_{11} = 6.41\ nm.$$

Since the radius of the hydrogen atom goes as n^2, the radius at $n = 11$ is 121 times the size of that atom at $n = 1$.

Example 223-5: Vibrational levels

A diatomic molecule (such as oxygen) has vibrational ground state energy of 0.11 eV.
  (a) What is the energy of the 4th excited state (n=4)?
  (b) What is the classical frequency of vibration associated with the ground state energy?

Solution:

Part (a) is a direct application of the vibrational energy level equation:

$$E_n = \frac{h\omega}{2\pi}\left(n + \frac{1}{2}\right).$$

If the ground state is 0.11 eV, we can find the value of the constants

$$\frac{h\omega}{2\pi} = \frac{E_n}{n + \frac{1}{2}} = \frac{0.11\ eV}{0 + \frac{1}{2}} = 0.22\ eV.$$

From this, we can find the value of the n=4 energy level

$$E_4 = \frac{h\omega}{2\pi}\left(n + \frac{1}{2}\right) = (0.22\ eV)\left(4 + \frac{1}{2}\right) = 0.99\ eV.$$

(b) Classically, $\omega = 2\pi f$, so the constants in the energy level equation can be rewritten as

$$\frac{h\omega}{2\pi} = 0.22\ eV = \frac{h2\pi f}{2\pi} = hf.$$

Solving for the frequency we see

$$f = \frac{0.22\ eV}{h} = \frac{0.22\ eV}{4.14 x 10^{-15}\ eV \cdot s} = 5.31 x 10^{13}\ Hz.$$

As you can see, the classical vibration frequency of diatomic molecules is extremely large!

224 – Atomic Physics

*In this unit, we explore how particles (such as electrons) can **move between** energy levels in bound systems. Specifically, we will explore each of the systems that were introduced in unit 223: the particle in a box, hydrogen atom, and harmonic oscillator. It is very important to remember that in unit 223 we found the energy equivalent of each level in those systems whereas unit 224 is all about **changing** energy levels.*

Integration of Ideas

Review energy quantization.
Review the particle in a box and Bohr model of the hydrogen atom, and the harmonic oscillator energy levels.
Review spin from our discussion of magnetism.

The Bare Essentials

- All fundamental particles exhibit the property of spin as discussed in the units on magnetism. Particles with an integer spin are known as **bosons** and particles with a half-integer spin are known as **fermions**.

- The **Pauli Exclusion Principle** states that no two fermions can be in the exact same quantum state in a bound system. Note that the exclusion principle does not apply to bosons.

- Electrons move between energy levels in bound systems by absorbing or emitting photons. The change in energy levels must be exactly equal to the energy of the photon involved.

Photon Emission and Absorption

$$\frac{hc}{\lambda_{nm}} = E_n - E_m$$

Description – This equation relates the wavelength of light absorbed or emitted (λ_{nm}) as an electron changes energy from state n (E_n) to or from state m (E_m) in a quantum system.

- The **spectra** of a given system is defined by all of the possible transitions created by exciting the system with a photon of a given energy. The system can be viewed using an **energy level diagram**.

224-1: Subatomic particles have "spin"

Consider: *What limits exist on particle energies discussed in unit 223?*
Are there rules for how many particles can be at one level?

ALL SUBATOMIC PARTICLES HAVE AN *intrinsic angular momentum* called *spin*. At a very basic level, you can think of each particle as a little bead that spins around some axis like a top. Although this is not too bad conceptually, it comes with a host of problems. First, if you work out how fast the particle is spinning, in a classical sense, its edges spin faster than the speed of light – oops. Second, the particles do not react to magnetic fields the way we would expect them too. Finally, the idea that a particle spins just because it is that type of particle does not fit with classical mechanics, that is, a classical top could be made to spin faster or slower if we choose, but an electron's spin has a fixed magnitude.

Particle spin is measured relative to a Planck's constant divided by 2π. Since this constant comes up quite often in particle physics, it has its own name: *h-bar*,

$$\hbar = \frac{h}{2\pi} = 1.055 \ x \ 10^{-34} J \cdot s = 6.582 \ x \ 10^{-16} eV \cdot s. \tag{224-1}$$

Electrons, protons and neutrons have *half-integer spin*, meaning that their intrinsic spin is $\hbar/2$. Photons, on the other hand, have a spin of $1\hbar$, which is called an *integer spin*, since it is a full integer multiple of $\hbar$. This allows us to put all subatomic particles into two categories

Fermions – particles with half-integer spin (relative to $\hbar$)

Bosons – particles with integer spin (relative to $\hbar$)

All quarks and leptons are fundamental fermions and the photon is a fundamental boson. Composite particles made from these subatomic particles are also characterized as either bosons or fermions. For example, the proton is a fermion (as noted above), because it is a composite of three quarks, and a composite of an odd number of fermions is a fermion itself (any odd number multiplied by a half-integer spin is a half-integer).

A free particle can have a spin that points in any direction – just as a top spinning in space could have it axis of rotation pointed in any direction. However, when we use a magnetic field to measure the spin, it becomes quantized. The process is similar to what we discussed in the last unit, since when we apply the magnetic field, we are placing the particle in a potential – and it becomes quantized. The possible values for the spin can be negative and are symmetric around zero. For example, an electron has a spin of ½, so when measured it will be either -½ or +½. A photon has a spin of 1, so when measured, it can be -1, 0, or 1.

So, why do we care? Well, bosons and fermions act differently when combined in a system such as a particle-in-a-box or an atom. Fermions must obey the *Pauli Exclusion Principle*, which states that no two fermions can be in the same state in a system. Bosons, however, do not follow this rule. Take, for example, two electrons in a particle-in-a-box setup. From the last unit, we know that the energy of a particle in this system is given by

$$E_n = \frac{\hbar^2 n^2}{8mL^2}, \tag{224-2}$$

where n is any positive integer. At first, you might think that only one electron could fit into the lowest energy level, one in n = 2, one in n = 3, etc., in order to follow the exclusion principle. However, since electrons have spin, we can fit an electron into the n=1 level with a spin of -½ (called spin-down) and one with spin +½ (called spin-up). Therefore, we can have two electrons in each level. If we try to put more than two electrons into the system, at least one of them will have to reside in the n = 2 level because if it were in the n = 1 level, it would have the same state as one of the electrons already there – which is prohibited by the exclusion principle.

Contrast this with photons (which are bosons). Since they do not follow the exclusion principle, as many photons as we desire can fit into the lowest energy level, with no restrictions. This has stark consequences for the overall energy level of a system, as can be seen in the example below.

Example 224-1: Comparing system energy

Compare the total energy of a particle in a box system with five electrons (which must follow the exclusion principle), to the same box with five magical bosons (which need not follow the exclusion principle). You may assume the bosons have the same mass as an electron..

$$E_1 = 2\frac{h^2 1^2}{8mL^2} = 2\frac{h^2}{8mL^2},$$

$$E_2 = 2\frac{h^2 2^2}{8mL^2} = 8\frac{h^2}{8mL^2},$$

$$E_3 = 2\frac{h^2 3^2}{8mL^2} = 18\frac{h^2}{8mL^2}.$$

Solution:

For the bosons, each of the five particles can be in the lowest energy state since there is no restrictions on how many can populate that level. Therefore

$$E_b = 5E_1 = 5\frac{h^2 n^2}{8mL^2} = 5\frac{h^2}{8mL^2},$$

since all the bosons can be in the $n = 1$ energy level. Note that I will leave all of the constants since they are the same for the electrons and bosons.

Now, for the electrons, we know that we can only have two electrons in the $n = 1$ level and two electrons in the $n = 2$ level. This leaves us one electron in $n = 3$. First, I will find the total energy in each level:

Adding this all together, we find that the total energy of the electrons is

$$E_e = 28\frac{h^2}{8mL^2}.$$

Comparing the two energies, we find

$$\frac{E_e}{E_b} = \frac{28}{5} = 5.6,$$

meaning that the system with electrons has 5.6 times the energy as the system with bosons.

For atoms, the situation is slightly more complicated. In addition to the intrinsic spin of the particle (usually, but not necessarily electrons), there can be orbital angular momentum of the particle around the nucleus. The particle's orbital angular momentum is important because since different orbital momentum states represent separate states, it allows for more particles in each energy level. What's more is that if the electron in an atom is placed into a magnetic field, the orbital angular momentum becomes quantized as well. So, taking the energy level, orbital angular momentum, quantization of the angular momentum in a magnetic field, and spin into account, we have four quantum numbers as shown in table 224-1.

Table 224-1. Description of the four quantum numbers for an electrons in an atom.

Symbol	Name of quantum number	Description
n	Principal	Energy level $n = 1,2,3\ldots$
l	Angular momentum	Orbital angular momentum $\ell = 0, 1, 2\ldots(n-1)$
m	magnetic	Quantization of l in B-field $m = -\ell \ldots \ell$
s	spin	Intrinsic angular momentum -s, -s+1,…s-1, s.

Example 224-2: How many electrons?

Compare how many electrons are allowed in the n = 3 levels of the particle-in-a-box potential versus that of an atom.

$$\ell = 0 \rightarrow m = 0 \qquad \text{two electrons}$$

$$\ell = 1 \rightarrow m = -1, 0, 1 \qquad \text{six electrons}$$

$$\ell = 2 \rightarrow m = -2, -1, 0, 1, 2 \qquad \text{ten electrons}$$

Solution:

The difference between the two systems is the additional orbital angular momentum in the atom. Since the particle in a box system does not have orbital angular momentum, the $n = 3$ level can support just two electrons (one with spin-up and one with spin-down).

In the $n = 3$ energy level of the atom, the following states are available: (remember that each state can also have spin-up **and** spin-down)

Therefore, the total number of electrons in this state is 18. So, the atom can have nine times as many electrons in the $n = 3$ level compared to the particle in a box. You can see that the addition of the orbital angular momentum states allows for many more electrons to fill in higher energy levels!

224-2: Absorption and emission of photons

Consider: *What happens when particles move between energy levels?*

We now know that particles in bound systems must have discrete energy levels, and that only a certain number of particles can be in each level if they are fermions. There is a mechanism by which particles can jump between energy levels that they are allowed to be in; however, in order to do this, the energy must either come from some place (if it wants to go to a higher energy level) or go somewhere (if it wants to go to a lower energy level).

Each of the three systems we studied in the last unit have different functional forms in regards to the energy level. For comparison, here are the energy-level equations again:

Particle in a box	$E_n = \frac{h^2 n^2}{8mL^2}$ n = 1, 2, 3, ...	(226-19)
Hydrogen atom	$E_n = \frac{-13.6\ eV}{n^2}$ n = 1, 2, 3, ...	(226-27)
Harmonic oscillator	$E_n = \frac{h\omega}{2\pi}(n + 1/2)$ n = 0, 1, 2, 3, ...	(226-29)

Notice that both the particle in a box and harmonic oscillator energy levels increase as the energy level increases, although the harmonic oscillator is linear and the particle in a box energy levels increase quadratically. The hydrogen atom potential energy, on the other hand, is negative and increases as $1/n^2$ towards zero as n approaches infinity. In the model we gave, both the particle in a box and harmonic oscillator functions are always bound, no matter how much energy is given to the particle contained in the potential. If an electron in the ground state of the hydrogen atom is given more than 13.6 eV it becomes unbound from the atom (called ionization). This is why the hydrogen atom is given as negative, whereas the other two energy functions are given as positive – there is no ionization energy in the particle in a box and harmonic oscillator potentials.

Figure 224-1 shows what are known as *energy level diagrams* for these three systems. Please note that for the hydrogen atom, energy levels get closer together, energy levels get farther apart for the particle in a box and the energy levels for the harmonic oscillator (vibrations) remain constant as n increases.

One of the most common mechanisms for particles to jump between energy levels is via the absorption or emission of photons, where the photon energy exactly matches the difference between the energy levels in question. *I cannot stress strongly enough that for a photon to be absorbed or emitted there must be a change in energy levels*. In unit 223, we calculated the energy of individual levels – this represents the energy of a particle while in a particular energy level. Emission and absorption are always about changes in levels.

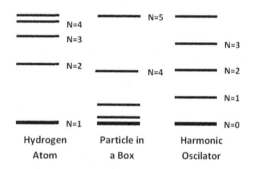

Figure 224-1. Energy-level diagrams for the Hydrogen atom, Particle in a box and Harmonic Oscillator potentials. Note that energy levels are scaled so that the $n = 5$ level of each coincides.

Photon Emission and Absorption

$$\frac{hc}{\lambda_{nm}} = E_n - E_m \qquad (224\text{-}3)$$

Description: This equation relates the wavelength of light absorbed or emitted (λ_{nm}) as an electron changes energy from state n (E_n) to or from state m (E_m) in a quantum system.

Note: The energy of the photon emitted is given by the difference in energy of the requisite levels, $E_P = E_n - E_m$

Spontaneous Emission

If a particle has an available energy level below where it currently resides, it will tend to drop down to the lower energy level on its own. This process is known as *spontaneous emission*, and is responsible for most of the energy release from excited atoms. The one exception (discussed in unit 222) is ***stimulated emission*** in lasers, where light already present forces electrons to drop to lower levels in excited atoms.

Example 224-3: Lower Hydrogen levels

What type of photon is released when an electron in an $n = 2$ level of a hydrogen atom drops down to the $n = 1$ state?

Solution:

This is a two-step process. First, must identify the system (hydrogen atom) and the energy levels involved, and then we must use our photon emission equation to determine the photon energy and/or wavelength.

First, I start with the equation for emission

$$\frac{hc}{\lambda_{nm}} = E_2 - E_1,$$

and place in the energy levels for the hydrogen atom, known that we need the 1st and 2nd energy levels

$$\frac{1240\ eV \cdot nm}{\lambda_{21}} = \frac{-13.6\ eV}{2^2} - \frac{-13.6\ eV}{1^2}.$$

Solving for the wavelength gives us

$$\lambda_{21} = 121.6\ nm.$$

This wavelength is in the ultraviolet portion of the electromagnetic spectrum. In fact, all of the hydrogen atom transitions to the $n = 1$ state are more energetic than any visible photons.

Photon absorption

A photon will only be absorbed by a system if its energy corresponds exactly to one of the energy transitions of the system and there is a particle that can move between the two energy levels. This means that there must be a particle in the lower energy level of the transition and an empty spot in the higher energy level.

Connection: Transparency

Materials that are transparent (so you can see through them) do not have any available energy level transitions in the visible part of the spectrum. Therefore, optical photons simply fly right through the material since there is no matter to interact with at that energy!

Example 224-4: Boxy absorption

What photon wavelength is required to excite an electron from the $n = 1$ to the $n = 3$ energy level of a box with length 7.2 nm?

Solution:

This is a direct application of the photon emission and absorption equation:

$$\frac{hc}{\lambda_{nm}} = E_n - E_m.$$

Inserting the equation for the particle in a box at $n = 3$ and $n = 1$, we get

$$\frac{hc}{\lambda_{13}} = \frac{h^2 3^2}{8mL^2} - \frac{h^2 1^2}{8mL^2} = 8\frac{h^2}{8mL^2} = \frac{h^2}{mL^2}.$$

This equation can be solved for the wavelength in a straightforward manner

$$\lambda_{13} = hc\frac{mL^2}{h^2} = \frac{mc^2 L^2}{hc} = \frac{(511{,}000\ eV)(7.2\ nm)^2}{1240\ eV \cdot nm}.$$

Simplifying this solution, we find

$$\lambda_{13} = 21{,}400\ nm = 21\ \mu m..$$

233

Example 224-5: Shaking it up

What type of photon is absorbed to excite diatomic oxygen from its vibrational ground state to the n = 5 level. Note that $\hbar\omega = 0.19\ eV$ for diatomic oxygen.

Solution:

Once again, we must employ our equation for energy level changes in quantum systems

$$\frac{hc}{\lambda_{nm}} = E_5 - E_0,$$

and insert the fact that we have vibrational levels using the harmonic oscillator equations:

$$\frac{hc}{\lambda_{nm}} = \hbar\omega\left(5 + \frac{1}{2}\right) - \hbar\omega\left(0 + \frac{1}{2}\right) = 5\hbar\omega.$$

We can now plug in our known values and solve for the wavelength

$$\lambda_{nm} = \frac{hc}{5\hbar\omega} = \frac{1240\ eVnm}{5(0.19\ eV)},$$

$$\lambda_{nm} = 1305\ nm.$$

This photon is in the infrared part of the EM spectrum. Infrared light is often associated with thermal transitions!

224-3: Emission and absorption spectra

> **Consider:** *What happens if instead of one photon, we use lots of photons (like in real life)?*

Above, we described how photons act as mediators for transitions in bound systems and calculated the wavelength of individual transitions. Real systems, however, are usually made up of a large number of individual quantum pieces. For example, if we want to excite atoms in a neon gas sample, we could be dealing with billions of neon atoms, each of which contain electrons with known energy levels. Similarly, liquid quantum dots, small glass particles that act like the particle in a box, could also contain huge numbers of small 'boxes.'

As an example, assume that we have a sample of hydrogen gas and we want to excite the gas to the n = 5 level. In order to do this, we would need to find the wavelength of light that would cause this excitation:

$$\frac{hc}{\lambda_{nm}} = E_2 - E_1 \quad \rightarrow \quad \frac{1240\ eV \cdot nm}{\lambda_{51}} = \frac{-13.6\ eV}{5^2} - \frac{-13.6\ eV}{1^2}, \tag{224-4}$$

which gives us

$$\lambda_{51} = 95\ nm.$$

Although 95 nm is in the ultraviolet part of the spectrum, we can easily excite the gas at this wavelength. Now, each photon that enters the gas has the ability to excite one atom. If we send one million photons into the gas, we can excite one million atoms, etc. Each of these atoms can then undergo spontaneous emission, but emission to what? Does the atom go from *n = 5* to *n = 1*? Does it go from *n = 5* to *n = 3*? The answer is yes. I know that answer seems vague, but each atom will choose to undergo a relaxation. Since all of the excited atoms are in the n = 5 level, this means each could drop to n = 4, 3, 2 or 1. Then, any atoms that dropped to the n = 4 level will later drop to n = 3, 2 or 1. Any that dropped to n = 3 will again drop to n = 2 or 1, etc. So, a gas of hydrogen that is excited to the n =5 energy level will emit a total of ten wavelengths of light as can be seen in figure 224-2. This is known as the **emission spectrum** for hydrogen gas (up to n = 5).

Emission spectra can be found for any bound system by exciting it to a specific energy level and detecting the wavelengths of light that are emitted from the system. Each of the systems we studied has a different spectrum. In addition, each atom or molecule also has a different spectrum, which makes the

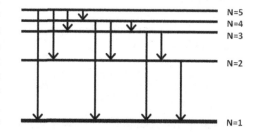

Figure 224-2. Emission spectra for hydrogen gas excited to the n = 5 energy level. Each vertical line represents an emitted photon.

technique just described very powerful for determining what particles constitute a system – be it gas, liquid or solid. The science of using spectra to determine what makes up a system is called **spectroscopy**. Spectroscopy is a very detailed and complex subject. Needless to say, we have just barely touched the surface and given you a brief introduction to the field.

For the example given above, we can write down each individual wavelength emitted by using our photon emission equation. Below are the 10 wavelengths emitted by this system. In writing this, I have used the notation $n \rightarrow m$ for a transition from level n to level m. For example, the transition from level 5 to level 1 is written $5 \rightarrow 1$.

$5 \rightarrow 4 \quad \lambda_{54} = 4052 \, nm$ $\qquad$ $5 \rightarrow 3 \quad \lambda_{53} = 1282 \, nm$ $\qquad$ $5 \rightarrow 2 \quad \lambda_{52} = 434 \, nm$

$5 \rightarrow 1 \quad \lambda_{51} = 95 \, nm$ $\qquad$ $4 \rightarrow 3 \quad \lambda_{43} = 1876 \, nm$ $\qquad$ $4 \rightarrow 2 \quad \lambda_{42} = 486 \, nm$

$4 \rightarrow 1 \quad \lambda_{41} = 97 \, nm$ $\qquad$ $3 \rightarrow 2 \quad \lambda_{32} = 656 \, nm$ $\qquad$ $3 \rightarrow 1 \quad \lambda_{31} = 103 \, nm$

$2 \rightarrow 1 \quad \lambda_{21} = 122 \, nm$

Inspecting these ten wavelengths, you can see that only three are in the visible part of the spectrum (656 nm, 434 nm and 486 nm). Four of the wavelengths are in the ultraviolet region and three are in the infrared region. This type of table can be made of each of our three systems if you know to which energy level the system is initially excited. As a practical note, using a spreadsheet program (such as Microsoft Excel) can be very useful in producing such a table.

Another way scientists use light to explore systems similar to what we're discussion is ***absorption spectroscopy***. As opposed to emission spectroscopy described above, in absorption spectroscopy, broadband light (light with a very wide wavelength range) is shined onto the sample. Wavelengths of light that correspond to energy level transitions are absorbed by the material (as electrons are excited to higher levels) and scientists look for the gaps in the light spectrum on the far side of the sample. Although a very different technique, absorption spectroscopy also allows us to get the specific energy levels of a system and complete a table similar to the one above for emission spectroscopy.

Example 224-6: Vibrational spectroscopy

Let's say you are doing an experiment on vibrational relaxation of an unknown gas. The two longest wavelengths that you measure as the gas relaxes are 2696 nm and 5391 nm (both in the infrared). What is the value of $\hbar\omega$ for this gas?

Solution:

Since we are dealing with vibrational relaxation, we can immediately use the harmonic oscillator function.

$$E_n = \hbar\omega(n + 1/2).$$

Now, using our photon emission equation, we find

$$\frac{hc}{\lambda_{nm}} = \hbar\omega(n_2 + 1/2) - \hbar\omega(n_1 + 1/2),$$

which simplifies to

$$\frac{hc}{\lambda_{nm}} = \hbar\omega(n_2 - n_1).$$

Remember that the spectrum for the harmonic oscillator is linear, meaning that any transition of adjacent energy levels has the same energy (and any transition of two energy levels has the same energy as well). Since we know the two longest wavelengths, this would correspond to a change of one energy level ($n_2 - n_1 = 1$) and two energy levels ($n_2 - n_1 = 2$).

Since we are looking for $\hbar\omega$, first let's solve for that value

$$\hbar\omega = \frac{hc}{\lambda_{nm}(n_2 - n_1)}.$$

Now we can use our known values for the two transitions

$$\hbar\omega = \frac{1240 \, eV \cdot nm}{(5391 \, nm)(1)} = 0.23 \, eV$$

$$\hbar\omega = \frac{1240 \, eV \cdot nm}{(2696 \, nm)(2)} = 0.23 \, eV$$

As you can see, each wavelength gives us the same information and we can conclude that $\hbar\omega = 0.23 \, eV$ for this gas.

225 – Nuclear Stability and Radioactivity

Nuclear physics is the study of the nuclei of atoms and their interactions. The strong nuclear interaction is responsible for holding nuclei together and the weak nuclear interaction is responsible for nuclear radioactivity. In these last few units we will discuss these seemingly esoteric interactions.

Integration of Ideas

Review the four fundamental interactions.

The Bare Essentials

- The nucleus is described by the total number of protons and neutrons contained within it.

Nuclear Symbols

$$^{A}_{Z}X$$

Description – The nuclear symbol for an atom with chemical symbol X is described by its number of protons (Z), number of neutrons (N) and the total number of nucleons (A).

Note 1: The chemical symbol (X) and the number of protons are redundant; however, in physics, both are often expressed for ease of use.

Note 2: The number of neutrons in a nuclei is given by the neutron number, $N = A - Z$.

- As fermions, protons and neutrons obey the Pauli Exclusion Principle and therefore form nuclear energy levels inside nuclei.

- For low nucleon number, the number of protons is approximately equal to the number of neutrons in a nucleus, but for large nuclei, the number of neutrons is greater than the number of protons to shield electrostatic repulsion.

- The weak nuclear interaction is responsible for the transformation of protons to neutrons and neutrons to protons leading to nuclear radiation.

- Alpha decay is the expulsion of a helium nucleus from a large nuclei in order to reduce electrostatic repulsion of protons.

Alpha Decay

$$^{A}_{Z}X \rightarrow ^{A-4}_{Z-2}Y + ^{4}_{2}He$$

Description – Alpha decay describes the release of a helium nucleus from a large nucleus creating a daughter nucleus with two less protons and two less neutrons.

Note: In order to be probable, $m_X > m_y + m_{He}$.

- Beta decay, which comes in three forms, is the release of an electron or positron following a proton $\rightleftarrows$ neutron transformation.

Beta Decay

$$^{A}_{Z}X \rightarrow ^{A}_{Z+1}Y + e^- + \bar{\nu} \qquad \text{Neutron Decay } (\beta^-)$$

$$^{A}_{Z}X \rightarrow ^{A}_{Z-1}Y + e^+ + \nu \qquad \text{Proton Decay } (\beta^+)$$

$$e^- + ^{A}_{Z}X \rightarrow ^{A}_{Z-1}Y + \nu \qquad \text{Electron Capture (EC)}$$

Description – Beta decay is characterized by the transformation of a proton to a neutron or a neutron to a proton with the creation/annihilation of an electron or positron to balance charge.

- Gamma decay is the release of a high energy photon following the energy level transformation of a nucleon inside a nucleus.

- Since the decay of a nucleus is a random process, and the average number of decays per second depends on the total number of nuclei present, nuclear radiation rates follow an exponential decay.

Decay Rate

$$N(t) = N_0 e^{-\lambda t}$$

Description: This equation relates the number of nuclei present at a certain time, t, to the original number of nuclei, N_0, in a sample. λ is the decay constant (with units of s⁻¹).

Note: The half-life of the material is related to the decay constant such that $t_{1/2} = \frac{\ln}{\lambda}$.

225-1: The atomic nucleus

Consider: *What makes up the nuclei of atoms? What symbols can be used to write this down efficiently?*

IN THE LAST YEARS OF THE 19TH CENTURY, little was known about atomic structure. The majority of physicists believed in the ***plum pudding*** model of the atom whereby negatively charged electrons (point particles) floated around in a positively charged fluid that filled the entire atom. In this model, the negative electrons are the plums and the positive charge is the pudding. Ironically, physicists had it very much backwards.

Physicists also knew that certain elements (later called radioactive elements) could spontaneously transmute themselves from one type to another by emitting particles. One example of this is radon gas, which slowly converts itself to polonium by ejecting a positively charged particle with about the same mass as helium atoms. These particles are known as ***alpha particles*** and we know that they are exactly a helium nucleus. Earnest Rutherford devised an experiment where he directed alpha particles from radon at thin sheets of metal to see how they deflected as they traversed the material. To his surprise, a vast majority of the alpha particles were deflected by a very small angle, which suggested that most of the atomic space of the metal sheets were empty. Also, some of the particles were deflected backwards as if they had bounced off of something positively charged, very small, and very massive.

We now know that the small, massive particles in the metal causing the alpha particles to rebound are the nuclei of the atoms and that these nuclei make up only about 1/100,000 the overall volume of the atom! Since the electrons that orbit the nuclei act as if they are point particles, this means that 99.9999% of an atom is empty space!

The nuclei of atoms contain ***protons*** and ***neutrons***, which are collectively known as ***nucleons***. The number of protons in a nucleus is called the atomic number (Z), and the number of neutrons is called the neutron number (N). The mass number (A) of a nucleus is given by the total number of nucleons:

$$A = Z + N. \tag{225-1}$$

The atomic number of a nucleus corresponds precisely with the chemical symbol of an atom found in the periodic table of the elements.   For completeness, we use the following symbol

$$_Z^A X, \tag{225-2}$$

where X is the chemical symbol that corresponds to the atomic number, Z. Note that the neutron number of the nucleus given by this notation is found as $N = A - Z$.

Nuclear Symbols

$$_Z^A X \qquad (225\text{-}2)$$

Description – The nuclear symbol for an atom with chemical symbol X is described by its number of protons (Z), number of neutrons (N) and the total number of nucleons (A).

Note 1: The chemical symbol (X) and the number of protons are redundant; however, in physics, both are often expressed for ease of use.

Note 2: The number of neutrons in a nuclei is given by the neutron number, $N = A - Z$.

The number of electrons needed to make an element with atomic number Z neutral is also Z electrons, since the positive electric charge of the nucleus is provided by the protons. An element (with a given Z) can have different number of neutrons, and therefore a different mass and mass number. Such elements are called ***isotopes***.   All known elements have multiple isotopes. For example, hydrogen ($_1^1 H$) has one proton and no neutrons, deuterium ($_1^2 H$) has one proton and one neutron and tritium ($_1^3 H$) has one proton and two neutrons. Hydrogen, deuterium and tritium are isotopes of each other. Isotopes which do not decay are known as ***stable*** and those that do decay are known as ***radionuclides***. In our example, hydrogen and deuterium are stable and tritium is an unstable radionuclide.

225-2: Three important properties

Consider: *What is the size of an atomic nucleus? How much energy in stored in the nucleus?*

Nuclear Radii

As mentioned above, nuclei are very small and tend to be measured in femtometers (fm), where $1\ fm = 10^{-15}\ m$. The *effective* radius of a hydrogen nucleus is approximately 1.2 fm. Since additional nucleons are filling a volume, we would expect the addition of nucleons to increase the radius by the cube-root of the mass number, A:

$$r = r_0 A^{1/3}, \tag{225-3}$$

where $r_0 = 1.2\ fm$ is the effective radius of a hydrogen nucleus (a proton).

Nuclear Binding Energy

Nuclei are held together by the ***strong nuclear force***. We will examine this force in greater detail in the next unit. For now, it is important to note that it takes a certain amount of energy to bind nucleons together inside of nuclei. As we discussed in the relativity section of Physics I, this potential energy, known as ***nuclear binding energy***, is realized as a change in mass of the nuclei relative to the parts that make it up. In the relativity section, we saw that the change in mass due to gravitational binding was miniscule; however, the strong nuclear force is so, well, strong, that the differences in mass can be measured. The nuclear binding energy can be found as

$$E_B = \Delta mc^2 = (Zm_P + Nm_N - m_{nucleus})c^2, \tag{225-4}$$

where m_P is the mass of the proton, m_N is the mass of the neutron, and $m_{nucleus}$ is the mass of the nucleus in question. Since nuclear masses are so small, they are often given in ***atomic mass units*** (u). In addition, they can also be given in relation to their rest-mass energy (in eV) by rearranging Einstein's famous equation

$$E = mc^2 \quad \rightarrow \quad m = E/_{c^2}. \tag{225-5}$$

Table 225-1. Masses of important particles

Conversion			
$1u = 1.6604\ x\ 10^{-27}kg = 931.48\ MeV/c^2$			
Particle	m (u)	m (MeV/c²)	m (kg)
Electron	0.0005486	0.5110	9.109 x 10⁻³¹
Proton	1.007277	938.27	1.6726 x 10⁻²⁷
Neutron	1.008665	939.57	1.6749 x 10⁻²⁷

The conversion between mass units, as well as some important nuclear masses, can be found in table 225-1. One of the benefits of using nuclear masses in MeV/c² is that when you apply them to the binding energy formula, the c² cancels, and you are left with energy in MeV. I strongly suggest you try to get used to these units as they greatly reduce calculation time.

Nuclear Stability

As with all natural systems, atomic nuclei like to move towards their most stable states – that is to say, the lowest energy state. When it comes to atomic nuclei, one of the ways to measure this is the ***binding energy per nucleon***. Note from above that the binding energy of a nucleus is the amount of energy that goes into holding the nucleus together. Inside of each nucleus, we have to worry about keeping each and every nucleon inside, which is why the energy per nucleon is a better measure than the overall binding energy in terms of stability. Figure 225-1 shows a plot of binding energy per nucleon up through uranium.

To calculate the binding energy per nucleon, we simply divide the binding energy of the nucleus by the mass number, A:

$$E_{B,N} = \frac{E_B}{A}. \tag{225-6}$$

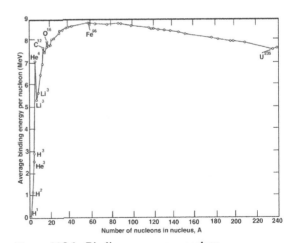

Figure 225-1. Binding energy per nucleon.

If we consider the most stable nuclei to be the one with the highest binding energy per nucleon, which is Nickel-62. This means that under nuclear processes, all nuclei below would like to combine to increase their binding energy per nucleon – this is called nuclear fusion, and that nuclei larger than iron would like to split to increase their stability – this is called fission. We will study fission and fusion in the next unit.

225-3: Radioactive Decay

Consider: *What is radioactive decay?*

Most of the known isotopes are not stable, and undergo *radioactive decay*, meaning that they randomly emit some form of fundamental or composite particle in order to create a more stable state. Figure 225-2 shows all of the known isotopes and their usual decay mechanism. We will describe each of these types of decay later in this section. First, we need to describe some of the math related to radiation.

As I just noted, radioactive decay is a *random process*. We do not know when an individual unstable nucleus will decay – it could be now, it could be in a million years. However, when we have a large number of the same nuclei together, there is an average time over which the process occurs – this is the definition of what is known as a *stochastic process* in science and engineering.

Consider again Figure 225-2. Note that at low atomic number, the number of protons and neutrons in a stable nucleus are about the same; however, as the atomic number increases, there must be more neutrons than protons in the nucleus for it to be stable. This is because of the electrostatic repulsion of the protons. Nuclei with a large number of protons must be buffered by neutrons – essentially increasing the distance between the protons and therefore reducing the electrostatic repulsion.

There are three main types of radioactive decay that we will consider, known as α (alpha), β (beta), and γ (gamma) decay. The types of radiation were named in order of their ability to penetrate matter, with α particles the least penetrating and γ particles the most penetrating. Before giving some detail on these three types of decay, it is important to note that these are not the only nuclear decay mechanisms, just the most common. As can be seen in Figure 225-2, spontaneous fission (see the next unit), and expulsion of a proton or a neutron are also mechanisms used by some nuclei.

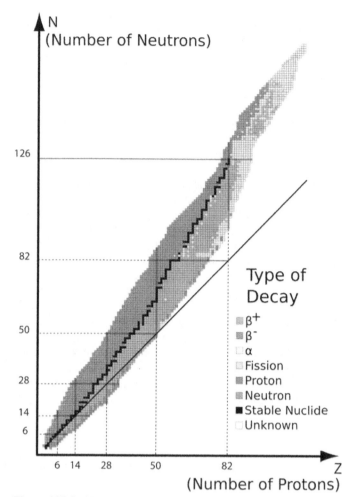

Figure 225-2. Known isotopes and their stability or decay mechanism.

Alpha (α) Decay

The nucleus of a helium atom is very stable, meaning that it has a much larger binding energy per nucleon when compared to other small nuclei (as can be seen in Figure 225-1). Large nuclei can become more stable and shed excess protons by releasing helium nuclei – a process known as *alpha decay*. In order for alpha decay to be feasible, the mass of the initial (parent) nuclei must be larger than the sum of the masses of the daughter nuclei and the alpha particle. Since the alpha particle carries away two protons and two neutrons, the daughter nuclei must change accordingly.

Alpha Decay

$$^A_Z X \rightarrow \,^{A-4}_{Z-2} Y + \,^4_2 He \qquad (225\text{-}7)$$

Description – Alpha decay describes the release of a helium nucleus from a large nucleus creating a daughter nucleus with two less protons and two less neutrons.

Note: In order to be probable, $m_X > m_y + m_{He}$.

Alpha particles are large and carry a charge equal to two fundamental units, so they do not penetrate matter very well. In fact, a single sheet of paper, or the dead outer layers of human skin are enough to stop most alpha particles.

Beta (β) Decay

Beta decay is the expulsion of an electron, or its positively charged cousin the positron, from a nucleus. This raises one important question - where do the electrons and positrons come from, since a nucleus is made up of only protons and neutrons? The answer is that they are **made** in the nucleus by the ***weak nuclear interaction***. The weak nuclear interaction can change a proton into a neutron, or vice versa. Conservation of charge says that if a neutron is converted to a proton, then there must be a particle with negative charge (electron) so that the overall reaction conserves charge. Likewise, if a proton is converted to a neutron, there must also be a positive charge created (positron) to conserve charge. The positron is the antiparticle of the electron, meaning that it has all the same properties of the electron, except that the sign of its charge is positive instead of negative.

Using nuclear symbols, proton decay and neutron decay can be written

$$p \rightarrow n + e^+ + \nu_e \qquad \text{(proton decay)}, \qquad (225\text{-}8)$$

$$n \rightarrow p + e^- + \bar{\nu}_e \qquad \text{(neutron decay)}. \qquad (225\text{-}9)$$

The particles ν_e and $\bar{\nu}_e$ are called the ***neutrino*** and ***anti-neutrino***, respectively. Neutrinos are electrically neutral fundamental particles (leptons) with an exceptionally low mass that are created in all reactions mediated by the weak nuclear force. Neutrinos were first theorized to balance conservation of energy and momentum in beta decay processes by Wolfgang Pauli in 1931, but not experimentally verified (because they have almost no interaction with normal matter) until 1956 by Clyde Cowan and Frederick Reines.

In beta decay processes, we write the nuclear equation in terms of the parent and daughter nuclei and not in terms of simple protons and neutrons. Also, proton decay is generally called β^+ decay since it releases a positively charged positron and neutron decay is called β^- decay because it releases a negatively charged electron.

In addition to beta decays where electrons and positrons are emitted, it is also possible for an electron in an inner shell of an atom to be combined with a proton in the nucleus to produce a neutron. This process is known as electron capture and has the fundamental nuclear equation

$$e^- + p \rightarrow n + \nu \qquad \text{(electron capture)}. \qquad (225\text{-}10)$$

Electron capture occurs in proton-rich nuclei where the expulsion of an electron or positron is not energetically favorable. In this case, the capture of an electron near the nucleus converts a proton to a neutron, reducing electrostatic repulsion.

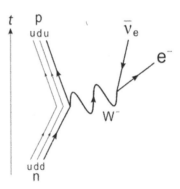

Figure 225-3. Feynman diagram for a β^- decay.

Beta Decay

$$^A_Z X \rightarrow _{Z+1}^{A}Y + e^- + \bar{\nu}$$ Neutron Decay (β⁻)   (225-8)

$$^A_Z X \rightarrow _{Z-1}^{A}Y + e^+ + \nu$$ Proton Decay (β⁺)   (225-9)

$$e^- + ^A_Z X \rightarrow _{Z-1}^{A}Y + \nu$$ Electron Capture (EC)   (225-10)

Description – Beta decay is characterized by the transformation of a proton to a neutron or a neutron to a proton with the creation/annihilation of an electron or positron to balance charge.

To be very precise, in β^+ (β^-) decay, one of the three quarks that make up a proton (neutron) is being converted from an up (down) quark to a down (up) quark. As the quark is converted, one of two gauge bosons, known as either a W⁻ or W⁺ particle, is emitted. Very quickly, the W-boson then decays into either a positron and neutrino, or an electron and an antineutrino. This process is shown in Figure 225-3, in what is called a Feynman Diagram.

Weak nuclear processes are very interesting, in that the force only acts over 10^{-18} m. Keep in mind that the effective size of a proton is on the order of 1 fm (10^{-15} m), which means that all weak nuclear interactions happen *inside* of protons and neutrons (on the scale of quarks). We never feel the weak nuclear force outside of this range – we can only see the electrons, positrons and neutrinos that are ejected as a result.

Gamma (γ) Decay

Daughter nuclei of radioactive decay often have one or more of their nucleons (protons or neutrons) in an excited energy state. Just as with electrons in atomic spectroscopy, as the nucleons drop down to their stable, lower state, they release the energy as a photon. The energy levels in a nucleus are spaced considerably farther apart than their atomic counterparts, which leads to emitted photons being in the gamma ray part of the electromagnetic spectrum. Quantitatively, gamma photons tend to be in the mega-electronvolt (MeV) range as opposed to photons emitted in atomic transitions which tend to be in the electronvolt (eV) range.

As very high-energy, uncharged photons, gamma rays tend to penetrate into matter farther than alpha or beta particles. It is typical for gamma particles to penetrate 20 cm through standard materials and even 5 or 6 cm through very dense lead.

Nuclear Stability Again

Now that we have discussed the three main types of nuclear decay processes, take a look at Figure 225-2 again. Note that whether it be alpha, beta or gamma decay, or even one of the less likely nuclear processes, each decay process brings the daughter nucleus closer to the line of stability than was the parent nucleus. Nature is always looking for the lowest-energy, most stable state, and for nuclei, this is where the binding energy per nucleon can be maximized for a given atomic number, Z.

225-4: Quantifying nuclear decay

Consider: *Is there a way to measure how fast nuclei decay?*

As noted above, the decay of an individual unstable nucleus is a completely random event – we have no way of knowing exactly when it will happen. However, if we have a large number of unstable atoms together, the average rate of radiation emission is quite stable. We will now take a look at how this works.

Let's say we have a sample of radioactive material – uranium-238 for example. We would expect that the amount of radiation given off by a sample would have to do with how large a sample we start with – meaning that if we have a 2 kg sample, we should expect twice as much radiation as a 1 kg sample. The other thing that is important to note is that as each uranium nucleus emits radiation, it will transmute and no longer be uranium – meaning that over time, we would expect the amount of uranium we have, and therefore the amount of radiation, to decline.

Mathematically, we can start by saying that the rate of change in the number of uranium nuclei (measured by the radiation given off) is proportional to the number of nuclei with which we start:

$$\frac{dN}{dt} = -\lambda_d N,$$ (225-11)

where λ_d is a constant of proportionality called the **decay constant**. Also note that I included a negative sign because the number of uranium nuclei is decreasing as time passes.

It is relatively straight forward to solve this equation for the number of uranium nuclei as a function of time. First, we need to isolate the N's on the left side by dividing the equation by N, and bring the dt to the right side by multiplying:

$$\frac{dN}{N} = -\lambda_d dt. \qquad (225\text{-}12)$$

Now, we can simply integrate both sides, assuming that we start at time t = 0 with N_0 nuclei and end at some undetermined time, t, with N(t) nuclei remaining:

$$\int_{N_0}^{N(t)} \frac{dN}{N} = \int_0^t -\lambda_d dt. \qquad (225\text{-}13)$$

Remembering that the integral of dN/N is the natural log of N, and that λ_d is a constant so it can be removed from the integral, we find

$$\ln N(t) - \ln N_0 = -\lambda_d (t - 0) \qquad \rightarrow \qquad \ln \left(\frac{N(t)}{N_0} \right) = -\lambda_d t, \qquad (225\text{-}14)$$

where I have also used the subtraction property for natural logs.

Finally, we can solve for N(t) by taking the exponential of both sides:

$$N(t) = N_0 e^{-\lambda_d t}. \qquad (225\text{-}15)$$

This equation shows that the number of radioactive nuclei, and therefore the radiation given off, decreases exponentially with time.

Decay Rate

$$N(t) = N_0 e^{-\lambda_d t} \qquad (225\text{-}15)$$

Description: This equation relates the number of nuclei present at a certain time, t, to the original number of nuclei, N_0, in a sample. λ_d is the decay constant (with units of s^{-1}).
Note: The half-life of the material is related to the decay constant such that $t_{1/2} = \frac{\ln 2}{\lambda_d}$.

A couple of notes about decay rates:

1. The decay constant, λ_d, **must have units of 1/time** for the argument of the exponential to be unitless. An important concept called the **activity** (A) is given by $A = \lambda N$. The activity measures the number of decays per second. The SI unit for activity is the becquerel (1 Bq = 1 disintegration per second). An older, English unit for activity is the curie (1 Ci = 3.7 x 10^{10} Bq). Note that activity follows the same decay curve as the number of particles, so

$$A(t) = A_0 e^{-\lambda_d t}. \qquad (225\text{-}16)$$

2. Another common way to measure the radioactivity of a substance is its **half-life**, or the time it takes half of a sample to radioactively decay. We can find the half-life by first substituting ½N_0 into our decay equation for N(t):

$$\frac{1}{2} N_0 = N_0 e^{-\lambda_d t_{1/2}} \qquad (225\text{-}17)$$

Then, we can solve this equation for time by dividing both sides by N_0 and taking the natural log of both sides:

$$\ln\left[\frac{1}{2}\right] = -\lambda_d t_{1/2}. \tag{225-18}$$

Finally, we can solve this for $t_{1/2}$:

$$t_{1/2} = \frac{\ln 2}{\lambda_d}, \tag{225-19}$$

where I have used the fact that $\ln[1/2] = -\ln[2]$. The half-life simply gives us a second way to describe the activity of a material. Note that an element with a short half-life emits more radiation than one with a long half-life. The half-life equation is noted in the decay rate box above.

Example 225-1: Nuclear decay

Cesium-137 is an unstable isotope of cesium that undergoes β^- decay with a half-life of 30.17 years. (a) Write the nuclear equation for cesium's β^- decay. (b) If we start with 22.1 g of cesium, how much will be left in 100 years?

Solution:

(a) Since cesium undergoes a β^- decay, we know that the atomic number of the nuclei will increase by one and the mass number will not change:

$$^{137}_{55}Cs \rightarrow ^{137}_{56}Ba + e^- + \bar{\nu}.$$

Note: I used a periodic table to find that barium has an atomic number of 56.

(b) In order to find the cesium left after 100 years, we must first find the decay constant from the half-life:

$$\lambda_d = \frac{ln(2)}{t_{1/2}} = \frac{ln(2)}{30.17\ y} = 0.0230\ y^{-1}.$$

We can leave this in years for consistency with the question. Now, we simply employ our decay equation – note that since N(t) represents the amount of material, we can use the mass as well:

$$N(t) = N_0 e^{-\lambda_d t} = 22.1\ g\ e^{-0.0230\ y^{-1}\ (100y)},$$

$$N(t) = 2.22\ g.$$

Approximately 10% of the cesium remains.

Example 225-2: Radiocarbon dating

Although the vast majority of carbon in the atmosphere at sea level is stable carbon-12 (98.9%) and carbon-13 (1.1%), a small trace is radioactive carbon-14. Plants and animals take in both stable carbon and carbon-14 while living; however, once they die, the carbon-14 will decay over time to carbon-12 reducing the relative abundance of the radioactive isotope. This fact can be used to determine the approximate age of fossils found that are up to approximately 60,000 years old. Here's how.

Solution:

Carbon-14 has a ½-life of 5730 years, meaning that if an archeological sample has an activity that is half of a living sample, we would estimate it to have died approximately 5730 years ago. If we found a sample with ¼ the activity of a living sample, we would estimate it died approximately 11,460 years ago.

We can find a general way of calculating the age of a sample by finding the ratio of the sample size to the half-

life size using our decay equation:

$$\frac{A(t)}{\frac{1}{2}A_0} = \frac{A_0 e^{-\lambda_d t}}{A_0 e^{-\lambda_d t_{1/2}}}.$$

Simplifying and taking the natural log of both sides to solve for t, we get

$$t = \left(\frac{\ln\frac{A(t)}{A_0}}{-\ln 2}\right) t_{1/2}.$$

So, let's say we have a sample where the activity is 2.45% that of a living sample, we would find the age to be

$$t = \frac{\ln(0.0245)}{-\ln(2)} 5730\ y = 30,660\ y.$$

That is we would estimate the sample was from approximately 30,700 years ago.

225-5: Measuring radiation and human factors

Consider: *How do we measure radiation in general and in terms of its effect on humans?*

As a biophysicist, I am always concerned with how the things we learn in physics affect our bodies and the environment. Although radioactivity is all around us every day, there are limits where excess radiation becomes devastating to humans. Let's take a look at how to measure radiation, what limits we need to place on radiation, and just how it affects our bodies.

The U.S. is one of the very few countries in the world that has not moved exclusively to the SI system for radiation units (as with many other units!). When reporting numbers throughout this section, I will use both the SI and English units with the SI units first and the English units in parentheses, such as for speed, I might say 10 m/s (22 mph).

As mentioned in the last section, the activity of a radioactive material is measured with the SI unit becquerels (Bq – one disintegration per second) or the English unit currie (Ci), where 1 Ci = 3.7 x 10^{10} Bq. However, the activity tells us only how many particles are coming from the source per second, and nothing at all about the energy that is carried away, or the effect of that radiation on material around the source, but we'll get to that in a minute.

The next important concept is the **radiation intensity**. The radiation from a given source is emitted in all directions around the object, so that the radiation that hits a given area decreases as we move away from the source, just as the ability of a light bulb to light a piece of paper decreases with distance. In order to find the intensity, we take the activity and divide it by the surface area of a sphere with a radius that is equal to how far we are from the source:

$$Radiation\ Intensity = \frac{activity}{4\pi r^2}.$$

(225-20)

Radiation intensity doesn't have its own symbol or official units; however, the concept is very important as we'll see that the effects of radiation drop off very quickly with distance from the source.

As the radiation travels from the source and impacts an object, the energy imparted to the object is quantified using the **Absorbed Dose**, which is energy imparted per mass. In the SI system we use the unit gray (Gy) which is a J/kg, and in the English system the unit is the rad (100 erg/g), where 1 rad = 0.01 Gy.

In order to find the absorbed dose, we must take the radiation intensity and multiply it by the area over which the intensity is absorbed. This will give us a measure, in counts per second, that radiation is absorbed by the given area. Also, we must convert this value to energy per count by including the energy of each particle. Finally, we must divide by the mass of the material absorbing the radiation to get an absorbed dose rate in J/kg/s = Gy/s:

$$\dot{D} = \frac{A}{4\pi r^2}(Area)(E)\frac{1}{m}$$

(225-21)

where $\dot{D}$ is the absorbed dose rate, A is the activity (in Bq), r is the distance from the source, E is the energy per nuclear particle (in Joules), and m is the mass of the material absorbing the radiation. This can be simplified by expanding the mass in this equation in terms of density and volume (where volume is the product of area, A, and thickness, d):

$$\dot{D} = \frac{A}{4\pi r^2}(\cancel{Area})(E)\frac{1}{\rho(\cancel{Area})(d)} = \frac{AE}{4\pi r^2 \rho d}.$$

(225-22)

Absorbed Dose Rate

$$\dot{D} = \frac{AE}{4\pi r^2 \rho d}$$

(225-22)

Description: This equation gives the absorbed dose rate ($\dot{D}$, in Gy/s) a distance r from a radioactive source with activity, A, emitting radiation with particle energy, E, into a material with density, ρ, and thickness, d.
Note: The absorbed dose is given by this dose rate multiplied by the time of exposure in units Gy (J/kg).

Example 225-3: Absorbed Dose

Radon gas is a natural byproduct of uranium decay in the ground. In the U.S., average radon gas activity in a basement is 48 Bq/m³, although this number varies widely. What is the absorbed dose of a person, 20 cm thick, standing in a basement due to 1 m³ of radon gas two meters away in 10 seconds? The alpha particle emitted from natural radon decay has an energy of 5.59 MeV.

Solution:

First, the energy per particle needs to be in Joules in order to use our dose rate equation:

$$5.59 \times 10^6 eV \cdot \frac{1.6 \times 10^{-19} J}{eV} = 8.94 \times 10^{-13} J.$$

Now, we can use our dose rate equation

$$\dot{D} = \frac{AE}{4\pi r^2 \rho d} = \frac{48\, Bq\,(8.94 \times 10^{-13} J)}{4\pi (2m)^2 \left(1000\, \frac{kg}{m^3}\right)(0.2\, m)}$$

$$= 4.27 \times 10^{-1}\, \frac{Gy}{s}.$$

Since we are asked for a 10 s exposure, the absorbed dose would then be $4.43 x 10^{-14}\, Gy$.

This is a very small absorbed dose; however, the example just given is for a tiny amount of gas over a short period of time. Radon gas exposure is a serious concern in many areas of the country.

The absorbed dose alone is not enough to tell us how radiation will affect the human body. There are two reasons for this: First, the ways that alpha, beta and gamma radiation interact with biological materials are all different. Second, radiation has different effects in different parts of the body.

In order to counter the first issue, we define the radiation weighting factor (W_R, also known as the relative biological effectiveness) for different types of radiation. The weighting scale is normalized so that photons (gamma, x-rays, etc.) and beta particles (electrons and positrons) have a W_R of 1, with other particles effectiveness defined around this value. Table 225-2 gives values of W_R for many nuclear radiation particles.

To get the effect of an absorbed dose of radiation on an entire body, we simply multiply the absorbed dose by the radiation weighting factor. This gives us a quantity known as the *effective dose*. The units of effective dose are *sieverts (Sv)* and *roentgen equivalent man (rem)* for SI and English units respectively. The effective dose is the most important quantity when we try to talk about the effect of radiation on the human body – the rest of the units and quantities we've discussed up until now are the ideas that we needed leading up to effective dose.

One last thing before we discuss ranges of radiation and their overall effect – what if radiation is only imparted on one small part of the body. This isn't important so much in, say, a nuclear reactor incident like Fukushima, but is very important when you get an x-ray of your arm. Since the radiation is only affecting part of your body, we would expect it to have less of an overall effect than if it hit your entire body. So, there is another weighting factor, called the tissue weighting factor, which can be used in this instance. If you want to find the effective dose on a section of body, you simply multiple the effective dose above by the tissue weighting factor for that tissue, which are given in table 225-3. Using these ideas, the overall effective dose is given by

$$D_{eff} = W_T \cdot W_R \cdot D, \tag{225-23}$$

where D_{eff} is the effective dose, W_T is the tissue weighting factor, W_R is the radiation weighting factor, and D is the absorbed dose.

Table 225-2. Radiation weighting factor for many types of radiation

Type	Energy	W_R
Photons	All	1
Electrons Positrons Muons	All	1
Neutrons	< 10 keV	5
	10 – 100 keV	10
	100 keV – 2 MeV	20
	2 – 20 MeV	10
	> 20 MeV	5
Protons	>2 MeV	5
Alpha particles	All	20
Fission fragments	All	20
Heavy nuclei	All	20

Table 225-3. Tissue weighting factors, W_T (2007 ICRP values)

Type	W_T
Gonads	0.08
Bone Marrow	0.12
Colon	0.12
Lung	0.12
Stomach	0.12
Breasts	0.12
Bladder	0.04
Liver	0.04
Esophagus	0.04
Thyroid	0.04
Skin	0.01
Bone Surface	0.01
Brain	0.01
Salivary Glands	0.01
Rest of Body	0.12
Whole Body	1

Takeaway

I realize that this section has introduced many new terms and units, so what are you really supposed to take away from this? As just mentioned, the most important quantity is the effective dose in Sv (rem). The rest of the section up until now gave us the physical background to understand what the effective dose means – it is a measure of how much energy is imparted on the human body, including how effective that type of radiation is at causing damage.

One Sv (100 rem) is a very large and dangerous amount of radiation. In the United States, the average person receives about 6 mSv (600 mrem) (note this is milli-Sv – 1/1000 of a Sv) of radiation from background sources in a given year. Of this, about 3 mSv (300 mrem) is naturally occurring radiation from radon, cosmic rays, and other natural radiation in the food we eat and water we drink. Americans get another 3 mSv (300 mrem) or so on average from medical tests and procedures, such as x-rays and radiation treatments for cancer. For college-aged people in the U.S., background radiation is usually closer to the 3 mSv (300 mrem), since you have less medical tests, etc., when compared to older people. Another source of artificial background radiation is what is left over from nuclear bomb tests back in the 1950's and 1960's, although this decreases every year as the fission products continue to decay.

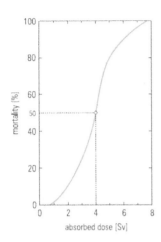

Figure 225-4. Plot of mortality versus effective dose.

The Nuclear Regulatory Commission and Environmental Protection Agency have set a limit for the general public of 1 mSv (100 mrem) of radiation exposure above background per year. Occupationally, limits can be increased to 5 mSv (500 mrem) for non-radiation workers and as high as 50 mSv (5000 mrem) for radiation workers. All of these limits are exceptionally conservative, as they should be.

A short-term exposure of 100 mSv (10 rem) is the lowest known exposure that leads to an increased risk of cancer (the mechanism is explained in the next section). At this exposure, however, the person exposed would have no short-term effects and might not even know they were exposed.

The dangerous situation begins at about 400 mSv (40 rem) of short-term exposure. At this level, the body undergoes a quick reaction to the radiation, known as *acute radiation sickness* as the radiation kills many cells in the body quickly. Up to about 1 Sv, most people survive the radiation sickness, although the mortality rate grows rapidly above 1 Sv until approximately 8 Sv, which is considered 100% fatal. The generally accepted mortality rate versus effective whole body dose is shown in Figure 225-4.

Some important levels of effective whole body dose are shown in Table 225-4.

Table 225-4. Effective doses of radiation

Effective Dose	Description
0.098 µSv	Banana Equivalent Dose – the approximate dose you get from eating a banana
10 µSv	Typical set of dental x-rays.
0.5 mSv	Typical dose for a mammogram.
1 mSv	NRC and EPA annual limit for a general member of the public
1.6 mSv	Typical annual dose for a flight attendant.
20 mSv	Typical dose for a full body CT scan.
50 mSv	NRC annual limit for radiation workers
69 mSv	Estimated dose to evacuated citizens near the Fukushima plant.
100 mSv	Lowest known short-term exposure that leads to an increase in cancer risk (approximately a 0.1% increase in cancer rate).
400 mSv	Onset of acute radiation sickness due to short term exposure
670 mSv	Largest dose for an emergency worker at Fukushima.
1 Sv	Fatal to approximately 1% in short-term exposure
4 Sv	Fatal to 50% - 100% in short-term exposure
5.1 Sv	Fatal dose to Harry Daghlin in a 1945 criticality incident (see Connection below)
8 Sv	Fatal to 100% in short-term exposure.
21 Sv	Fatal dose to Louis Slotin in a 1946 criticality incident (see Connection below)
64 Sv	Non-lethal 21 year dose to Albert Stevens due to a 1945 Plutonium injection experiment. (See Connection below)

Example 225-4:   Radiation in Our Physics Labs

Many of our physics sections will do radiation experiments or demonstrations as part of this unit. What doses of radiation might be expected during such an experiment?

Solution:

One of the sources we use contains 1 µCi of Cesium-137, with an activity of 3.7×10^4 Bq (disintegrations per second). If your fingers are 5 mm from the source (the approximate distance if you are holding the disk correctly), the radiation intensity is

$$Rad.Int. = \frac{3.7 \times 10^4 \, Bq}{4\pi(0.005 \, m)^2} = 1.18 \times 10^8 \, \frac{Bq}{m^2}.$$

One disintegration of Cesium-137 releases 1.176 MeV of energy, which we must convert to joules:

$$\left(\frac{1.176 \times 10^6 \, eV}{d}\right)\left(\frac{1.6 \times 10^{-19} \, J}{eV}\right) = 1.88 \times 10^{-13} \, J,$$

We are now ready to use our general equation for effective dose

$$D_{eff} = W_T \cdot W_R \cdot D,$$

where we will use 0.01 for W_T (skin), 1 for W_R (gamma radiation), and D is the effective dose.

In order to find the absorbed dose rate, I will assume that we are holding the disk with two fingers, each with a thickness of 2 cm, for a total thickness of 4 cm:

$$\dot{D} = \frac{AE}{4\pi r^2 \rho d} = \frac{3.7 \times 10^4 \, Bq(1.88 \times 10^{-13} \, J)}{4\pi(0.005m)^2 \left(1000 \, \frac{kg}{m^3}\right)(0.04 \, m)}$$

$$= 5.5 \times 10^{-7} \, \frac{Gy}{s}.$$

If we assume it takes two seconds to put the disk in place and another two seconds to remove it, for a total time of irradiation of 4 seconds, we find a total absorbed dose of

$$D = \dot{D}t = 5.5 \times 10^{-7} \, \frac{Gy}{s}(4 \, s) = 2.2 \times 10^{-6} \, Gy.$$

Finally, we put all of this information together in our equation for effective dose to find

$$D_{eff} = (0.01)(1)(2.2 \times 10^{-6} Gy) = 2.2 \times 10^{-8} \, Sv.$$

This value of 22 nSv is well below any dangerous levels and is the equivalent dose of eating ¼ of a banana or 1/100th the dose of a dental x-ray. We consider our radiation experiments and demos to be quite safe.

Connection: United States Radiation Deaths

There have been 21 known acute deaths due to radiation accidents in the United States. Of these, 17 were due to miscalculation of medical radiotherapy doses in Columbus, OH and Houston, TX between 1974 and 1980. The other four deaths are due to "criticality incidents," where a critical mass of nuclear materials was unexpectedly created (more on critical mass in the next chapter). When criticality is reached, a tremendous number of neutrons and gamma rays are created, often accompanied by a blue glow known as Cherenkov radiation.

The first two of these criticality incidents are of particular interest because they involved the same piece of radioactive material. In two separate incidences, Harry K. Daghlian, Jr and Loius Slotin, both eminent physicists, were working with a sphere of Plutonium. In each case, the researchers were surrounding the plutonium with a neutron reflecting material when an accident occurred – Daghlian dropped the final brick that was to encase the plutonium over the core and Slotin was using a screwdriver to keep two hemispheres of neutron reflectors separated when the screwdriver slipped. Once encased, the plutonium went critical. Daghlian received 5.1 Sv of radiation and Slotin received 21 Sv. Both died horribly of acute radiation poisoning. The plutonium core they were using was dubbed the "Demon Core" and was later detonated in the Abel nuclear bomb test of Project Crossroads.

225-6: Biological effects of ionizing radiation (optional)

Consider: *Exactly how does radiation cause damage to human tissue?*

So, just how does nuclear radiation cause such sickness and death? It all comes down to your DNA. Deoxyribonucleic acid, or DNA, is a double-stranded set of genetic instructions in the shape of a double helix (twisted ladder). DNA is made up of just four "bases", Adenine (A), Guanine (G), Cytosine (C) and Thymine (T). The bases are always paired on opposite sides of the ladder shape with Adenine paired with Thymine and Cytosine paired with Guanine. Therefore, each strand contains the same information, even though the order of the bases may look quite different.

direct DNA-damage

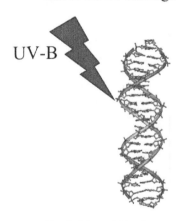

UV-B

Figure 225-5. Direct DNA damage via a UV photon.

Damage to DNA can be done in two different ways. First, direct damage occurs when radiation is incident directly upon one of the DNA bases and ionizes the base. Since the bases are bonded together by a subtle equilibrium in their electron clouds, this ionization causes a break in one of the two strands of DNA. This is called a single-strand break. Indirect damage occurs when ionizing radiation strikes a water molecule in one of your cells creating highly oxidizing free radicals (oxygen radicals and hydroxyl radicals) that can then interact with a DNA base to chemically cause a single-stranded break. The mechanism of direct damage is shown in Figure 225-5, and that of indirect damage in Figure 225-6.

Your body is amazingly good at fixing single-stranded breaks. First of all, if only one strand of DNA is damaged, the information is still encoded in the other strand to correctly repair the molecule. Single-stranded breaks happen regularly in your body (up to 20,000 times a day per cell) even in the absence of radiation, so your body's defenses are adept at this type of repair, although these defenses are not always perfect. There are three types of DNA repair mechanisms for single-stranded breaks – Base Excision repair, Nucleotide Excision Repair, and Mismatch Repair.

Therefore at low levels of nuclear radiation, the body can clear any damage pretty effectively. In fact, some recent studies have suggested that low levels of radiation may, in fact, strengthen your body's ability to use these processes and actually increase overall health. Note that this is certainly not a fact at this point, just a supposition.

At high levels of radiation, multiple single-stranded breaks, or even double stranded breaks can occur in DNA if the radiation is causing damage faster than the body can repair the damage. Double-stranded breaks are especially a problem, since your body lacks the roadmap to recreate the DNA as it does for single-stranded breaks. This can lead to mismatches in the DNA – mutations. More often than not, a DNA mutation will cause the cell to kill itself off by *apoptosis* (programmed cell death). At low-to-medium levels of radiation, a few cells die off gracefully and are eventually replaced by your body. No problem.

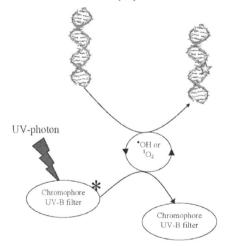

UV-photon

•OH or ¹O₂

Chromophore UV-B filter

Chromophore UV-B filter

Figure 225-6. Indirect DNA damage via a UV photon.

At higher levels of radiation, however, large numbers of cells in an area of your body may die in a short period of time – leading to sickness or death depending on where in the body the damage occurs and to what extent it occurs. This is why low levels of radiation cause little problem, medium levels cause radiation sickness and high levels cause sickness and death.

Also, at high levels of radiation, not only is the DNA in your cells damaged, but also many proteins may be damaged by the radiation. In this case, the cell may undergo a catastrophic death known as necrosis, whereby the cell ruptures and its contents are released into the extracellular environment. The cell's contents are highly inflammatory when in the extracellular matrix, which can lead to further damage.

Radiation sickness occurs when the damage to cells in your body is to an extent that tissues and organs stop functioning correctly. Your body will continue to attempt to repair the damage, which is why some people experiencing radiation sickness recover. Since damage to DNA is the main culprit in radiation sickness, it is tissues that replicate quickly that are most affected – such as your bone marrow. Stem cell treatment for bone marrow replacement has been shown to be very effective at increasing survivability to radiation doses in the 2-6 Sv range – a range that would have certainly been fatal in the past.

Finally, radiation doses that cause DNA mutations, but not high levels of cell death carry an increased risk of cancer. Without regulation, the cells in our bodies would replicate out-of-control. Under normal circumstances a regulatory system is in place to slow down cell replication to the necessary rate. This process is mediated by *tumor suppressor genes* and *tumor suppressor proteins*. If a mutation happens in one of the DNA genes for this system, the cell may very well survive, but lose its ability to regulate its own replication. This is how tumors form. Without the regulation pathways, the initial cell and its progeny divide unregulated and become a ***tumor***. If the tumor is capable of spreading to other parts of the body, it is called a ***malignant tumor***, and this is cancer.

Tumorigenesis, or the production of tumors, is actually very rare – more often than not, your body either repairs the cell, or the cell dies. However, we simply do not notice this happening. It is only the very rare occurrence where the cell's ability to regulate itself is hampered that we see the results – cancer.

Connection: The Atomic Man

On May 14th, 1945, Albert Stevens was injected with 3.55 µCi of Plutonium without his knowledge or consent. Stevens, a house painter living in California, had recently been diagnosed with terminal stomach cancer. He was unwittingly put into one of the medical programs of the Manhattan Project designed to test the effect of radiation on the human body. There was only one problem – Stevens didn't have cancer. He had been misdiagnosed, and when doctors finally performed surgery to remove the tumor, four days *after* the plutonium injection, they found no sign of cancer. All they found was a very bad ulcer.

Over the course of the next 20 years, Stevens absorbed an effective dose of over 64 Sv from the plutonium in his system. He died January 9th, 1966 of a heart attack, apparently unconnected to the plutonium he carried all those years. What's amazing is that the U.S. Government never told Stevens or his family that he had been injected with plutonium; he spent the last 20 years of his life thinking that he had made a remarkable recovery from terminal stomach cancer. It wasn't until 1993 that investigative journalist Eileen Welsome published a series of articles related to the plutonium injection experiments that Stevens' story came out. We now know that there were 18 patients in the experiment. Albert Stevens received by far the largest effective dose, which is now considered eight times the lethal limit of plutonium. The fate of the other 17 people in the experiment are not all known; however, it is believe that they died of the pre-existing conditions that placed them, unknowingly, in the study in the first place.

226 – Nuclear Physics and Fusion

Our final unit extends unit 225 to include nuclear fission, nuclear fusion and nuclear technology. Along the way, we will learn about natural nuclear reactors, how stars produce so much energy and how splitting and combining atoms has been harvested for destruction and energy. As time goes on, it is likely that the technologies discussed in this unit will take on greater importance, and may, someday, solve our energy crisis.

The Bare Essentials

- The strong nuclear interaction is responsible for holding quarks together inside of nucleons and nucleons together into nuclei.

- Large nuclei are less stable because the strong nuclear force has a very short range, and therefore electrostatic repulsion plays a larger role.

- The stability of each nucleus is measured by its binding energy per nucleon. Nickel-62 has the largest binding energy per nucleon and is therefore the most stable nuclei known.

- The energy released in a nuclear reaction is given by the difference in binding energies between the reactants and products.

- Nuclear weapons (fission weapons) were first developed during WWII and utilize the fission of Uranium or Plutonium to create explosive power many times larger than what was seen before.

- Thermonuclear weapons (fusion weapons) utilize a nuclear (fission) explosion to create the high energy densities required to initiate fusion.

- Stars, such as the sun, utilize nuclear fusion to create the energy to power the solar system.

Energy Released in Nuclear Reactions

$$E = \sum E_{B,products} - \sum E_{B,initial}$$

Description – This equation states that the energy released in a nuclear reaction is equal to the difference in binding energy between the product nuclei and the initial nuclei.

Note 1: The excess energy is usually released in the form of kinetic energy of one or more of the reaction products (often neutrons).

- Nuclei that are larger than iron tend to release energy when they break apart, or undergo *fission*. For very large nuclei, the energy released per nuclei can be substantial.

- Nuclei that are smaller than iron tend to release energy when the combine with another small nuclei, or undergo *fusion*. The energy released per nuclei can be substantial for very small nuclei.

226-1: The strong nuclear interaction

Consider: *How does the strong nuclear interaction work?*

If atomic nuclei were governed primarily by the electromagnetic force, every atom in the universe would instantly explode. Although neutrons inside of nuclei are immune to electrostatic effects, every proton in every nucleus is constantly pushing its neighbors away. With no negative charges inside the nucleus, there would be nothing to offset this constant repulsion. In steps the **strong nuclear interaction** (aka, strong interaction). The strong nuclear interaction can be viewed in two ways:

1) On the order of 1 – 3 fm, the strong nuclear interaction is responsible for holding nucleons together inside of nuclei,
2) On the order of about 0.8 fm, the strong nuclear interaction is responsible for holding quarks together to form nuclei.

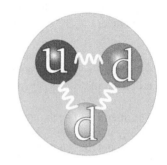

The strong interaction is like nothing we have studied before – it is always attractive, meaning that it always acts to pull particles together. Furthermore, it has a limit of just a couple of femtometers, and outside of this range the strong interaction has no effect. (Compare this to electromagnetism and gravity which have an infinite range.) In fact, and this is weird, as two particles separate from 0.8 fm to 1.0 fm, the strong force between them *actually increases up to about 25,000 N*, trying to pull them back together. Between 1 and 3 fm, the force decreases quickly to almost zero.

The leading theory for the strong interaction is called **quantum chromodynamics (QCD)**. In this theory, interaction bosons called **gluons** are transferred between quarks, which bind the quarks together. The quark-gluon interaction acts similar to small springs between quarks (as can be seen in Figure 226-1). Quark-gluon interaction also happens between quarks of two different nuclei; however this interaction has a slightly different flavor than the intra-nucleon interaction.

Figure 226-1 shows each quark as having a **color charge** – either red, green, or blue. Don't confuse the word color here with our traditional colors of light. The term *color charge* is used because it comes in three types and has been named after colors of the rainbow. Stable nucleons have one quark of each color. As a gluon moves between two

Figure 226-1. Depiction of the strong interaction where the gluons are represented as squiggle lines between quarks.

quarks, it carries color charge with it and changes the colors of its two quarks. For example, a blue quark may emit a blue-antigreen gluon, which turns the blue quark to a green quark. Then, when the blue-antigreen gluon reaches a green quark, it converts it to a blue quark. Except for the transit times of gluons, all stable nucleons are **color neutral**, meaning that they contain one quark of each color. Any non-symmetric color exchange will cause a nuclear structure to become unstable, leading to many of the effects we are going to explore in the rest of this unit.

226-2: Nuclear fission

Consider: *What causes large nuclei to split apart?*

Large nuclei have a tendency to be unstable because the strong nuclear force binding the nucleus together does not extend much beyond neighboring nucleons. Therefore protons that are relatively far apart in a nucleus have a lower attraction due to the strong interaction, but yet are still repelled by the electrostatic interaction with other protons. This is one of the reasons that nuclei that are heavier than iron-56 tend to have a lower binding energy per nucleon than iron.

Although most of these nuclei can maintain stability when undisturbed, some will spontaneously break into two or more pieces if induced by a passing neutron. This process of nuclei splitting is known as **nuclear fission**. Take uranium-235 for example. When a passing neutron is absorbed by the uranium, it becomes uranium-236. However, excess energy from the collision causes the uranium nucleus to form lobes like a dumbbell. Since protons in one lobe are now slightly farther away from protons in the other lobe, the attractive strong force is greatly decreased, while the repulsive electrostatic force does not change by much. This causes the uranium-236 nucleus to split into two uneven daughter nuclei, usually accompanied by a couple of free neutrons. This process is depicted in Figure 226-2. When uranium-235 undergoes fission, there are a number of possible reactions that can take place. Three common nuclear equations for such fission are

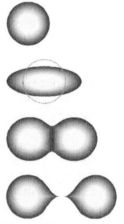

Figure 226-2. A model for nuclear fission.

$$^{235}_{92}U + {}^{1}_{0}n \rightarrow {}^{236}_{92}U \rightarrow {}^{89}_{36}Kr + {}^{144}_{56}Ba + 3{}^{1}_{0}n + 173\ MeV. \qquad (226\text{-}1)$$

$$^{235}_{92}U + {}^{1}_{0}n \rightarrow {}^{236}_{92}U \rightarrow {}^{137}_{55}Cs + {}^{95}_{37}Rb + 4{}^{1}_{0}n + 191\ MeV. \qquad (226\text{-}2)$$

$$^{235}_{92}U + ^1_0n \rightarrow ^{236}_{92}U \rightarrow ^{140}_{54}Xe + ^{94}_{38}Sr + 2^1_0n + 200 \; MeV. \qquad (226\text{-}3)$$

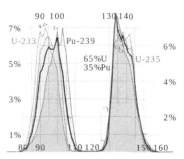

Although there are many fission products from uranium-235, they do tend to come in unequal sizes. The spectrum of fission products from uranium-235 can be seen in Figure 226-3. Each of these fissions have one thing in common through – they release energy. You can see on the right side of each of the nuclear equation that between 177 and 200 MeV of energy is released per fission. Although this is a relatively small amount of energy, remember that this is for just one uranium nucleus, and that even a 1 g block of uranium would contain 10^{21} nuclei! Also, compare the above energy released for one fission with the energy released by the combustion of one ethane molecule – 16.2 eV; *over ten million times less than one nuclear fission*.

Figure 226-3. Spectrum of fission products from $^{235}_{92}U$.

The energy released in a nuclear fission is given by the difference in binding energy of the product nuclei and initial nuclei. The excess energy goes into kinetic energy, with most of the kinetic energy given to released neutrons.

Energy Released in Nuclear Reactions

$$E = \sum E_{B,products} - \sum E_{B,initial} \qquad (226\text{-}4)$$

Description – This equation states that the energy released in a
nuclear reaction is equal to the difference in binding energy
between the product nuclei and the initial nuclei
Note 1: The excess energy is usually released in the form of kinetic
energy of one or more of the reaction products (often neutrons).

Example 226-1-1: Fission energy

Determine the energy released during the fission of one $^{235}_{92}U$ nuclei to $^{89}_{36}Kr$ and $^{144}_{56}Ba$ as shown above.

Solution:

The energy is given by our energy released equation above:

$$E = \sum E_{B,products} - E_{B,initial}$$

$$E = (E_{B,Kr} + E_{B,Ba} - E_{B,U})$$

We must look up the binding energy for each of our isotopes and substitute them into this equation:

$$E = 766.643 \; MeV + 1190.014 \; MeV - 1783.870 \; MeV,$$

$$E = 172.787 \; MeV.$$

Rounding to three significant figures, this gives us 173 MeV per fission, consistent with what is shown above.

226-3: Critical mass

Consider: *Is there a certain mass needed to maintain nuclear fission?*

The fission discussed in section 226-2 required a free neutron to strike a $^{235}_{92}U$ nuclei in order to start the fission process. This is generally true of fission reactions; they require some form of initiation, which is often a neutron striking a large nuclei. Many of the fission reactions produce additional free neutrons, which can then either strike another nuclei or escape the material. Consider Figure 226-4. On the left, the Figure shows a small piece of material, where an initial fission produces two neutrons. One of these product neutrons then strikes another nuclei, while the second leaves the material. When the new nuclei is struck by this neutron, it fissions producing an additional two neutrons - which then leave the material. Since the material has no excess neutrons following this second fission, the fissioning of nuclei stops and the material settles down. Compare this with the diagram on the right of Figure 226-4. In this case, the material is large enough that for each fission more than one neutron (on average) hits another nuclei and causes a fission. In this case, although some neutrons may leave the material, on average the neutrons released by a fission create more than one fission. This is a ***chain reaction***. The mass at which a given material can start to produce a self-sustaining chain reaction is called the ***critical mass***. A mass below critical is called a ***subcritical mass*** and one above is called a ***supercritical mass***.

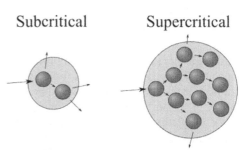

Figure 226-4. Diagram depicting why a certain mass is required to set up a fission chain reaction. Note: arrows are neutrons, spheres are nuclei.

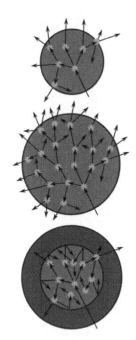

It turns out that the ability of a material to go critical depends on more factors than just the mass. The critical mass is the smallest possible free mass that can cause a spontaneous chain reaction. One of the largest factors determining criticality is the shape of the material – and specifically the ratio of the surface to the volume. A large surface area allows many neutrons to escape, reducing the ability of the material to form a chain reaction. Remember that it is excess neutrons from a fission creating more fissions which keeps a chain reaction going, so losing neutrons through the surface reduces its ability to fission.

Another way to increase the ability for a material to sustain a chain reaction is to surround it with a ***neutron reflector***. With a neutron reflective material surrounding material that can undergo fissions, neutrons that would escape the material are reflected back into it, increasing the number of free neutrons eligible to create a fission. Increasing a material's ability to fission by increasing its mass and surrounding it with a neutron reflector are shown in Figure 226-5. It turns out that hydrogen is a very good neutron reflector – therefore water and all organic material tend to act as very good neutron reflectors.

Figure 226-5. Why increasing mass or surrounding with a neutron reflector increases fission.

Connection: Daghlian and Slotin again

In one of the Connection boxes in unit 224, I described how two Los Alomos physicists, Harry K. Daghlian, Jr and Loius Slotin, died in criticality accidents in the 1940's. The scientists were playing with neutron reflectors around a plutonium core, exploring the limits for nuclear weapons testing. In each case, they completely surrounded the plutonium with the neutron reflector by accident, causing it to go supercritical. The radiation released by the unexpected chain reaction is what killed them. They had, unfortunately and unexpectedly, created an *unshielded nuclear reactor*. This is one of the reasons that we have to be so careful to control and shield nuclear power plants, as we're about to see.

226-4: Controlled and uncontrolled chain reactions.

Consider: *How have we, as humans, harnessed nuclear fission?*

We have harnessed nuclear fission in two ways – the controlled reactions inside of a nuclear power plants and the uncontrolled reactions of nuclear weapons. The histories of these two technologies are tightly entwined. Hungarian physicist Leo Szilard realized that chain reactions of the recently discovered uranium fission (discovered in 1938) could be used to create both power production and powerful *atomic* weapons. Concerned that German scientists would come to the same conclusion, Szilard wrote a letter to President Roosevelt expressing his concern, and he asked his colleague and friend Albert Einstein to sign the letter to give it more weight. Einstein admitted during that meeting that he had not thought of the possibility of nuclear bombs created through fission, and he signed the letter. The letter was delivered by Szilard to President Roosevelt on October 11th, 1939. Roosevelt was immediately concerned with the consequences outlined in the letter and two major initiatives resulted. First, Szilard was given funding by the government to build a carbon moderated nuclear reactor (then called a *nuclear pile*) with Enrico Fermi. Second, a commission was formed that would eventually lead to the development of the **Manhattan Project** – the United States' secret project to develop the first nuclear weapon.

The development of both nuclear energy and nuclear weapons were hampered by the lack of ***fissile material***. Not all large elements will readily fission when struck with a neutron, and in fact, only a very small percentage of uranium as mined is fissile. Natural uranium ore is a mixture of uranium-238 (99.27%) and uranium-235 (0.72%). Uranium-238 does not readily fission and so natural uranium is not directly adequate for nuclear technology. To be useful for power production, *highly-enriched uranium*, with >20% uranium-235 must be produced. Nuclear weapons require >85% uranium-235 (known as *weapons-grade uranium*). One of the major thrusts of the Manhattan Project was to produce the necessary uranium concentrations through a process known as *gaseous diffusion*. The physics of gaseous diffusion is beyond the scope of this text, but I introduce the term for those that may be interested.

Another important fissile element, especially for weapons development, is *plutonium-239*. There is just one problem – before the Manhattan Project, there was close to no plutonium on earth! Even so, what little plutonium that is left over from the formation of the earth is plutonium-244, a non-fissile isotope of plutonium. Plutonium-239 is created by neutron activation of uranium-238 inside of nuclear reactors. Said another way, uranium-238 will absorb a neutron, which then, in time, decays via two beta decays to plutonium-239:

$$^{238}_{92}U + ^{1}_{0}n \longrightarrow ^{239}_{92}U \xrightarrow{\beta(23.5\ min)} ^{239}_{93}Np \xrightarrow{\beta(2.35\ days)} ^{239}_{94}Pu.$$

In the above equation, the time in parentheses for a given process represents that half-life for that process. You can see that within a few days, activated uranium-238 will substantially decay to plutonium-239. Therefore, a nuclear reactor designed to produce plutonium would use uranium that has a high percentage of uranium-238 as opposed to reactors used for energy production that use a high percentage of uranium-235. A nuclear reactor designed to produce more fissile material than it consumes (i.e., produce plutonium) is called a *breeder reactor*.

> **Connection**: Plutonium
>
> Plutonium is an interesting beast. Since there is almost no natural plutonium in the environment, we have developed *very* little tolerance to it in our systems - a mere 20 milligrams of plutonium inhaled is enough to kill you in a short period of time. Plutonium is an alpha-emitter. A 5 kg mass of plutonium gives off the equivalent of 9.68 Watts of power due to its internal radiation, and the internal deceleration of alpha particles (friction) make a sphere of plutonium warm to the touch, even in cold environments!

Nuclear Power

Today, controlled nuclear reactions are performed all around the world - nuclear power makes up 19.4% of electricity generation in the U.S, and a whopping 75% of electricity in France, for example. In many ways, nuclear power plants work the same way as coal power plants that have been existence since 1882. In both designs, water is heated until it turns to steam.

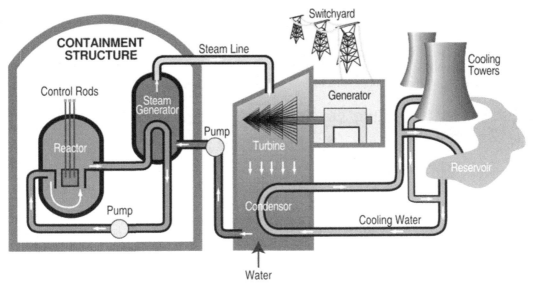

Figure 226-6. Basic schematic of a nuclear power plant (pressurized water reactor).

As the steam rises, it turns a turbine connected to an electric generator, and voila!, we have electricity. The main difference between the two types of plants is that nuclear fission is used to produce heat in a nuclear power plant as opposed to the burning of coal in a coal power plant. The important components of a nuclear power plant are shown in Figure 226-6.

The reactor vessel is where the nuclear reactions take place. The fuel used in modern nuclear reactors consists of enriched uranium dioxide pellets made into fuel rods when clad by the metal alloy zircaloy. The fuel rods are placed vertically in the reactor vessel in a geometry that encourages fission. High pressure water is also pumped through the vessel for two reasons. First, the water acts as a moderator, slowing down neutrons to increase their effectiveness at causing a fission when they intersect a uranium nuclei. Second, the water will be heated by the nuclear reactions, and the thermal energy will be used later

255

to create steam, while at the same time keeping the reactor fuel at operating temperature. The water that flows through the reactor vessel and the steam generator create the ***primary cooling loop***.

Another important feature of the reactor vessel is the ***control rods***. The control rods are made of neutron absorbing materials, so that when inserted between the fuel rods, they drastically reduce the number of neutrons available to create fission. These rods are used to control the rate of fission and therefore the overall power production of the power plant. In addition, the control rods are part of the ***scram system***, whereby in an emergency, a single button will cut power to an electromagnet holding the control rods, thereby causing the control rods to drop completely into place between the fuel rods (using only gravity), ending the fission reactions in the vessel. This is a very important and robust safety feature built into most U.S. nuclear power plants.

> **Connection**: Natural Nuclear Reactor
>
> Although we think of nuclear reactors as manmade (with good reason), it turns out that there was a natural nuclear reactor about two billion years ago in Gabon, West Africa. Analysis of ore taken from the Oklo Mine had a reduced level of $^{235}_{92}U$. Further examination of the ore found many isotopes of Xenon, an element often found in long-term fission products. The conclusion was stark – water had permeated into uranium-rich rock and moderated a chain reaction for hundreds of thousands of years.

As can be seen in Figure 226-6, water in the primary coolant loop and secondary loops are kept separate from each other. Water in the primary coolant loop may become radioactive via neutron activation in the reactor vessel. Since the water in the primary and secondary loops does not come into contact with each other, this greatly reduces the chance of water in the secondary loop from being contaminated.

Once in the steam generator, the very hot, high pressure water in the primary coolant loop creates steam from water in the secondary coolant loop. This steam is then forced through a turbine, which cranks a generator producing electricity. Finally, cool water from an external source is used to condense the steam in the secondary coolant loop back to water, which returns to the steam generator.

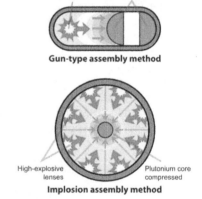

Figure 226-7. Diagrams of two nuclear weapons designs.

Again, the overall design of a nuclear power plant is similar to many fossil fuel power plants; however, traditional plants to do not require the use of separate primary and secondary loops since the creation of radioactive water is not a concern.

Nuclear power plants are large – both physically and in terms of electricity production. The large containment domes that many of us use to instantly recognize a nuclear plant take quite a bit of space and add to the physical infrastructure (and safety) of such plants. In terms of electricity, modern nuclear plants are rated to produce between 800 and 1,000 megawatts of power. Although this is a tremendous amount of energy, I should note that modern coal-fired power plants can produce 2,000 – 3,000 megawatts; however, their pollution signature is well known.

In the wake of nuclear power accidents such as Three Mile Island, Chernobyl and

Table 226-1. Mortality rates related to different types of electricity production.

Source	Deaths (per billion kWh)	% of Electricity
Coal (global)	170	41%
Coal (China)	(280)	(75%)
Coal (U.S.)	(15)	(44%)
Oil	36	5.5%
Biofuel	24	1.3%
Natural Gas	4	21%
Hydro	1.4	16%
Solar	0.44	0.064%
Wind	0.15	1.1%
Nuclear	0.09	14%

Fukushima, many people also question the safety of nuclear power. However, a detailed look at deaths related to different power sources shows that nuclear power is actually the safest form of energy production currently known. Table 226-1 compares deaths per billion kilowatt-hour of energy produced for many types of power production. Please note that this table *does* include deaths due to both the Chernobyl and Fukushima accidents (there are no known deaths related to Three Mile Island).

Nuclear Weapons

Although our nuclear power industry is a direct result of the Manhattan Project, the main thrust of the project was to produce the world's first atomic bomb (nuclear fission bomb). In the end, the Project produced three bombs – one used as a test (known as the Trinity Test) and the two used by the United States to end the Second World War. As noted early, simply bringing a critical mass of a fissile material together *does* cause a chain reaction; however, simply doing this essentially creates a nuclear reactor and not a nuclear bomb. The bomb makers at Los Alamos needed to create a design that would contain the fission energy long enough to create an explosion.

In the end, physicists and engineers at Los Alamos created two bomb designs: an implosion device using plutonium (used at the Trinity test and over Nagasaki) and a *gun-type* device (used over Hiroshima). Brief diagrams of each device type are shown in Figure 226-7.

In the gun-type device, two subcritical pieces of weapons-grade uranium (~80% uranium-235) are separated at each end of the bomb. A conventional explosive is used to propel one of the pieces (called the bullet) towards the other (called the target). When the two pieces combine they form a supercritical mass that undergoes a spontaneous fission chain reaction. This design was first used in the bombing of Hiroshima, Japan on August 6th, 1945. The blast of this bomb was the equivalent of 16,000 tons of TNT (called 16 kT). This is the equivalent of 67 TJ of energy released (67,000,000,000,000 J). The radius of total destruction was approximately one mile, and fires completely consumed an area of 4.4 square miles. Believe it or not, the bomb dropped on Hiroshima was incredibly inefficient with only about 1.7% of its possible uranium fissioning.

The implosion-type device is far more subtle than the gun-type device. A subcritical mass of plutonium is placed at the center of a spherical distribution of conventional explosives. The shell of explosives is actually a combination of fast and slow explosives that is designed to create a perfectly spherical, inward moving shockwave (called an explosive lens) when the bomb detonates. This shockwave compresses the plutonium core, increasing its density. As the density of the plutonium increases, the effectiveness of free neutrons also increases until a runaway fission chain reaction is initiated. This device also has the added bonus (from a bomb maker's perspective) that the inward moving shockwave also holds the core together for an extra amount of time, increasing the overall blast energy.

The implosion bombs used in the Trinity test on July 16th, 1945 and over Nagasaki, Japan on August 9th, 1945 each had similar explosive yields to that of the device used over Hiroshima (17 kT for Trinity and 21 kT over Nagasaki). Again, this represents an enormous energy release of over 70 TJ (70,000,000,000,000). To this date, the bombs used at Hiroshima and Nagasaki are the only nuclear weapons ever detonated in war. As powerful as these nuclear fission weapons are, their introduction into the world arsenal led to another weapon, the hydrogen bomb, developed less than 10 years later with almost 1,000 times the destructive power as the fission weapons just described.

226-4: Nuclear fusion

> **Consider:** *What causes small nuclei to combine? How does the energy of this nuclear fusion compare to nuclear fission?*

Just as energy is released when very large nuclei split under nuclear fission, small nuclei release energy when forced together, a process called nuclear fusion. If you refer to the plot of binding energy per nucleon in unit 224, you will see that small nuclei such as hydrogen, helium and lithium have a much lower binding energy per nucleon than does the most stable element – nickel-62. So, when these nuclei combine, energy is released.

The sun (and most stars) are powered by nuclear fusion. One of the main mechanisms used by the sun is called the ***proton cycle***. In this cycle, six hydrogen nuclei will combine to form a helium nucleus and two hydrogen nuclei. This process is summarized in table 226-2.

Table 226-2. Energy released in the proton-proton cycle.

Reaction	Energy Released
$^1_1H + {}^1_1H \rightarrow {}^2_1H + e^+ + \nu_e$	1.44 MeV *
$^2_1H + {}^1_1H \rightarrow {}^3_2He + \gamma$	5.49 MeV
$^3_2He + {}^3_2He \rightarrow {}^4_2He + {}^1_1H + {}^1_1H + \gamma.$	12.86 MeV

* Note that 0.42 MeV is released in the formation of ^2_1H, and an additional 1.02 MeV comes from the annihilation of the positron with an electron.

To find the total energy released, first note that each of the ^3_2He nuclei in the third equation is produced by the processes represented in the first two equations, so the total energy released is 2(1.44 MeV) + 2(5.49 MeV) + 12.86 MeV = 26.7 MeV. Note that fission processes release more energy per fission (around 200 MeV), but the energy released *per nucleon* is larger in fusion processes. Also, I'd like to point out that there is an implied β^+ decay in the first equation of the proton-proton cycle as one of the protons decays to a neutron.

Example 226-1-1: The second step

Verify that the second step in the proton-proton cycle releases 5.49 MeV.

Solution:

The energy is given by our energy released equation above:

$$E = \sum E_{B,products} - E_{B,initial},$$

$$E = \left(E_{B,{}^3_2He} - E_{B,{}^1_1H} - E_{B,{}^2_1H} \right).$$

We must look up the binding energy for each of our isotopes and substitute them into this equation:

$$E = 7.718058 \ MeV - 0 - 2.224573 \ MeV,$$

$$E = 5.493485 \ MeV.$$

Rounding to three significant Figures, this gives us 5.49 MeV per fusion, consistent with what is shown above.

Fusion in astronomical processes was imperative for the formation of all elements larger than hydrogen. The early universe was made almost entirely of hydrogen and helium. As stars coalesced to the point that fusion was possible, more helium was produced as hydrogen was used as fuel. Then, as stars exhaust their hydrogen fuel, they can fuse helium and later lithium, etc. A star about the size of our sun will not be able to fuse elements beyond carbon and oxygen – it just does not have enough energy. Very large stars (blue giants) can fuse elements up to iron. However, very large elements, such as uranium and plutonium discussed before, are only formed during supernovae. Our sun is a second-generation star, meaning that it and our solar system formed from the remnants of a supernova – which is good for us because it allows us to have elements larger than iron!

Humans have not yet been able to harness the power of fusion in a meaningful and controlled fashion - we have been told that fusion energy is about 20 years away for the past 50 years or so. This is not to say that physicists and engineers have not been working diligently on the problem the entire time. In fact, the technology has come a long way – fusion reactors that produce more energy than they consume have been constructed. Unfortunately, these reactors are still vulnerable to many subtle instabilities that make commercial application still beyond our reach.

Thermonuclear Weapons

A thermonuclear weapon uses the heat generated during a fission reaction to ignite a fusion reaction. That is to say, a fission bomb is used to produce a fusion bomb. The first bombs designed to use this principle were similar to the plutonium implosion devices described above; however, their cores also contain isotopes of hydrogen that could fuse to release additional energy. Such weapons are called *fusion-boosted fission weapons*, because although they were primarily fission bombs, the small fusion reaction greatly enhanced yield. Using this basic principle, the yield of boosted weapons easily reached between 50 and 100 kT (compare to the ~20 kT for fission-only weapons).

The real advance to a full thermonuclear device (fusion device) was the development of the *Teller-Ulam design*, pictured in Figure 226-8. In this design, the energy from an implosion-type fission bomb (the primary) is funneled to a cylindrical secondary. The secondary is a multilayered cylinder with more plutonium fuel on the inside of the cylinder, a uranium *tamper* on the outside and fusion fuel in the meat of the cylinder. The energy from the fission bomb enters the center of the secondary setting off fission in the plutonium cylinder (called the spark plug). The energy from this plutonium fission then sets up fusion reactions

Figure 226-8. Teller-Ulam design.

in the fusion fuel, which creates neutrons with incredibly high energies (called fusion neutrons). When the fusion neutrons reach the uranium tamper, the uranium fissions, creating an inward moving shockwave that contains the fusion fuel until a large percentage has completed the fusion process. Once the shockwave from the uranium fission has subsided, the entire energy of the fusion reactions is released.

The yield of true thermonuclear weapons was unexpected. The first full test of an operational Teller-Ulam device[1] – the Bravo shot of Operation Castle on March 1st, 1954 - was expected to have a yield 6 MT (MT = megatons – 1000 kT) of TNT; however, when detonated, 15 MT of energy were released – nearly 1,000 times the energy of the Trinity test. The test shot was performed at Bikini Atoll in the Marshall Islands – a relatively secluded area. However, the detonation produced such a high yield that fallout fell on inhabited areas thought to be out of range. The inhabitants were evacuated three days later when

---

[1] The *Ivy Mike* test of a Teller-Ulam device preceded Castle Bravo; however, *Ivy Mike* used super-cooled fuel that could never have been used operationally.

some started to show the effects of radiation sickness. In addition, the Japanese fishing vessel *Daigo Fukuryū Maru* ("Lucky Dragon No. 5") was just outside the anticipated blast zone. A number of the crew members became ill with radiation sickness and one died.

The Castle Bravo device, at 15 MT, was the largest detonation performed by the U.S. and is considered the largest operational weapon exploded. The Soviet Union tested a device called the *Tsar Bomba* (Emperor Bomb) on October 30th, 1961, with an astounding yield of 50 MT. The Tsar Bomba was 26 ft long, 7 ft wide and 60,000 pounds – it could never have been used as an operational device, but does stand as the largest thermonuclear weapon ever tested. No thermonuclear device has ever been detonated in a hostile act.

The proliferation of nuclear weapons in the 1960's, 70's and 80's was extreme. In 1985, it is estimated that there were 68,000 active thermonuclear warheads around the world with an average yield of around 1 MT each. The United States and Soviet Union were responsible for the vast majority of these weapons, although the U.K, France, China and Israel were all known to have a nuclear arsenal. Thankfully, the number of active thermonuclear warheads has decreased worldwide to around 4,000, with another 12,000 in storage. I don't know about you, but this still sounds like a large number to me.

A very serious concern is what happened to the nuclear materials of the more than 40,000 dismantled nuclear warheads. Although the U.S has, generally, maintained very tight control over nuclear materials, records and regulations during the breakup of the Soviet Union were shoddy to say the least. Oh, and did you know that even the U.S. has lost eight thermonuclear weapons that are completely unaccounted for? *Eight*, just lost…

226-5: Nuclear Terrorism - Radiological Dispersion Devices (RDDs) and Improvised Nuclear Devices (INDs)

Consider: *Why should we be so worried about nuclear energy and terrorists?*

So, why should you care about radioactivity and nuclear processes as future Coast Guard Officers? Although the threat of full scale nuclear war has decreased over the last couple of decades, the risks of terrorist action involving radiological material has increased. In 2010, President Barak Obama said that nuclear terrorism is "the single most important national security threat that we face." What's more is that both the U.S and Russia are known to have produced nuclear weapons small enough to fit in a backpack (so called suitcase nukes) – eliminating the need to drop a large nuclear bomb or fire a ballistic missile to devastate many city blocks. Security experts estimate the risk of a terrorist attack using radiologic material at between 10% and 50% over the foreseeable future.

The most likely terrorist use of nuclear material are ***radiological dispersion devices*** (***RDDs*** – also called ***dirty bombs***) and ***improvised nuclear devices*** (***INDs***). RDDs use conventional explosives to spread radioactive material. This type of device would be relatively easy to construct and may be the weapon of choice for a terrorist group that cannot acquire enough fissile material for a nuclear detonation. Although long considered an important threat, recent studies by the U.S. Department of Energy (DOE) suggest that any mortality from the explosion of a dirty bomb would come from the conventional explosion and that the spread of the radioactive material would have considerably smaller effects than previously thought. In fact, tests done by the DOE suggest that even if the area around a dirty bomb explosion were not cleared, those living in the area for full year would experience a high, but not lethal, level of radiation. However, the psychological effects of an RDD explosion on U.S. soil (or anywhere in the world for that matter) would be extreme. Since terrorism is as much psychological as physical, RDDs are still considered a serious terrorist threat.

INDs are nuclear weapons fabricated by non-state entities, most likely a terrorist group. Unlike RDDs, INDs would contain enough fissile material to create a fission bomb. An IND would most likely use the gun-type fission weapon design described above because of its relative simplicity when compared to the implosion-type bomb. In order to ensure that the IND undergoes fission, the nuclear core could very possibly be *overengineered*, meaning that there will be more fissile material than is necessary for a nuclear detonation. This will cause excess radioactive material to be spread throughout the fallout area. The International Atomic Energy Agency (IAEA) estimates that only 25 kg of highly-enriched uranium or 8 kg of plutonium would be needed to produce an effective IND. Such devices would have a potential explosive yield in the 10 – 20 kT range – similar to the Trinity test and the bombs dropped on Hiroshima and Nagasaki. Even a poorly constructed bomb that *fizzles* (only fissions a small percentage of its capability) could reach a yield 1 kT – devastating many city blocks.

259

Figure References

201-1. "CPT-sound-physical-manifestation" by Pluke - Own work. Licensed under Creative Commons Zero, Public Domain Dedication via Wikimedia Commons - http://commons.wikimedia.org/wiki/File:CPT-sound-physical-manifestation.svg

201-3. "Simple sine wave". Licensed under Creative Commons Attribution-Share Alike 3.0-2.5-2.0-1.0 via Wikimedia Commons - http://commons.wikimedia.org/wiki/File:Simple_sine_wave.svg

201-4. (top) "Wave new sine" by Kraaiennest. Originally created as a cosine wave, by User:Pelegs, as File:Wave_new.svg - Own work. Licensed under Creative Commons Attribution-Share Alike 3.0 via Wikimedia Commons - http://commons.wikimedia.org/wiki/File:Wave_new_sine.svg. (Bottom) Modification of the top image made by E. Page.

201-5. "Fourier Series" by Jim.belk - Own work. Licensed under Public domain via Wikimedia Commons - http://commons.wikimedia.org/wiki/File:Fourier_Series.svg

202-1. "Nodo2" by Josell7 - Own work. Licensed under Creative Commons Attribution-Share Alike 3.0-2.5-2.0-1.0 via Wikimedia Commons - http://commons.wikimedia.org/wiki/File:Nodo2.svg

202-2. Modified by E.Page from "Beating Frequency" by Ansgar Hellwig - created with gnuplot and Corel Draw. Licensed under Creative Commons Attribution 2.5 via Wikimedia Commons - http://commons.wikimedia.org/wiki/File:Beating_Frequency.svg

202-3. "Inverted Reflection". Licensed under Creative Commons Attribution-Share Alike 3.0 via Wikimedia Commons - http://commons.wikimedia.org/wiki/File:Inverted_Reflection.PNG

202-4. "Erect Reflection". Licensed under Creative Commons Attribution-Share Alike 3.0 via Wikimedia Commons - http://commons.wikimedia.org/wiki/File:Erect_Reflection.PNG

202-6. By Vegar Ottesen - Own work, CC BY-SA 3.0, https://commons.wikimedia.org/w/index.php?curid=26296568

202-7. Modified by E.Page from "Schwingende Saiten". Licensed under Creative Commons Attribution 2.5 via Wikimedia Commons - http://commons.wikimedia.org/wiki/File:Schwingende_Saiten.svg

202-8. "Schwingende Saiten". Licensed under Creative Commons Attribution 2.5 via Wikimedia Commons - http://commons.wikimedia.org/wiki/File:Schwingende_Saiten.svg

202-9. Modified by E.Page from "Overtones closed pipe". Licensed under Public domain via Wikimedia Commons - http://commons.wikimedia.org/wiki/File:Overtones_closed_pipe.png

203-2. Modified by E.Page from "Longitudinalwelle Transversalwelle" by Debianux - Own work. Licensed under Creative Commons Attribution-Share Alike 3.0 via Wikimedia Commons - http://commons.wikimedia.org/wiki/File:Longitudinalwelle_Transversalwelle.png

203-3. "Ear-anatomy-text-small-en" by Iain at en.wikipedia, SVG conversion by User:Surachit - Made by Iain 05:39 29 Jun 2003 (UTC)Transferred from en.wikipedia; transferred to Commons by User:Papa November using CommonsHelper.. Licensed under Creative Commons Attribution-Share Alike 3.0 via Wikimedia Commons - http://commons.wikimedia.org/wiki/File:Ear-anatomy-text-small-en.svg

203-4. "Cochlea" by Cochlea.png: Original uploader was Dicklyon at en.wikipediaderivative work: Fred the Oyster - Cochlea.png. Licensed under Public domain via Wikimedia Commons - http://commons.wikimedia.org/wiki/File:Cochlea.svg

203-5. "HearingLoss" by Thomas.haslwanter - Own work. Licensed under Creative Commons Attribution-Share Alike 3.0 via Wikimedia Commons - http://commons.wikimedia.org/wiki/File:HearingLoss.svg

203-6. "Plates tect2 en" by USGS - http://pubs.usgs.gov/publications/text/slabs.html. Licensed under Public domain via Wikimedia Commons - http://commons.wikimedia.org/wiki/File:Plates_tect2_en.svg

203-7. Modified by E.Page from "Pswaves" by No author given. PD beacause from http://earthquake.usgs.gov/ - http://earthquake.usgs.gov/learn/glossary/images/PSWAVES.JPG or: http://earthquake.usgs.gov/learn/glossary/?term=seismic%20wave. Licensed under Public domain via Wikimedia Commons - http://commons.wikimedia.org/wiki/File:Pswaves.jpg

203-8. "Earth-crust-cutaway-english" by Surachit - Self-made, based on the public domain image File:Earth-crust-cutaway-english.png by Jeremy Kemp. This vector image was created with Inkscape.. Licensed under Creative Commons Attribution-Share Alike 3.0-2.5-2.0-1.0 via Wikimedia Commons - http://commons.wikimedia.org/wiki/File:Earth-crust-cutaway-english.svg

203-9. "Earthquake wave shadow zone". Licensed under Creative Commons Attribution-Share Alike 3.0 via Wikimedia Commons - http://commons.wikimedia.org/wiki/File:Earthquake_wave_shadow_zone.svg

203-10. Modified by E.Page from "Pswaves" by No author given. PD beacause from http://earthquake.usgs.gov/ - http://earthquake.usgs.gov/learn/glossary/images/PSWAVES.JPG or: http://earthquake.usgs.gov/learn/glossary/?term=seismic%20wave. Licensed under Public domain via Wikimedia Commons - http://commons.wikimedia.org/wiki/File:Pswaves.jpg

204-1. "Electromagnetic wave" by P.wormer - Own work. Licensed under Creative Commons Attribution-Share Alike 3.0 via Wikimedia Commons - http://commons.wikimedia.org/wiki/File:Electromagnetic_wave.png

204-2. "EM Spectrum Properties edit" by Inductiveload, NASA - self-made, information by NASABased off of File:EM_Spectrum3-new.jpg by NASA. The butterfly icon is from the P icon set, P biology.svg. The humans are from the Pioneer plaque, Human.svg. The buildings are the Petronas towers and the Empire State Buildings, both from Skyscrapercompare.svg. Licensed under Creative Commons Attribution-Share Alike 3.0 via Wikimedia Commons - http://commons.wikimedia.org/wiki/File:EM_Spectrum_Properties_edit.svg

204-3. "Blackbody emission" by Ant Beck - Own work. Licensed under Creative Commons Attribution 3.0 via Wikimedia Commons - http://commons.wikimedia.org/wiki/File:Blackbody_emission.svg

206-1. "Gravity field lines" by Sjlegg - Own work. Licensed under Public domain via Wikimedia Commons - http://commons.wikimedia.org/wiki/File:Gravity_field_lines.svg

206-2. "VFPt minus thumb" by Geek3 - Own work. This plot was created with VectorFieldPlot. Licensed under Creative Commons Attribution-Share Alike 3.0 via Wikimedia Commons - http://commons.wikimedia.org/wiki/File:VFPt_minus_thumb.svg

206-3. "VFPt plus thumb" by Geek3 - Own workThis plot was created with VectorFieldPlot. Licensed under Creative Commons Attribution-Share Alike 3.0 via Wikimedia Commons - http://commons.wikimedia.org/wiki/File:VFPt_plus_thumb.svg

206-4. (Left) "VFPt charges plus minus" by Geek3 - Own work. This plot was created with VectorFieldPlot. Licensed under Creative Commons Attribution-Share Alike 3.0 via Wikimedia Commons - http://commons.wikimedia.org/wiki/File:VFPt_charges_plus_minus.svg. (Right) "VFPt charges plus plus" by Geek3 - Own workThis plot was created with VectorFieldPlot. Licensed under Creative Commons Attribution-Share Alike 3.0 via Wikimedia Commons - http://commons.wikimedia.org/wiki/File:VFPt_charges_plus_plus.svg

206-5. "Non-symmetrical charge field" by derivative work by: PLATO Learning - VFPt metal ball grounded transparent.svghttp://commons.wikimedia.org/wiki/File:VFPt_metal_ball_grounded_transparent.svg. Licensed under Creative Commons Attribution-Share Alike 3.0 via Wikimedia Commons - http://commons.wikimedia.org/wiki/File:Non-symmetrical_charge_field.jpg

207-1. "VFPt charges plus minus thumb" by Geek3 - Own workThis plot was created with VectorFieldPlot. Licensed under Creative Commons Attribution-Share Alike 3.0 via Wikimedia Commons - http://commons.wikimedia.org/wiki/File:VFPt_charges_plus_minus_thumb.svg

207-2. "Dipole vector plus minus" by Geek3 - Own work. Licensed under Creative Commons Attribution 3.0 via Wikimedia Commons - http://commons.wikimedia.org/wiki/File:Dipole_vector_plus_minus.svg

207-4. "Miri2" by J:136401 - Own work. Licensed under Creative Commons Attribution-Share Alike 3.0 via Wikimedia Commons - http://commons.wikimedia.org/wiki/File:Miri2.jpg

207-5. "Electric dipole torque uniform field" by Maschen - Own work. Licensed under Creative Commons Zero, Public Domain Dedication via Wikimedia Commons - http://commons.wikimedia.org/wiki/File:Electric_dipole_torque_uniform_field.svg

208-2. "VFPt dipole magnetic1" by Geek3 - Own workThis plot was created with VectorFieldPlot. Licensed under Creative Commons Attribution-Share Alike 3.0 via Wikimedia Commons - http://commons.wikimedia.org/wiki/File:VFPt_dipole_magnetic1.svg

208-4. Modified by E.Page from "Dominios" by Original uploader was 4lex at es.wikipedia - Originally from es.wikipedia; description page is/was here.. Licensed under Creative Commons Attribution-Share Alike 3.0 via Wikimedia Commons - http://commons.wikimedia.org/wiki/File:Dominios.png

208-6. "Barmagnet1" by P.Sumanth Naik - Own work. Licensed under Creative Commons Attribution-Share Alike 3.0 via Wikimedia Commons - http://commons.wikimedia.org/wiki/File:Barmagnet1.png

208-7. Modified by E.Page from "Geomagnetisme" by JrPol - Own work. Licensed under Creative Commons Attribution-Share Alike 3.0-2.5-2.0-1.0 via Wikimedia Commons - http://commons.wikimedia.org/wiki/File:Geomagnetisme.svg

208-8. "Magnetic North Pole Positions" by Tentotwo - Own work. Observed pole positions taken from M. Mandea and E. Dormy, "Asymmetric behavior of magnetic dip poles", Earth Planets Space, 55, 153–157, 2003.Observed position in 2007 taken from Newitt et al., "Location of the North Magnetic Pole in April 2007", Earth Planets Space, 61, 703–710, 2009Modelled pole positions taken from the National Geophysical Data Center, "Wandering of the Geomagnetic Poles"Shoreline and borders from Natural Earth. Licensed under Creative Commons Attribution-Share Alike 3.0 via Wikimedia Commons - http://commons.wikimedia.org/wiki/File:Magnetic_North_Pole_Positions.svg

209-2. "FuerzaCentripetaLorentzN2" by Jfmelero - Own work. Licensed under Creative Commons Attribution-Share Alike 3.0 via Wikimedia Commons - http://commons.wikimedia.org/wiki/File:FuerzaCentripetaLorentzN2.svg

209-3. Modified by E.Page from "Magnetic deflection helical path" by Maschen - Own work. Licensed under Public domain via Wikimedia Commons - http://commons.wikimedia.org/wiki/File:Magnetic_deflection_helical_path.svg

Example 209-2. "Wienscher geschwindigkeitsfilter massenspektroskopie" by Sgbeer - Own work. Licensed under Creative Commons Attribution 3.0 via Wikimedia Commons - http://commons.wikimedia.org/wiki/File:Wienscher_geschwindigkeitsfilter_massenspektroskopie.svg

209-4. (left) "VFPt charges plus minus" by Geek3 - Own work This plot was created with VectorFieldPlot. Licensed under Creative Commons Attribution-Share Alike 3.0 via Wikimedia Commons - http://commons.wikimedia.org/wiki/File:VFPt_charges_plus_minus.svg. (right) "VFPt dipole magnetic3" by Geek3 - Own work. This plot was created with VectorFieldPlot. Licensed under Creative Commons Attribution-Share Alike 3.0 via Wikimedia Commons - http://commons.wikimedia.org/wiki/File:VFPt_dipole_magnetic3.svg

209-5. "Torque of a magnetic dipole" by Original uploader was 老陳 at zh.wikipedia - Transferred from zh.wikipedia; transferred to Commons by User:Shizhao using CommonsHelper.. Licensed under GNU Free Documentation License via Wikimedia Commons - http://commons.wikimedia.org/wiki/File:Torque_of_a_magnetic_dipole.png

210-3. "GaussLaw1" by LeyGauss1.jpg: Original uploader was Gonfer at es.wikipedia derivative work: Nicoguaro (talk) - LeyGauss1.jpg. Licensed under Creative Commons Attribution-Share Alike 3.0 via Wikimedia Commons - http://commons.wikimedia.org/wiki/File:GaussLaw1.svg

210-4. "GaussLaw2" by LeyGauss2.jpg: Original uploader was Gonfer at es.wikipedia derivative work: Nicoguaro (talk) - LeyGauss2.jpg. Licensed under Creative Commons Attribution-Share Alike 3.0 via Wikimedia Commons - http://commons.wikimedia.org/wiki/File:GaussLaw2.svg

210-6. "Electromagnetism". Licensed under Creative Commons Attribution-Share Alike 3.0 via Wikimedia Commons - http://commons.wikimedia.org/wiki/File:Electromagnetism.png

211-3. "Contour map (PSF)" by Pearson Scott Foresman - Pearson Scott Foresman, donated to the Wikimedia Foundation. Licensed under Public domain via Wikimedia Commons - http://commons.wikimedia.org/wiki/File:Contour_map_(PSF).png

211-4. "Electric field point lines equipotentials" by Sjlegg - Own work. Licensed under Public domain via Wikimedia Commons - http://commons.wikimedia.org/wiki/File:Electric_field_point_lines_equipotentials.svg

211-5. "VFPt plus thumb" by Geek3 - Own workThis plot was created with VectorFieldPlot. Licensed under Creative Commons Attribution-Share Alike 3.0 via Wikimedia Commons - http://commons.wikimedia.org/wiki/File:VFPt_plus_thumb.svg

211-6. "Field lines equipotentials parallel plates" by Sjlegg - Own work. Licensed under Public domain via Wikimedia Commons - http://commons.wikimedia.org/wiki/File:Field_lines_equipotentials_parallel_plates.svg

212-1. "Field lines parallel plates" by Sjlegg - Own work. Licensed under Public domain via Wikimedia Commons - http://commons.wikimedia.org/wiki/File:Field_lines_parallel_plates.svg

Example 212-1. "Cylindrical Capacitor" by Fabian R. Licensed under Public domain via Wikimedia Commons - http://commons.wikimedia.org/wiki/File:Cylindrical_Capacitor.svg

212-4. "Dielectric notext" by Smack (talk) - Own work. Licensed under Public domain via Wikimedia Commons - http://commons.wikimedia.org/wiki/File:Dielectric_notext.png

214-7. "Capacitor resistor series" by Sjlegg - Own work. Licensed under Public domain via Wikimedia Commons - http://commons.wikimedia.org/wiki/File:Capacitor_resistor_series.svg

214-8. "Multimeter". Licensed under Creative Commons Attribution-Share Alike 3.0 via Wikimedia Commons - http://commons.wikimedia.org/wiki/File:Multimeter.JPG

216-1. "Lenz's Law02" by User:老陳 - Own work. Licensed under Creative Commons Attribution-Share Alike 3.0-2.5-2.0-1.0 via Wikimedia Commons - http://commons.wikimedia.org/wiki/File:Lenz%27s_Law02.jpg

216-2. "Right hand rule simple" by Original uploader was Schorschi2 at de.wikipedia - Transferred from de.wikipedia(Original text : Eigene Zeichnung). Licensed under Public domain via Wikimedia Commons - http://commons.wikimedia.org/wiki/File:Right_hand_rule_simple.png

216-3. Modified by E.Page from "Solenoid, air core, insulated, 20 turns, (shaded)" by Inductiveload - Own work, created with Solid Edge and Inkscape. Licensed under Public domain via Wikimedia Commons - http://commons.wikimedia.org/wiki/File:Solenoid,_air_core,_insulated,_20_turns,_(shaded).svg

216-5. "Current continuity in capacitor" by Brews ohare - Own work. Licensed under Creative Commons Attribution-Share Alike 3.0 via Wikimedia Commons - http://commons.wikimedia.org/wiki/File:Current_continuity_in_capacitor.JPG

217-4. "Hydroelectric dam". Licensed under Public domain via Wikimedia Commons - http://commons.wikimedia.org/wiki/File:Hydroelectric_dam.png

217-5. "Transformer3d col3" by BillC at en.wikipedia - Own workTransferred from en.wikipedia. Licensed under Creative Commons Attribution-Share Alike 3.0 via Wikimedia Commons - http://commons.wikimedia.org/wiki/File:Transformer3d_col3.svg

217-6. "Electricity grid simple- North America" by United States Department of Energy, SVG version by User:J JMesserly - http://www.ferc.gov/industries/electric/indus-act/reliability/blackout/ch1-3.pdf Page 13 Title:"Final Report on the August 14, 2003 Blackout in the United States and Canada" Dated April 2004. Accessed on 2010-12-25. Licensed under Public domain via Wikimedia Commons - http://commons.wikimedia.org/wiki/File:Electricity_grid_simple-_North_America.svg

218-13. Left - "Klaudia 801" by Mohylek - Own work. Licensed under Creative Commons Attribution-Share Alike 3.0 via Wikimedia Commons - https://commons.wikimedia.org/wiki/File:Klaudia_801.JPG. Center - "SafirVehicle4" by M-ATF, from military.ir and iranmilitaryforum.net - http://gallery.military.ir/albums/userpics/10187/_DSC1376.jpg. Licensed under Creative Commons Attribution-Share Alike 3.0 via Wikimedia Commons - http://commons.wikimedia.org/wiki/File:SafirVehicle4.jpg. Right - "Recreational Walkie Talkies" by Original uploader was Wtshymanski at en.wikipedia - Transferred from en.wikipedia; transferred to Commons by User:Premeditated Chaos using CommonsHelper.. Licensed under Public domain via Wikimedia Commons - http://commons.wikimedia.org/wiki/File:Recreational_Walkie_Talkies.jpg

218-14. "Elem-doub-rad-pat-pers". Licensed under Creative Commons Attribution-Share Alike 3.0 via Wikimedia Commons - https://commons.wikimedia.org/wiki/File:Elem-doub-rad-pat-pers.jpg

218-15. "Vpol dual band blade antenna blade L1 3D" by Cwru53 - Own work. Licensed under Creative Commons Attribution-Share Alike 3.0-2.5-2.0-1.0 via Wikimedia Commons - http://commons.wikimedia.org/wiki/File:Vpol_dual_band_blade_antenna_blade_L1_3D.jpg

218-16. "Rabbit-ears dipole antenna with UHF loop 20090204" by Mark Wagner (User:Carnildo) - Own work. Licensed under Creative Commons Attribution 2.5 via Wikimedia Commons - https://commons.wikimedia.org/wiki/File:Rabbit-ears_dipole_antenna_with_UHF_loop_20090204.jpg

219-2. "Reflection angles". Licensed under Creative Commons Attribution-Share Alike 3.0 via Wikimedia Commons - https://commons.wikimedia.org/wiki/File:Reflection_angles.svg

219-3: "Diffuse reflection" by Original uploader was Theresa knott at en.wikipedia - Transfered from en.wikipedia Transfer was stated to be made by User:gretana.. Licensed under Creative Commons Attribution-Share Alike 3.0 via Wikimedia Commons - https://commons.wikimedia.org/wiki/File:Diffuse_reflection.PNG

219-4: "Snells law" by Original uploader was Cristan at en.wikipedia Later version(s) were uploaded by Dicklyon at en.wikipedia. - Transfered from en.wikipedia. Licensed under Public domain via Wikimedia Commons - https://commons.wikimedia.org/wiki/File:Snells_law.svg

219-5: "Road to nowhere" by Brocken Inaglory - Own work. Licensed under Creative Commons Attribution-Share Alike 3.0 via Wikimedia Commons - https://commons.wikimedia.org/wiki/File:Road_to_nowhere.JPG

219-6: "RefractionReflextion" by Josell7 - Own work. Licensed under Creative Commons Attribution-Share Alike 3.0 via Wikimedia Commons - https://commons.wikimedia.org/wiki/File:RefractionReflextion.svg

220-1. "Lens3" by w:en:DrBob - w:en:File:Lens3.svg. Licensed under GNU Free Documentation License via Wikimedia Commons - https://commons.wikimedia.org/wiki/File:Lens3.svg

220-2. "Concavo 1". Licensed under Creative Commons Attribution-Share Alike 3.0 via Wikimedia Commons - https://commons.wikimedia.org/wiki/File:Concavo_1.png

220-3: Figure 25.40: OpenStax College. (2014, July 30). *College Physics*. Retrieved from the OpenStax-CNX Web site: http://cnx.org/content/m42955/1.9/

220-4. "Aberration de sphéricité d'un miroir sphérique concave" by Jean-Jacques MILAN 17:15, 5 June 2008 (UTC) - Own work. Licensed under Creative Commons Attribution-Share Alike 3.0-2.5-2.0-1.0 via Wikimedia Commons - https://commons.wikimedia.org/wiki/File:Aberration_de_sph%C3%A9ricit%C3%A9_d%27un_miroir_sph%C3%A9rique_concave.svg

Table 220-1. "Convexmirror raydiagram" by Cronholm144 - Own work. Licensed under Creative Commons Attribution-Share Alike 3.0 via Wikimedia Commons - https://commons.wikimedia.org/wiki/File:Convexmirror_raydiagram.svg

Table 220-2(a) "Concavemirror raydiagram F" by Cronholm144 - Own work. Licensed under Creative Commons Attribution-Share Alike 3.0 via Wikimedia Commons - https://commons.wikimedia.org/wiki/File:Concavemirror_raydiagram_F.svg

Table 220-2(b). "Concavemirror raydiagram FE" by Cronholm144 - Own work. Licensed under Creative Commons Attribution-Share Alike 3.0 via Wikimedia Commons - https://commons.wikimedia.org/wiki/File:Concavemirror_raydiagram_FE.svg

Table 220-2(c). "Concavemirror raydiagram 2FE" by Cronholm144 - Own work. Licensed under Creative Commons Attribution-Share Alike 3.0 via Wikimedia Commons - https://commons.wikimedia.org/wiki/File:Concavemirror_raydiagram_2FE.svg

220-5. "Lenses en" by ElfQrin - Own work. Licensed under Creative Commons Attribution-Share Alike 3.0 via Wikimedia Commons - https://commons.wikimedia.org/wiki/File:Lenses_en.svg

Table 220-3(a) "Lens1" by Original uploader was DrBob at en.wikipedia - Originally from en.wikipedia.SVG version of image:lens1.png by DrBob. Licensed under Creative Commons Attribution-Share Alike 3.0 via Wikimedia Commons - https://commons.wikimedia.org/wiki/File:Lens1.svg

Table 220-3(b) "Large convex lens" by User Fir0002 on en.wikipedia - http://en.wikipedia.org/wiki/Image:Large_convex_lens.jpg. Licensed under Creative Commons Attribution-Share Alike 3.0 via Wikimedia Commons - https://commons.wikimedia.org/wiki/File:Large_convex_lens.jpg

Table 220-3(c) "Lens1b" by Original uploader was DrBob at en.wikipedia - Originally from en.wikipedia; description page is/was here.. Licensed under Creative Commons Attribution-Share Alike 3.0 via Wikimedia Commons - https://commons.wikimedia.org/wiki/File:Lens1b.svg

Table 220-3(d) "Concave lens" by User Fir0002 on en.wikipedia - This file is lacking source information.Please edit this file's description and provide a source.. Licensed under Creative Commons Attribution-Share Alike 3.0 via Wikimedia Commons - https://commons.wikimedia.org/wiki/File:Concave_lens.jpg

220-6. "Lens3b". Licensed under Creative Commons Attribution-Share Alike 3.0 via Wikimedia Commons - https://commons.wikimedia.org/wiki/File:Lens3b.svg

Table 220-4(a). "Lens3b". Licensed under Creative Commons Attribution-Share Alike 3.0 via Wikimedia Commons - https://commons.wikimedia.org/wiki/File:Lens3b.svg

Table 220-4 (b). Same as 223-1.

Table 220-4 (c). "Lens4" by w:en:DrBob - w:en:File:Lens4.svg. Licensed under GNU Free Documentation License via Wikimedia Commons - https://commons.wikimedia.org/wiki/File:Lens4.svg

Example 220-6. "Opticke zobrazeni dalekohled kepleruv" by Pajs - Own work. Licensed under Public domain via Wikimedia Commons - http://commons.wikimedia.org/wiki/File:Opticke_zobrazeni_dalekohled_kepleruv.svg

220-7. "Schematic diagram of the human eye en" by Rhcastilhos - Schematic_diagram_of_the_human_eye_with_English_annotations.svg. Licensed under Public domain via Wikimedia Commons - https://commons.wikimedia.org/wiki/File:Schematic_diagram_of_the_human_eye_en.svg

220-8. Adapted by E. Page from "Myopia color" by Gumenyuk I.S. - Original artwork made specially for Wikipedia. Licensed under Creative Commons Attribution-Share Alike 3.0 via Wikimedia Commons - https://commons.wikimedia.org/wiki/File:Myopia_color.png

220-9. "Myopia color" by Gumenyuk I.S. - Original artwork made specially for Wikipedia. Licensed under Creative Commons Attribution-Share Alike 3.0 via Wikimedia Commons - https://commons.wikimedia.org/wiki/File:Myopia_color.png

220-10. "Hypermetropia color" by Gumenyuk I.S. - Own work. Licensed under Creative Commons Attribution-Share Alike 4.0 via Wikimedia Commons - https://commons.wikimedia.org/wiki/File:Hypermetropia_color.png

220-11. "Astigmatismus-Sonne" by Bautsch - Own work. Licensed under Creative Commons Zero, Public Domain Dedication via Wikimedia Commons - https://commons.wikimedia.org/wiki/File:Astigmatismus-Sonne.png

221-1. "Polarisation state - Linear polarization parallel to x axis" by Cepheiden - Own work. Licensed under Public domain via Wikimedia Commons - https://commons.wikimedia.org/wiki/File:Polarisation_state_-_Linear_polarization_parallel_to_x_axis.svg

"Polarisation state - Linear polarization oriented at +45deg" by Cepheiden - Own work. Licensed under Public domain via Wikimedia Commons - https://commons.wikimedia.org/wiki/File:Polarisation_state_-_Linear_polarization_oriented_at_%2B45deg.svg

"Polarisation state - Right-circular polarization" by Cepheiden - Own work. Licensed under Public domain via Wikimedia Commons - https://commons.wikimedia.org/wiki/File:Polarisation_state_-_Right-circular_polarization.svg

"Polarisation state - Left-circular polarization" by Cepheiden - Own work. Licensed under Public domain via Wikimedia Commons - https://commons.wikimedia.org/wiki/File:Polarisation_state_-_Left-circular_polarization.svg

"Polarisation state - Right-elliptical polarization A" by Cepheiden - Own work. Licensed under Public domain via Wikimedia Commons - https://commons.wikimedia.org/wiki/File:Polarisation_state_-_Right-elliptical_polarization_A.svg

221-2. "Circular.Polarization.Circularly.Polarized.Light Circular.Polarizer Passing.Left.Handed.Helix.View" by Dave3457 - Own work. Licensed under Public domain via Wikimedia Commons - https://commons.wikimedia.org/wiki/File:Circular.Polarization.Circularly.Polarized.Light_Circular.Polarizer_Passing.Left.Handed.Helix.View.svg

221-3. "Ebohr1 IP" by Stannered - File:Ebohr1.svg. Licensed under Creative Commons Attribution-Share Alike 3.0 via Wikimedia Commons - https://commons.wikimedia.org/wiki/File:Ebohr1_IP.svg

221-4: Figure 27.11: OpenStax College. (2014, July 30). *College Physics*. Retrieved from the OpenStax-CNX Web site: http://cnx.org/content/m42955/1.9/

221-5: Figure 27.14: OpenStax College. (2014, July 30). *College Physics*. Retrieved from the OpenStax-CNX Web site: http://cnx.org/content/m42955/1.9/

221-6: "Single Slit Diffraction" by jkrieger - File:Beugungsspalt fuer schlitzblende2.svg. Licensed under Creative Commons Attribution-Share Alike 3.0 via Wikimedia Commons - https://commons.wikimedia.org/wiki/File:Single_Slit_Diffraction.svg

221-9: "Airy disk spacing near Rayleigh criterion" by Spencer Bliven - Own work. Licensed under Public domain via Wikimedia Commons - https://commons.wikimedia.org/wiki/File:Airy_disk_spacing_near_Rayleigh_criterion.png

221-10: "Dispersion-curve" by Original uploader was DrBob at en.wikipedia - Transferred from en.wikipedia. Licensed under Creative Commons Attribution-Share Alike 3.0 via Wikimedia Commons - https://commons.wikimedia.org/wiki/File:Dispersion-curve.png

221-11: "Prism rainbow schema". Licensed under Creative Commons Attribution-Share Alike 3.0 via Wikimedia Commons - https://commons.wikimedia.org/wiki/File:Prism_rainbow_schema.png

221-12: "Rainbow1" by KES47 - Own work. Licensed under Public domain via Wikimedia Commons - https://commons.wikimedia.org/wiki/File:Rainbow1.svg

221-13: "Refraction - Huygens-Fresnel principle" by Arne Nordmann (norro) - Own illustration, based on Image:Wellen-Brechung.png and Image:Huygens_brechung.png. Licensed under Creative Commons Attribution-Share Alike 3.0 via Wikimedia Commons - https://commons.wikimedia.org/wiki/File:Refraction_-_Huygens-Fresnel_principle.svg

222-1. "Stimulated Emission" by V1adis1av - Own work. Licensed under Creative Commons Attribution-Share Alike 3.0-2.5-2.0-1.0 via Wikimedia Commons - https://commons.wikimedia.org/wiki/File:Stimulated_Emission.svg

222-2. "Hene-2" by DrBob at en.wikipedia - DrBob (talk) (Uploads)Transferred from en.wikipedia by SreeBot. Licensed under Creative Commons Attribution-Share Alike 3.0 via Wikimedia Commons - https://commons.wikimedia.org/wiki/File:Hene-2.png

222-3. "Photoelectric effect" by Wolfmankurd - en:Inkscape. Licensed under Creative Commons Attribution-Share Alike 3.0 via Wikimedia Commons - https://commons.wikimedia.org/wiki/File:Photoelectric_effect.svg

223-1. "Infinite potential well-en" by Infinite_potential_well.svg: Bdeshamderivative work: Papa November (talk) - Infinite_potential_well.svg. Licensed under Creative Commons Attribution-Share Alike 3.0 via Wikimedia Commons - https://commons.wikimedia.org/wiki/File:Infinite_potential_well-en.svg

225-1. "AvgBindingEnergyPerNucleon". Licensed under Public domain via Wikimedia Commons - http://commons.wikimedia.org/wiki/File:AvgBindingEnergyPerNucleon.jpg

225-2. "Table isotopes en" by Table_isotopes.svg: Napy1kenobiderivative work: Sjlegg (talk) - Table_isotopes.svg. Licensed under Creative Commons Attribution-Share Alike 3.0-2.5-2.0-1.0 via Wikimedia Commons - http://commons.wikimedia.org/wiki/File:Table_isotopes_en.svg

225-3. "Beta Negative Decay" by Joel Holdsworth (Joelholdsworth) - Own workThis vector image was created with Inkscape.. Licensed under Public domain via Wikimedia Commons - http://commons.wikimedia.org/wiki/File:Beta_Negative_Decay.svg

225-5. Modified be E.Page from "Direct DNA damage" by Gerriet41 - Own work. Licensed under Public domain via Wikimedia Commons - http://commons.wikimedia.org/wiki/File:Direct_DNA_damage.png

225-6. Modified by E.Page from "Indirect DNA damage" by Gerriet41 - Own work. Licensed under Public domain via Wikimedia Commons - http://commons.wikimedia.org/wiki/File:Indirect_DNA_damage.png

226-1. "Quark structure neutron". Licensed under Creative Commons Attribution-Share Alike 2.5 via Wikimedia Commons - http://commons.wikimedia.org/wiki/File:Quark_structure_neutron.svg

226-2. "Stdef2" by Hullernuc - Own work. Licensed under Creative Commons Attribution-Share Alike 3.0 via Wikimedia Commons - http://commons.wikimedia.org/wiki/File:Stdef2.png

226-3. "ThermalFissionYield" by JWB at en.wikipedia - Transferred from en.wikipedia by SreeBot. Licensed under Creative Commons Attribution 3.0 via Wikimedia Commons - http://commons.wikimedia.org/wiki/File:ThermalFissionYield.svg

226-5. "Critical mass". Licensed under Public domain via Wikimedia Commons - http://commons.wikimedia.org/wiki/File:Critical_mass.svg

226-6. "PWR nuclear power plant diagram" by Tennessee Valley Authority - tva.com. Licensed under Public domain via Wikimedia Commons - http://commons.wikimedia.org/wiki/File:PWR_nuclear_power_plant_diagram.svg

226-7. "Fission bomb assembly methods" by Fastfission - Own work. Licensed under Public domain via Wikimedia Commons - http://commons.wikimedia.org/wiki/File:Fission_bomb_assembly_methods.svg

226-8. "Teller-Ulam design" by User:Fastfission modified by User:HowardMorland - http://commons.wikimedia.org/wiki/File:Teller-Ulam_device_3D.svg. Licensed under Public domain via Wikimedia Commons - http://commons.wikimedia.org/wiki/File:Teller-Ulam_design.png

SI Prefixes

Prefix	Power	Symbol
yocto	10^{-24}	y
zepto	10^{-21}	z
atto	10^{-18}	a
femto	10^{-15}	f
pico	10^{-12}	p
nano	10^{-9}	n
micro	10^{-6}	μ
milli	10^{-3}	m
centi	10^{-2}	c
deci	10^{-1}	d
deka	10^{1}	da
hecto	10^{2}	h
kilo	10^{3}	k
mega	10^{6}	M
giga	10^{9}	G
tera	10^{12}	T
peta	10^{15}	P
exa	10^{18}	E
zetta	10^{21}	Z
yotta	10^{24}	Y

Important Physical Constants

Name	Symbol	Value
Atomic mass unit	amu	$1\,u = 1.6605 \times 10^{-27}\,kg$
Avogadro's number	N_A	$6.0221 \times 10^{23}\,\dfrac{particles}{mole}$
Boltzmann constant	$k = \dfrac{R}{N_A}$	$1.3806 \times 10^{-23}\,J/K$ $8.6173 \times 10^{-5}\,eV/K$
Coulomb Constant	$k = \dfrac{1}{4\pi\epsilon_0}$	$8.9875 \times 10^{9}\,N \cdot m^2/C^2$
Fundamental Charge	e	$1.6022 \times 10^{-19}\,C$
Gas Constant	R	$8.3145\,J/(mol \cdot K)$
Universal Gravitational Constant	G	$6.6738 \times 10^{-11}\,N \cdot m^2/kg^2$
Permeability of Free Space	μ_0	$4\pi \times 10^{-7}\,N/A$
Permittivity of Free Space	ϵ_0	$8.8542 \times 10^{-12}\,C^2/(N \cdot m^2)$
Planck's Constant	h $\hbar = h/2\pi$	$6.626 \times 10^{-34}\,J \cdot s = 4.135 \times 10^{-15}\,eV \cdot s$ $1.055 \times 10^{-34}\,J \cdot s = 6.582 \times 10^{-16}\,eV \cdot s$
Speed of light	c	$2.9979 \times 10^{8}\,m/s^2$
Stephan-Boltzmann constant	σ	$5.67037 \times 10^{-8}\,W/(m^2K^4)$

Fundamental Masses

Name	Mass (kg)	Mass (u)	Mass (MeV/c²)
Electron	9.109×10^{-31}	0.0005486	0.5110
Proton	1.6726×10^{-27}	1.007277	938.27
Neutron	1.6749×10^{-27}	1.008665	939.57
Muon	1.8835×10^{-28}	0.113429	105.66

Terrestrial and Astronomical Data

Name	Symbol	Value
Gravitational Field Strength at earth's surface	g	$9.81\,N/kg = 9.81\,m/s^2$
Radius of Earth	R_E	$6.38 \times 10^{6}\,m$
Mass of Earth	M_E	$5.98 \times 10^{24}\,kg$
Mass of Sun	M_S	$1.99 \times 10^{30}\,kg$
Mass of the moon	M_m	$7.35 \times 10^{22}\,kg$
Earth-moon distance (mean)		$3.84 \times 10^{8}\,m$
Earth-sun distance (mean)		$1.50 \times 10^{11}\,m$

Some Conversion Factors

Length
1 m = 39.37 in = 3.21 ft = 1.094 yard
1 km = 0.6214 mi
1 mi = 5280 ft = 1.602 km
1 inch = 2.54 cm

Volume
1 L = 10^3 cm³ = 10^{-3} m³ = 1.057 qt

Time
1 h = 3600 s
1 y = 3.156×10^7 s

Speed
1 km/h = 0.278 m/s = 0.9214 mi/h
1 ft/s = 0.3048 m/s = 0.6818 mi/h

Angular Variables
1 rev = 2π rad = 360 degrees
1 rad = 57.30 degrees

Mass
1 kg = 2.205 lb (at the surface of the earth)

Energy
1 J = 10^7 erg
1 kWh = 3.6 MJ
1 cal = 4.186 J
1 eV = 1.602×10^{-19} J

Made in the USA
Middletown, DE
28 August 2021